Burials of War of 1812 Veterans in the Commonwealth of Virginia

First Addendum

by the

Society of the War of 1812 in the Commonwealth of Virginia

Myron E. Lyman, Sr., Compiler, Editor

William J. Simmons, Assistant Editor

HERITAGE BOOKS
2018

HERITAGE BOOKS

AN IMPRINT OF HERITAGE BOOKS, INC.

Books, CDs, and more—Worldwide

For our listing of thousands of titles see our website
at
www.HeritageBooks.com

Published 2018 by
HERITAGE BOOKS, INC.
Publishing Division
5810 Ruatan Street
Berwyn Heights, Md. 20740

Heritage Books by the author:
Burials of War of 1812: Veterans in the Commonwealth of Virginia
Burials of War of 1812: Veterans in the Commonwealth of Virginia: First Addendum

International Standard Book Number
Paperbound: 978-0-7884-5823-1

TABLE OF CONTENTS

ACKNOWLEDGMENTS

As with the original burial book, a lot of the members of the Society of the War of 1812 in the Commonwealth of Virginia have contributed with their time, material, and financial support to produce this addendum. Others outside of the Society have provided a lot of material support as well. The presidents of the Society, C. Dennis Fritts and Charles (Chuck) Poland, and the officers of the Society have given me encouragement and provided financial support to complete this project.

My assistant editor William (Billy) Simmons has worked with me diligently, especially for preparing some of the paragraphs and all of the appendices. The publication date would be considerably extended without his valuable assistance.

Another key individual who assisted is author and President-Elect of the Society, Stuart L. Butler. His outstanding book *A Guide to Virginia Militia Units in the War of 1812* was essential in determining the military service for most all of the veterans identified in this publication.

Contributing immensely to the preparation of the manuscript has been Karen Hart who provided technical and technological assistance in formatting the text and appendices in the preparation of the final copy for the publisher.

Myron E. Lyman, Sr.
Compiler/Editor
January 2018

INTRODUCTION

This addendum provides over 900 new War of 1812 veterans with known burial locations in Virginia. It provides more information on approximately 185 others as identified in the original publication by the War of 1812 Society in the Commonwealth of Virginia, *Burials of War of 1812 Veterans in the Commonwealth of Virginia,* published by Heritage Books Inc. in 2012. It also lists fifteen deletions. These additions make a total of more than 5325 identified veteran burials in Virginia.

A large portion of the new burials is the result of contributors of memorials in Findagrave.com. The contributors also provided considerably more genealogical data from their research and from inscriptions on family gravestones. Also, the publishing on Fold3.com of pension and bounty land records from the National Archives and Records Administration (NARA) has contributed significant genealogical information on the widows of the veterans. The new second edition to author Stuart Butler's *Guide to Virginia Militia Units in the War of 1812,* published in 2011 by New Papyrus Publishing (Athens, GA), has greatly increased the determination of the correct service and unit commanders for the interred veteran.

The reader will need to use the original *Burials* alongside this addendum for a thorough understanding of the work and for a full listing of abbreviations, references, and codes. This addendum lists alphabetically the additions, updates, and deletions to the veteran information. This addendum also features supplements to the original appendices and illustrations. New codes to service and burial sources can be found in appendices C and D to expand on those listed in the original book. New photographs show cemetery plaques and monuments placed by the Society since the first book's publication.

The paragraphs of veteran information follow the same format as the original book. A key to the bolded category abbreviations is located at the bottom of each page. Readers should also take note of the additional explanations below.

Birth and Death Dates and Places
If the birth place or death place is not given, it means that it is the same as the county or independent city in Virginia in which the burial is listed.

Rank
The rank listed is the highest obtained during the war period. In some cases the veteran obtained a higher rank after the war ended in February 1815. If so this may be listed in the paragraph under VI (other veteran information).

Unit (for militia soldiers)
The Virginia Militia Regiment (VMR) is given for the county or independent city in which the veteran resided. This is followed, if known, by the company commander's name, the county or independent city, and to whom the unit was attached when arriving at its duty location. This information is in Butler's *Guide* (see above) and the reader should refer to that publication for more details about the service. If the unit is not listed for the veteran, readers may be able to obtain service records from NARA. An example of a service record from NARA is illustrated in *Burials* (page xiii).

Cemetery
The cemetery name is followed by the county or independent city, GPS reading if known, and address or general directions.

LIST OF SUPPLEMENTAL ILLUSTRATIONS

Additional Cemetery Plaques and Monuments
placed by Society of the War of 1812 in the Commonwealth of Virginia

ADDENDUM TO
BURIALS OF WAR OF 1812 VETERANS IN THE COMMONWEALTH OF VIRGINIA

(Listings updated January 27, 2018)

Key: *Additional veteran entry   **Corrected veteran entry   *** Deleted veteran entry

*ABBOTT, William; b 1784, d 18 May 1823 **RU:** Private, 31st, 51st or 122nd VMR a company of Frederick Co, attached to 4th or 5th VMR **CEM:** Mount Hebron; Winchester; 305 E Boscawen St **GS:** Y **SP:** No spouse info **VI:** No further data **P:** No **BLW:** No **PH:** No **SS:** A rec 411 or 412 **BS:** 245

*ABERNATHY, Raleigh H; b 1789, d 30 Jun 1858 **RU:** Corporal, 66th or 96th VMR Capt John B Rice's Co., attached to 1st VMR [Byrne] **CEM:** Bethel Methodist Church; Brunswick; Cemetery Rd, Alberta **GS:** Yes **SP:** Mar 1) Martha T Avery 2) Susan Peterson (__-4 Jun 1880) **VI:** His pension # 15399-40-50, Susan's 68099-20-55 **P:** Both **BLW:** Yes **PH:** No **SS:** B pg 49; M pg 1 **BS:** 245

*ABY, Jonas, b PA, d 25 Feb 1849 **RU:** Private, 31st VMR, Capt Thomas Robert's company of Riflemen, Frederick Co, attached to 4th VMR (Beatty) **CEM:** Ridings Family; Frederick; Middletown **GS:** Unk **SP:** Mar (Winchester Co, 28 Jul 1808) Margaret Hulett (__-aft 1871) **VI:** Probably disabled in service as he is listed at NARA in Old War files #12008 **P:** Spouse #3165 **BLW:** No **PH:** No **SS:** B pg 79; BD pg 3 **BS:** 31

*ADAMS, James; b 3 Sep 1765, d 14 Dec 1849 **RU:** Private 57th VMR in unknown company of Loudoun County **CEM:** Sharon; Loudoun; Jay & Federal Sts, Middleburg **GS:** Yes **SP:** No spouse info **VI:** It is possible that this service belongs to another person this name from Loudoun Co **P:** No **BLW:** No **PH:** No **SS:** A rec 1627 **BS:** 245

*ADAMS, William; b 1796, d 20 Sep 1846 **RU:** Private, 56th & 57th VMR, companies in Loudoun County commanded by Lt Colonels Timothy Taylor and Armistead Mason **CEM:** Ebenezer Baptist Church; Loudoun; Bluemont **GS:** Yes **SP:** Nancy Robertson (__-1845 **VI:** No further info **P:** No **BLW:** No **PH:** No **SS:** A rec 2332, 2333; B pgs 119-121 **BS:** 245

*ALBRIGHT, John; b c1791, d 8 Oct 1851 **RU:** Private, 4th or 5th VMR, in a company of these regiments at Norfolk **CEM:** McDowell Family; Rockbridge; Rt 11 nr jct with Rt 712, Fairfield **GS:** Yes **SP:** No spouse info **VI:** No further data **P:** No **BLW:** No **PH:** No **SS:** A recs 1325; 1324 **BS:** 245 GPS 37.86860, -79.31080

*ALDRIDGE, John; b 1769, d 1837 **RU:** Private; 31st, 51st, or 122nd VMR; A company of Frederick Co, attached to 1st VMR (Taylor) **CEM:** Goose Creek; Loudoun; Rt 722 S of jct with Rt 723 **GS:** Unk **SP:** Harriet West (1765, MD-1857) **VI:** No further info **P:** No **BLW:** No **PH:** No **SS:** A rec 1681; B pgs 78-80 **BS:** 245

ALEXANDER, James B; b 1780, d 17 Feb 1835 **RU: Sergeant, 2nd Corps D'Elite (Green) **CEM:** Timber Ridge; Rockbridge; Rt 11, 6.5 mi N of Lexington **GS:** Yes **SP:** Mary Ashcraft **VI:** Son of Joseph Alexander (1742-1813) and Sarah Reid **P:** No **BLW:** No **PH:** No **SS:** K pg 222 **BS:** 193

*ALEXANDER, Richard Brown; b 1792, d 25 Sep 1856, Alexandria **RU:** Private, Col Whann's 8th Regiment of Cavalry, MD Militia **CEM:** Pohick Episcopal Church; Fairfax; Jct Rts 1 & 611, Lorton **GS:** Yes **SP:** No spousal info **VI:** Son of Charles Alexander (1737-1806) and Frances Brown (1751-1823). Reinterred fr Preston Cem in Alexandria. Was a doctor **P:** No **BLW:** No **PH:** No **SS:** O- Serv Index Card, fold3 **BS:** 245

Key: *Additional veteran entry   **Corrected veteran entry   ***Deleted veteran entry
RU=Rank/Unit   CEM=Cemetery   GS=Gravestone   SP=Spousal Information   VI=Other Veteran Info   P=Pension
BLW=Bounty/Land Warrant   PH=Photo   SS=Service Source   BS=Burial Source   VMR= VA Military Regiment
LNR= Last Known Residence

1

*ALEXANDER, William Reid; b 12 May 1775, d 12 Mar 1825 **RU**: Sergeant; 8th VMR, Capt Alexander Campbell's Co, Rockbridge Co, attached to 2d Corps De Elite (Green) **CEM**: Timber Ridge; Rockbridge; GPS-37.84200,-79.35800, off Rt 11, just E of I-81 **GS**: Yes **SP**: Elizabeth Newell Campbell (27 Jan 1780-19 Aug 1850), daug of Duncan Campbell and Margaret Newell **VI**: Son of Joseph Alexander and Sarah Reid **P**: No **BLW**: No **PH**: No **SS**: A rec 2254; B pg 179; K pg 209 **BS**: 245

*ALLAN, John: b c1780 Ayreshire, Scotland, d 27 Mar 1834 **RU**: Private, 74th VMR, Capt Benjamin Pollard's Co, Hanover County, attached to 9th VMR (Sharp) **CEM**: Shockoe Hill; Richmond City; 100 Hospital St; **GS**: Yes **SP**: Mar 1) (Feb 1803) Frances Keeling, Valentine (1784-28 Feb 1829) 2) Louisa Gabriella Patterson (1800-1881) **VI**: Merchant **P**: No **BLW**: No **PH**: Yes **SS**: A rec 293; B pgs 94; 95; 233 **BS**: 31; 245

ALLEN, David Hume; b 24 Jul 1781, Shenandoah Co; d 31 Oct 1854 **RU: Private, 16th VMR (Waller), Spotsylvania Co **CEM**: Grace Episcopal; Clarke; 110 N Church St, Berryville **GS**: Yes **SP**: Mar (Frederick Co) Sarah Griffin Taylor (__-at Clifton 11 Aug 1874) **VI**: Son of Thomas Allen and Abigail Milar **P**: No **BLW**: No **PH**: No **SS**: A rec 702 **BS**: 92 pg 34; 245

*ALLEN, Israel, Sr; b 28 Aug1779, d 18 Nov 1831 **RU**: Private, 119th VMR, company of Harrison Co **CEM**: Union Church; Shenandoah; Mount Jackson **GS**: Yes **SP**: Sarah Phifer (1783-1849) **VI**: No further data **P**: No **BLW**: No **PH**: No **SS**: A rec 1003; B pg 97 **BS**: 245

*ALLEN, James; b 15 Oct 1769, d 9 Jun 1845 **RU**: Adjutant, 56th or 57th VMR, Loudoun Co company attached to and adjutant of 6th VMR **CEM**: Allen Family; Loudoun; Watson **GS**: Yes **SP**: Mar Elizabeth Ann Lee (1775-1820) **VI**: Son of William Allen and Sarah (__) **P**: No **BLW**: No **PH**: No **SS**: A rec 1082 **BS**: 245

*ALLEN, John; b 1790, d 16 May 1856 **RU**: Private, 70th or 105th VMR, Washington Co company attached to Bradley's command **CEM**: Kelly's Chapel; Washington; Vic jct Rts 736 & 608 **GS**: Yes **SP**: No spouse info **VI**: No further info **P**: No **BLW**: No **PH**: No **SS**: A rec 1162 **BS**: 245

*ALLEN, John C: b 1781; Prince Edward Co, d 30 Nov 1834 **RU**: Private, 63rd VMR, Capt Anthony Woodson's Artillery Co, Prince Edward County, attached to Battalion of Artillery **CEM**: Lakeview; Nottoway; Blackstone **GS**: No **SP**: Mar (Prince William Co, 20 Oct 1808) Nancy Watson **VI**: Son of Charles Allen (1754-1814); burial from family records **P**: No **BLW**: No **PH**: No **SS**: A rec 1154; B pg 168 **BS**: 245

*ALLEN, Joseph; b 5 Aug 1796; d 27 Mar 1862 **RU**: Private, 9th VMR, Capt Thomas Hoskins Company and Capt William Hutchison's Company, King & Queen Co, with service at Alexander's Field, Georgetown, DC **CEM**: Hollywood; Richmond City; 412 S Cherry St **GS**: Yes **SP**: Mary Ann Stetson (25 Aug 1806-24 Dec 1880) **VI**: No further info **P**: No **BLW**: No **PH**: No **SS**: A rec 1388; B pg 113; L pg 448, 468 **BS**: 245

*ALLEN, Reuben; b 21 Mar 1796, d 21 Feb 1874; **RU**: Private, 5th or 34th VMR Capt Reuben Moore's Co, Culpeper Co attached to 6th VMR (Coleman) **CEM**: Emanuel Lutheran Church; Shenandoah; 127 East High St, Woodstock **GS**: Unk **SP**: Not married **VI**: Recd pen and BLW of 160 acres **P**: Yes **BLW**: Yes **PH**: No **SS**: A rec 1602; B pg 62; O- pen files, Fold3 **BS**: 245

*ALLEN, Robert Millar; b 16 Jan 1790, Shenandoah Co, d 14 May 1819 **RU**: Private, 13th or 97th VMR, Captain Peter Hays, Shenandoah Co, attached to 1st VMR (Yancey) **CEM**: Prospect Hill; Shenandoah; Front Royal **GS**: Unk **SP**: Mar (4 Apr 1816) Rebecca Branson **VI**: Son of Thomas Allen (1724-1822) and Abigail Millar (1747-1823) **P**: No **BLW**: No **PH**: No **SS**: A rec 1656; B pg 242 **BS**: 245

*ALLEN, William C; b 6 Jan 1794, d 6 Sep 1874 **RU**: Private, 23rd VMR, Capt Alexander Gibbs, Chesterfield Co, attached to 2d VMR (Ambler/Brown) **CEM**: Hollywood; Richmond City; 412 S Cherry St **GS**: Yes **SP**: No spouse info **VI**: No further data **P**: No **BLW**: No **PH**: No **SS**: B pg 60; L pg 354 **BS**: 245

ALLEN, William T; b 1784, d 07 May 1843 **RU: Private, 4th VMR **CEM**: Grace Episcopal; Clarke; GPS: 39.15810,-77.97690, Green Hill, 110 N Church St, jct Mosby Blvd and Rt 340 Berryville **GS**: Yes **SP**: No spouse info **VI**: No further data **P**: No **BLW**: No **PH**: No **SS**: A rec 1951 **BS**: 86 pg 14; 245

Key: *Additional veteran entry   **Corrected veteran entry   ***Deleted veteran entry
RU=Rank/Unit   CEM=Cemetery   GS=Gravestone   SP=Spousal Information   VI=Other Veteran Info   P=Pension
BLW=Bounty/Land Warrant   PH=Photo   SS=Service Source   BS=Burial Source   VMR= VA Military Regiment
LNR= Last Known Residence

2

*ALLEN, William, Jr: b 1788, d 17 Oct 1876 **RU**: Private, 54th VMR Capt Julian Magagnos Co, Norfolk Borough Norfolk, attached to 3rd VMR (Preston) **CEM**: Jones Family; Franklin; 2930 Runnett Bag Creek Rd, Endicott **GS**: No **SP**: Mar 1) (Franklin Co, 10 Mar 1812) Susannah Smith, 2) (Floyd Co, 15 Dec 1872) Martha Jane (Turner) Boyd (__-1 Oct 1910). Recd pen **VI**: Son of William Allen and Rachel Elizabeth Kendall **P**: Both **BLW**: Yes **PH**: No **SS**: B pg 145; 229; M pg 5 **BS**: 245

*ALLISON, Henry; b 1176, d 1851 **RU**: Corporal, 13th or 97th VMR in Company Shenandoah Co, attached to 4th VMR **CEM**: Saint Pauls Lutheran Church; Shenandoah; Strasburg **GS**: No **SP**: No spouse info **VI**: No further data **P**: No **BLW**: No **PH**: No **SS**: A rec 2380 **BS**: 245

*ALLISON, John; b 1774, Prince George County, d 18 Jan 1859 **RU**: Private, 39th VMR, Petersburg **CEM**: Blandford; Petersburg; 319 S Crater Rd **GS**: No **SP**: Mar (Lancaster Co, 28 Nov 1796) Frances Hill Currie **VI**: Son of William Allison & Alice (__) **P**: No **BLW**: No **PH**: No **SS**: A rec 2413; B pg 159 **BS**: 245

ALLISON, Robert, Jr; b 1787; d 5 Sep 1814 **RU: Private, 9th VMR, Captain William Hutchinson's Co, King & Queen County, attached to First DC Militia Regt at Alexander's Field, Georgetown DC **CEM**: Old Presbyterian Meeting House; Alexandria; Wilkes & Hamilton **GS**: Yes **SP**: No spouse info **VI**: Killed in battle at "The White House" at now Ft Belvoir in Fairfax Co at age 27, Inscription tombstone reads, "In memory of Robert Allison, Jr. who fell In battle on 5 September 1814 at the white house gallantly defending his country, aged 27 years. Our lives belong to God & our country. He was a dutiful son. An affectionate brother. Conciliating of manners. Beloved by all. Erected by his kinsman James M. Stewart" **P**: No **BLW**: No **PH**: No **SS**: A rec 2437; B pg 113 **BS**: 32 pg 9

*AMBLER, Edward; b May 1783, Jamestown, d 10 Jul 1846, Rappahannock Co **RU**: Lieutenant, 4th VMR Commanded a Calvary unit at Norfolk **CEM**: Shockoe Hill; Richmond City; 100 Hospital St **GS**: Unk **SP**: Mar (15 May 1807) Sarah Taylor Holcombe (1787, Amelia Co-1867) **VI**: Son of John Ambler (1762-1836) and Katherine Bush (1773-1846), obtained rank of Colonel, served also in Navy before war as Midshipman under Commodore Rogers. Served in VA Legislature **P**: No **BLW**: No **PH**: No **SS**: O-Fold3 service rec **BS**: 245

*AMISS, John S; b 4 Jul 1790, Culpeper Co, d 9 Nov 1885 **RU**: Corporal, 47th or 88th VMR, Capt Samuel Leake's Co, Albemarle Co, attached to 2d VMR (Ballowe) **CEM**: New Bethel; Albemarle; GPS: 37.90060,-78.43610; 4899 Rolling Rd; Woodridge **GS**: Yes **SP**: Mar (1 Aug 1812) Mary Taylor (Hattan) Shifflett **VI**: No further data **P**: Both **BLW**: Yes **PH**: N **SS**: A rec 846; B pg 36; M pg 36 **BS**: 245

*ANDERSON, James; b 11 Aug 1777, d 11 Oct 1827 **RU**: Sergeant, 58th VMR, Captain Adam Hornberger's Co attached to 1st VMR (Trueheart's) **CEM**: Cooks Creek Presbyterian Church; Rockbridge; Harrisonburg **GS**: Yes **SP**: No spouse info **VI**: No further data **P**: No **BLW**: No **PH**: No **SS**: B pg 181 **BS**: 245

*ANDERSON, James Baker; b 1 May 1787, d Apr 1856 **RU**: Sergeant, 17th VMR, Captain Allen's Co attached to 1st VMR (Trueheart) **CEM**: James Anderson Family; Cumberland; Reeds **GS**: No **SP**: Charlotte Hazelgrove (1800-1874) **VI**: Son of Thomas (1754-1804) and Sarah Anderson (1760-1814) **P**: No **BLW**: No **PH**: No **SS**: K pg 43 **BS**: 245

*ANDERSON, Jesse; b 1767; d Apr 1818 **RU**: Private, 43rd VMR, a company in Franklin Co **CEM**: Anderson Family; Bedford; GPS 37.38970, -79.39045 2370; Citax Rd, Goode **GS**: Yes **SP**: Mar (21 Jun 1791) Elizabeth West Jones (Bedford Co, 1775 - Sep 1849), daug of John Jones and Frances Barber **VI**: Son of Nelson Anderson (1735-1820) and Frances Jackson (1741-1818) **P**: No **BLW**: No **PH**: No **SS**: A rec 1632 **BS**: 245

*ANDERSON, Joseph B; b Unk, d 3 Apr 1873 **RU**: Sergeant Major, Squadron of Calvary, Headquarters Staff, Major John T Woodford Commanding, Amelia Co **CEM**: Anderson Family; Amelia; Rt 620, N of Rodaphil **GS**: Unk **SP**: 1) Sally Meriwether 2) Jane S. Archer (__-20 Oct 1885) **VI**: Was a doctor after war **P**: Widow **BLW**: Yes **PH**: No **SS**: A rec 1875; B pg 8; L pg 47 **BS**: 245

Key: *Additional veteran entry   **Corrected veteran entry   ***Deleted veteran entry
RU=Rank/Unit   CEM=Cemetery   GS=Gravestone   SP=Spousal Information   VI=Other Veteran Info   P=Pension
BLW=Bounty/Land Warrant   PH=Photo   SS=Service Source   BS=Burial Source   VMR= VA Military Regiment
LNR= Last Known Residence

3

***ANDERSON**, Joseph Edward, Jr; b 30 Dec 1788, d 28 Dec 1855 **RU**: Private, 5th VMR (McDowell) **CEM**: Anderson Family; Clarke; "Springfield" Claggett Farm **GS**: Yes **SP**: Jane Ross (1802- 24 Aug 1842; Aged 40 yrs, 10 mos, 13 days) "Wife of Joseph Anderson" (stone.) **VI**: Died age 66 yrs, 11 mos, 28 days. Son of Joseph Edward Anderson, Sr (1746-1825) and Hannah D Blue (1754-1843) **P**: No **BLW**: No **PH**: No **SS**: A rec 1865 **BS**: 92 pg 1; 245

***ANDERSON**, Richard; b 1779, d Mar 1857 **RU**: Private, 19th VMR, Richmond City **CEM**: Shockoe Hill; Richmond; 100 Hospital St **GS**: Unk **SP**: No spouse info **VI**: No further data **P**: No **BLW**: No **PH**: No **SS**: B pg 74; L pg 198 **BS**: 245

***ANDERSON**, Samuel; b 1 Jan 1796, d 28 Mar 1819 **RU**: Drum Major 13th VMR, Capt George Shum's Co, Shenandoah Co, attached to Maj Nathaniel Perkin's Garrison **CEM**: Emanuel Lutheran Church; Shenandoah; 127 E High St, Woodstock **GS**: Unk **SP**: No spousal info **VI**: No further data **P**: No **BLW**: No **PH**: No **SS**: A rec 2061; B pg 185; K pg 227 **BS**: 245

***ANDERSON**, Thomas Bates; b 14 Jan 1792, d 3 May 1872 **RU**: Private, 30th VMR, Capt Daniel DuVal's or Capt James Gate's Co, Caroline Co, attached to 9th VMR (Boyd) **CEM**: Topping Castle; Caroline; GPS: 37.92332,-77.54283; Nr S end unmarked Rd; E of Jct Rts 700 and 658 **GS**: Unk **SP**: Harriet McLaughlin **VI**: Was a doctor **P**: No **BLW**: No **PH**: No **SS**: A rec 2135; B pg 55 **BS**: 245

***ANDERSON**, William; b 1787, d 1871 **RU**: Private, 58th or 116th VMR of Rockingham Co, attached to 6th VMR (Coleman) **CEM**: Woodbine; Harrisonburg City; 381 East Market St (Rt 33) **GS**: Yes **SP**: No spouse info **VI**: No further data **P**: No **BLW**: No **PH**: No **SS**: A rec 2247 **BS**: 245

***ANDREWS**, Anderson; b 20 Jul 1785, d 1821 **RU**: Corporal 39th VMR and 98th VMR As a private was in Capt Richard McRae's Co in 1812 in Petersburg Volunteers 39th VMR and as Corporal in 98th VMR in one of several companies, Mecklenburg County, attached to LTC Grief Green **CEM**: William Andrews Family; Mecklenburg; GPS: -36.78894, -78.13494, Whittles Mill Rd, RT 654, past Bridge Rd to end **GS**: Unk **SP**: Mar (28 Nov 1816) Sarah Gee (c1796- 19 Aug 1853) **VI**: Son of Varney Andrews (25 Jul 1755- and Lucy Green (1765-___) **P**: No **BLW**: No **PH**: No **SS**: A rec 177; B pgs 130; 131; 159; G pg 138 **BS**: 245

***ANDREWS**, Bullard R; b 1782, d 1841 **RU**: Private, 23rd VMR Capt Lawson Burfoot's Co, Chesterfield County, attached to 1st VMR (Yancey) **CEM**: Andrews Family; Petersburg City; River Rd on private property, lower district **GS**: Unk **SP**: Elizabeth Eppes Belcher (1792-___) **VI**: Son of Bullard Andrews (1745-1828) **P**: No **BLW**: No **PH**: No **SS**: A rec 212; B pg 59 **BS**: 245

***ANDREWS**, Erasmus Granger; b 1793; d 7 Jun 1875 **RU**: Private, 23rd VMR, Capt Littleberry West's company, Chesterfield County **CEM**: Hollywood; Richmond City; 412 S Cherry St **GS**: Unk **SP**: Mar (20 Aug 1845) Rebecca B Scott (___-c1879) **VI**: Son of Bullard Andrews (20 Jul 1745- 7 Dec 1828), Recd pen and two BLW - 40 acres 1850, 120 acres 1855 **P**: Both **BLW**: Yes **PH**: N **SS**: B pg 61, L pg 825; Open Files, fold 3 **BS**: 245

*****ANDREWS**, George: deleted as service depends upon movement to Frederick Co that was not found

***ANNIS**, Custis; b 1789, d May 1851 or b 1779, d Mar 1857 **RU**: Private, 99th VMR, Capt George Ewell's Co, Accomack County **CEM**: Tehern; Accomack; Guilford **GS**: Unk **SP**: No spouse info **VI**: No further data **P**: No **BLW**: No **PH**: No **SS**: B pg 32; L pg 311 **BS**: 245

***ARBOGAST**, Michael Jr; b 1768, Augusta Co, d aft 1855, Highland Co **RU**: Private 46th VMR, Capt Peter Hienkle's Co, Pendleton County, attached to 5th VMR (Mason-Preston) **CEM**: Arbogast Family; Highland; Crab Bottom Pauper Farm (see property records for loc) **GS**: Unk **SP**: Mar (1792) Barbara Buzzard (1770-1824) **VI**: Son of Michael Arbogast (1734-1812) and Mary Elizabeth Samuels (1736-___), Recd BLW 1855 **P**: No **BLW**: Yes **PH**: No **SS**: A rec 1341; B pg 158 **BS**: 245

***ARCHER**, Richard Thompson; b 16 Aug 1796; d c1853 **RU**: Private, 19th VMR, Capt William Murphy, Light Infantry Blues, Richmond City, attached to 1st Corps d'Elite (Randolph) **CEM**: Red Lodge; Amelia; Rt 625, 0.6 mi from Rt 609 **GS**: Yes **SP**: Ann Maria Barnes (1805-1882) **VI**: Son of Colonel William Archer **P**: No **BLW**: No **PH**: No **SS**: K pg 259; B pg 175 **BS**: 266 pg 251

Key: *Additional veteran entry   **Corrected veteran entry   ***Deleted veteran entry
RU=Rank/Unit   CEM=Cemetery   GS=Gravestone   SP=Spousal Information   VI=Other Veteran Info   P=Pension
BLW=Bounty/Land Warrant   PH=Photo   SS=Service Source   BS=Burial Source   VMR= VA Military Regiment
LNR= Last Known Residence

4

ARCHER, Robert; b 28 Aug 1794, Norfolk, d 19 May 1877 **RU: Surgeon's Mate, 6th VMR (Reade) **CEM:** Hollywood; Richmond City; 412 S Cherry St **GS:** Yes **SP:** No spouse info **VI:** Physician at Fort Nelson and later at Fort Monroe, and was an iron manufacturer in Richmond that made ordnance for Confederacy **P:** No **BLW:** No **PH:** Yes **SS:** A rec 1481 **BS:** 31

ARMISTEAD, John Baylor; b 1765; d Unk **RU: Captain, 52d VMR, Company commander, Charles City Co, attached to 2d VMR (Ambler & Brown) **CEM:** Upperville United Methodist; Fauquier; Delaplane Rd, Upperville **GS:** Yes Govt **SP:** Anne B Carter **VI:** Son of John Armistead & Lucy Baylor **P:** No **BLW:** No **PH:** No **SS:** A rec 132; B pg 143 **BS:** 4 pg 204; 49

*ARMSTRONG, William J; b 1762, d 28 Jun 1814 **RU:** Private, 8th VMR Capt John Mullin's Co, Rockbridge County, attached to 4th VMR (Boyd's) **CEM:** Old Providence Presbyterian Church; Rockbridge; GPS: 37.95170,-79.30250, 1208 New Providence Rd, Raphine **GS:** Yes **SP:** Mary P. (__) (1802-22 Apr 1837) **VI:** No further data **P:** No **BLW:** No **PH:** No **SS:** A rec 770; B pgs179; 180 **BS:** 245

ARMSTRONG, William Morris; b 30 Jan 1789, Shelby Co, KY, d 27 Jun 1861 **RU:** Midshipman, commissioned 30 Nov 1814 U.S. Navy **CEM:** Cedar Grove; Norfolk; Cnr Salter St & E. Princess Anne Rd **GS:** Yes **SP:** Adelaide Tyler (17 Sep 1806-15 Sep 1881), daug of Samuel Tyler (brother of President Tyler) and Eliza Bray **VI:** Became Capt U.S.N, (24 Mar 1855); Gr St lists service; Obit in Norfolk Newspaper **P:** Unk **BLW:** Unk **PH:** No **SS:** AL General Naval Register; pg 28 **BS:** 245

*ARRINGTON, Richard; b c1787, d 5 Dec 1849 **RU:** Private, 36th VMR (Reno), Prince William Co **CEM:** Arrington Family; Prince William; GPS 38.75520,-77.43390, betw Mineda Ct and Arrington Family Ct **GS:** Yes **SP:** No spouse info **VI:** No further data **P:** No **BLW:** No **PH:** No **SS:** A rec 1453 **BS:** 245

*ARTHUR, John A Jr; b 7 Dec 1791, d 21 Aug 1844 **RU:** Sergeant, 43rd & 110th VMR, Capt William Jones Co, Franklin County, attached to 5th VMR (Mason & Preston) **CEM:** Arthur Family; Franklin; GPS 37.15580,-80.01823, John Arthur Rd, 500 yds behind old home place **GS:** Yes **SP:** Mar (Charlotte Co, 7 Sep 1818) Ann Smith (1798-1882), daug of Shelby Smith **VI:** No further info **P:** Widow **BLW:** Yes **PH:** No **SS:** B pg 77; BD pg 55; M pg 12 **BS:** 245

*ASHBY, James; b c1764, d 10 Jan 1857 **RU:** Private USA 25th Infantry Capt John T Chunn's Co, and/or USA 37th Inf Capt John Brown's Co **CEM:** Heron Hill; Accomack; Nr jct Rts 615 & 614, Locustville **GS:** Unk **SP:** 1) Elizabeth (__) (1755-14 Jul 1824), 2) Sarah Bayley(__-Jan 1833), daug of Charles Bayley **VI:** Recd BLW 160 acres, 1855 **P:** No **BLW:** Yes **PH:** No **SS:** O- Fold3 BLW files **BS:** 245

*ATKINS, John; b 1784, d 14 May 1859 **RU:** Sergeant, 9th VMR, Lieutenant William Newsome's Co, King & Queen County **CEM:** Shackelford's Chapel UMC; King & Queen; Plain View **GS:** Yes **SP:** Mar (19 Dec 1812) Nancy Taylor (1786-3 Dec 1876) **VI:** No further data **P:** Both **BLW:** Yes **PH:** No **SS:** B pgs 113,114; BD pg 59; L pg162; M pg 13 **BS:** 245

*ATKINS, John; b Unk, Rappahannock Co, d 13 Apr 1876 **RU:** Private; 5th VMR, Capt John Thom's Co, Culpeper County, attached to 2d VMR (Ballowe) **CEM:** Atkins; Rappahannock; Rt 211, W of Sperryville, just inside Shenandoah National Park **GS:** Yes **SP:** Mar (16 Mar 1837) Francis Lucy Campbell (__-12 Dec 1892), daug of Owen Campbell a Rev War veteran and Jemina Lear, daug of John Lear, her LNR Sperryville 1878 **VI:** Son of Ambrose Atkins a Rev War veteran **P:** Both **BLW:** Yes **PH** No **SS:** A rec 329; B pg 63; BD pg 59; M pg 13 **BS** 31

*ATKINSON, William Mayo; b 22 Apr 1796, Chesterfield Co; d 24 Feb 1849 **RU:** Private, 19th VMR, company of Richmond City, attached to 2d VMR (Ballowe) **CEM:** Hebron; Winchester; GPS 39.18170, -78.1572, 305 E Boscawen St **GS:** Unk **SP:** No spouse info **VI:** Son of Robert Atkinson and Mary Tabb Mayo **P:** No **BLW:** No **PH:** No **SS:** A rec 558 **BS** 245

*AULD, Hugh, Jr; b 20 Jul 1767, d 3 Nov 1820 **RU:** Lieutenant Colonel; Commander, 26th Regiment, MD Militia **CEM:** Arlington National; Arlington; off George Washington Parkway **GS:** Yes, buried in section 2 **SP:** Mar (Baltimore, 24 Jul 1793) Ziporah Wilson **VI:** In Battle of St. Michaels. **P:** No **BLW:** No **PH:** No **SS:** A rec 1024; AL pg 206 **BS:** 53 pg 19

Key: *Additional veteran entry   **Corrected veteran entry   ***Deleted veteran entry
RU= Rank/Unit   CEM=Cemetery   GS=Gravestone   SP=Spousal Information   VI=Other Veteran Info   P=Pension
BLW=Bounty/Land Warrant   PH=Photo   SS=Service Source   BS=Burial Source   VMR= VA Military Regiment
LNR= Last Known Residence

*AULICK, Frederick A; b 1792, d 28 Feb 1872 **RU:** Private; 31ˢᵗ VMR, Capt Michael Coyle's Co, Frederick County, attached to 1ˢᵗ VMR (Taylors) **CEM:** Mt Hebron; Winchester; GPS 39.18170, -78.1572, 305 E Boscawen St **GS:** Yes **SP:** Mar (Frederick Co, 13 Jun 1817) Elizabeth Smith (__-26 Feb 1813) **VI:** Son of Charles Aulick **P:** Yes **BLW:** Yes **PH:** No **SS:** A rec 1051; B pg 79; BD pg 62: M pg 14 **BS:** 245

*AYLETT, John; b 11 Dec 1795, d 7 Oct 1796 **RU:** Private, 74ᵗʰ VMR, Capt Mallory's Co and/or Capt Thomas Sorrilles Co, Hanover County, attached to Robert Crutchfield's Detachment **CEM:** Aylett Graves; Fairfield Plantation; King William; Aylett **GS:** Yes **SP:** No spousal info **VI:** Son of Phillip Aylett and Elizabeth (__) **P:** No **BLW:** No **PH:** No **SS:** B pg 95; O Serv index card; Fold 3 **BS:** 245

*AYRES, John Grigsby Sr.; b 10 Jun 1786, d 1 Sep 1876 **RU:** Private, 8ᵗʰ VMR, Rockbridge Co, attached to 5ᵗʰ VMR (McDowell's) **CEM:** Collierstown Presbyterian; Rockbridge; Collierstown **GS:** Yes **SP:** Mar (16 Sep 1817) Rachel Entsminger **VI:** Son of William Ayres (___-1832) and Nancy Grigsby **P:** Both **BLW:** Yes **PH:** No **SS:** A rec 2424, B pg 179, M pg 14 **BS:** 245

*BAGBY, Bennett, M; b 3 May 1796, d 30 Nov 1884, Powhatan Co **RU:** Private, 23ʳᵈ VMR, Capt Lawson Burfoot's Co, Chesterfield County, attached to 1ˢᵗ VMR (Yancey) **CEM:** Forest lawn; Richmond; GPS 37.59390, -77.43530, 4000 Pilots Lane **GS:** Unk **SP:** Mar 4) (Prince Edward Co, 11 Dec 1860) Louisa B Flippen (__- 1 Jul 1903) **VI:** Recd BLW of 80 acres, 1850 **P:** Both **BLW:** Yes **PH:** No **SS:** A rec 1096; B pg 59, 242; G pg 15; O- pen files Fold3 **BS:** 245

*BAGBY, John; b 1791; d 22 Jun 1880 **RU:** Private, 9ᵗʰ VMR, Ensign Thomas Gresham's Co, King & Queen County **CEM:** Bruington Baptist Church; King & Queen; GPS: 37.77775, -76.98140, 4784 The Trail **GS:** Yes **SP:** Mar (King & Queen Co) Elizabeth Motley (1799-11 Feb 1880) **VI:** Originally buried Society Hill **P:** Yes **BLW:** Yes **PH:** No **SS:** B pg 113; G pg 15; L pg 383 **BS:** 245

*BAGEANT, William; b 15 Aug 1784, d 23 Sep 1863 **RU:** Private, Frederick County company attached to 1ˢᵗ VMR (Taylor) **CEM:** Bageant Family; Frederick; Cross Junction **GS:** Yes **SP:** Mar (19 Nov 1813) Elizabeth Dick (8 Dec 1789-19 Feb 1859) **VI:** Son of John Dick and Catherine Whitnak **P:** No **BLW:** No **PH:** No **SS:** A rec 1140 **BS:** 56 pg 115; 245

BAILEY, James; b 11 Aug 1775, d 07 Feb 1833 **RU: Fifer, 64th VMR, Henry Co **CEM:** Old City Cemetery; Lynchburg; 401 Taylor St **GS:** Yes **SP:** Nancy Harvey (1785-1849) **VI:** No further data **P:** No **BLW:** No **PH:** No **SS:** A rec 1721 **BS:** 87 pg 85

*BAILEY, Jeremiah; b 2 Jun 1784, d 9 Apr 1825 **RU:** Private: 1st VMR, Capt John C Hill's Co, Amelia County attached to 1st VMR (Trueheart) **CEM:** Bailey Family; Amelia; GPS-37.29721,-77.97475; on Poorhouse Rd, beyond jct with Rt 701 at W.C. Golden farm **GS:** Yes **SP:** Mary Watkins Anderson (9 Mar 1786-4 Jun 1874), daug of Jordan Anderson and Margaret Easter. She applied for pen, that was rejected for proof of 60 day service, however company was on duty more than 60 days **VI:** Son of Benjamin Bailey (1755-1813) **P:** No **BLW:** No **PH:** No **SS:** B pg 37; M pg 16 **BS:** 245

*BAILEY, John; b 1790, d 1870 **RU:** Private, 13ᵗʰ or 97ᵗʰ VMR, Capt Peter Hay's Co, Shenandoah Co, attached to 1ˢᵗ VMR (Yancey) **CEM:** John Bailey Family; Page; Overlook Drive, Rileyville **GS:** Yes **SP:** Lucretia Bryan **VI:** No further data **P:** No **BLW:** No **PH:** No **SS:** B pg 242; O Serv Index Card, Fold3 **BS:** 245

BAILEY, Philip; b c1773 (assume age 21 on first marriage) d Unk **RU: Private, 1st VMR (Allen) **CEM:** Bailey Family; Sussex; Sussex **GS:** Yes **SP:** Mar 1) (Sussex Co, 2 Jan 1794) Betty Maglamie, 2) (Sussex Co, 17 Oct 1796) Susannah Cotton **VI:** No further data **P:** No **BLW:** No **PH:** No **SS:** A rec 1948 **BS:** 31

*BAILIE, Robert Bruce; b 27 Mar 1776, Ireland, d 21 Feb 1854 **RU:** Private, Capt Lander's Co, TN Militia **CEM:** Enoch P. Lane Family; Scott; GPS: 36.65250,-82.62889; off Rt 665; Manville Rd; Nr Smith Chapel & William's Cem **GS:** Yes **SP:** Mar (1803) Eleanora Ellen Carnahan (26 Oct 1783, Ireland-Nov 1875) **VI:** Son of William Law Bailie (1750-1824, Abingdon, Washington Co) and Mary Wright; plaque in cem lists War of 1812 service **P:** Widow **BLW:** Both **PH:** No **SS:** Gr St **BS:** 245

Key: *Additional veteran entry   **Corrected veteran entry   ***Deleted veteran entry
RU=Rank/Unit   CEM=Cemetery   GS=Gravestone   SP=Spousal Information   VI=Other Veteran Info   P=Pension
BLW=Bounty/Land Warrant   PH=Photo   SS=Service Source   BS=Burial Source   VMR= VA Military Regiment
LNR= Last Known Residence

*BAIRD, Peter; b 1795, d aft 4 Jul 1851 **RU**: Private, 62nd VMR, volunteered 4 Aug 1812 for 6 mos in Capt Allen Temple's Co, Prince George County, attached to 4th VMR (Lucas & Wills); discharged 7 Feb 1813 **CEM**: Peter Baird Family; Surry; off Rt 40 on field with fence, Spring Grove **GS**: Yes **SP**: Mar (1817) Eliza M Bingham **VI**: Held rank of Colonel 4 Jul 1851 **P**: Yes **BLW**: Yes **PH**: No **SS**: M pg 17; O-BLW file; Fold3 **BS**: 245

*BAKER, Jacob; b 25 Jul 1799, d 10 Mar 1875 **RU**: Private, 31st VMR, Capt Thomas Robert's Co, Frederick County, attached to 4th VMR (Beatty) **CEM**: St Stephens; Shenandoah; Strasburg Junction **GS**: Yes **SP**: No spouse info **VI**: Son of Philip Peter Baker (1759-1837) and E Dorothea Volkner (1765-1842) **P**: No **BLW**: No **PH**: No **SS**: A rec 484; B pg 79; S pg 106 **BS**: 245

*BAKER, John; b 31 Jan 1784, d 2 Apr 1849, Mt Crawford **RU**: Private, 58th VMR, Capt Robert Hooke's Co, Rockingham County, attached to McDowell's Flying Camp **CEM**: St Jacobs- Spaders Lutheran Church; Rockingham; Mount Crawford **GS**: Yes **SP**: Mar (1 Nov 1810) Mary Spader (1788-3 Jun 1879) **VI**: No further data **P**: Spouse appl **BLW**: Yes **SS**: B pg 181; BD pg 84 **BS**: 245

BAKER, John B, b 12 Nov 1767, bur 14 May 1856 **RU: Private, 74th VMR, Capt James Mallory's Co, Hanover County **CEM**: Hollywood; Richmond City; 412 S Cherry St, Sec I, lot 2 **GS**: Unk **SP**: Mar 1) Mary (__), (1786-1 Nov 1871) age 75, Sec I, lot 14; 2) (Frederick Co, 5 Dec 1815) Alcinda Louisa Tapscott, may have also have married Adeline (__) (Germany, 1784-21 Apr 1856) **VI**: No further data **P**: No **BLW**: No **PH**: No **SS**: K pg 176 **BS**: 237; 260

*BAKER, Henry; b 12 Jul 1795, d 11 Jun 1846 **RU**: Private; Frederick Co VMR, company of which was attached to either Flying Camp (McDowell) or 5th VMR (Mason & Preston) **CEM**: Mt Hebron; Winchester; 305 E Boscawen St **GS**: Yes **SP**: No spouse info **VI**: Nearby Hampshire Co, WVA (then VA) had a private Henry Baker in the war so perhaps him. He is buried in the Centenary Reformed UCC portion of Mt Hebron Cemetery. A Rev War veteran, Henry Baker is buried in the cemetery, perhaps his father **P**: No **BLW**: No **PH**: No **SS**: A rec 407, 420; B pg 78-80 **BS**: 245

*BAKER, Isaac; b 2 Dec 1787, d 26 Jun 1861 **RU**: Private, 97th VMR, Capt Samuel Colville's Co, Shenandoah County, attached to 6th VMR (Coleman) **CEM**: St. Stephens; Shenandoah; Strasburg Junction **GS**: Yes **SP**: Savilla (__) (__-1879) **VI**: No further data **P**: Widow applied pen (#WO-31010), but died before receiving it **BLW**: No **PH**: No **SS**: B pg 184; BD pg 83 **BS**: 245

*BALDWIN, Robert Thomas; b 4 May 1793, d 11 Sep 1863 **RU**: Surgeon's Mate, 31th VMR, commanded by Lt Col Henry Beatty, Frederick Co, attached to his 4th VMR at Norfolk **CEM**: Mount Hebron; Winchester; 305 E Boscawen St **GS**: Yes **SP**: 1) Sally Makay 2) (22 Feb 1830) Portia Lee Hopkins 3) (Aug 1807-14 Dec 1885) **VI**: No further data **P**: Widow Portia recd pen 15 Jun 1878, under Act of Mar 1878, # 23601 **BLW**: Yes **PH**: No **SS**: A rec 1689; B pg 78; M pg 18 **BS**: 245

BALL, George Washington; b 20 Mar 1789; d 1815 **RU: Captain, 57th VMR, Company Commander, Troop of Cavalry, Loudoun Co, attached to Green's Regiment **CEM**: Ball Family; Loudoun; Leesburg **GS**: Yes **SP**: Mary Randolph **VI**: Son of Colonel Burgess Ball and Frances Ann Washington **P**: No **BLW**: No **PH**: No **SS**: A rec 2106 **BS**: 245

*BALL, Moses III; b 12 Dec 1768, d 9 Jul 1840 **RU**: Private, 94th VMR, Lt John Graham's Co, Lee County, attached to 4th VMR (McDowell, Koontz, Chilton) **CEM**: Chadwell Station Baptist Church; Lee; 11 mi E of Cumberland Gap, RT 58, top of hill at church site **GS**: Unk **SP**: Mar (17 Feb 1789) Elizabeth Yeary, daug of Henry Yeary and Elizabeth Croxstall **VI**: Son of George Ball and Keziah Ann Hanson **P**: No **BLW**: No **PH**: No **SS**: A rec 2225; B pg 118 **BS**: 245

*BANKS, John; b 1787, d 1872 **RU**: Private, 54th VMR, Captain John West's Co, Norfolk Borough, attached to 8th VMR (Magnien) **CEM**: Elmwood; Norfolk City; 228 E Princess Anne Rd **GS**: No **SP**: Mar (20 Oct 1825) Sarah K Martin (1791-16 Jun 1831) **VI**: No further data **P**: Yes **BLW**: No **PH**: No **SS**: A rec 402; B pg 146; BD pg 146 **BS**: 245

*BANKS, William Tunstall; b 17 Feb 1788, d Jan 1859 **RU**: Surgeon Master, 6th VMR (Ritchie's) **CEM**: Piedmont Episcopal: Madison; 214 Church St **GS**: Yes **SP**: Pamela Harris (17 Mar 1794-10 Aug 1824) **VI**: A physician **P**: No **BLW**: No; **PH**: No **SS**: A rec 451 **BS**: 30

Key: *Additional veteran entry   **Corrected veteran entry   ***Deleted veteran entry
RU=Rank/Unit   CEM=Cemetery   GS=Gravestone   SP=Spousal Information   VI=Other Veteran Info   P=Pension
BLW=Bounty/Land Warrant   PH=Photo   SS=Service Source   BS=Burial Source   VMR= VA Military Regiment
LNR= Last Known Residence

*BARKSDALE, Claiborne; b 25 Aug 1783, d 1839 **RU**: Second Lieutenant, 26th VMR, Capt Grief Barksdale's Co, Charlotte County, attached to the 4th VMR (Greenhill) **CEM**: South Isle; Charlotte; 2471 Ridgeway Rd; Brookneal **GS**: Yes **SP**: No spouse info **VI**: No further data **P**: No **BLW**: No **PH**: No **BS**: A rec 2105; B pg 57 **BS**: 245

*BARE, Jacob; b 30 Dec 1772 d 21 Jul 1841 **RU**: Private, Captain William Harrison's Company 116th VMR, Rockingham Co., attached to 1st VMR (Truehart) **CEM**: Radar Lutheran Church; Rockingham; Timberville **GS**: Yes **SP**: No spouse info **VI**: No further data **P**: No **BLW**: No **PH**: No **SS**: B pg 181; K pg 51 **BS**: 31; 245

*BARLEY, Frederick; b c1773, d Feb 1847 **RU**: Private, Frederick Co militia company attached to 1st VMR (Taylor) **CEM**: Mt Hebron; Winchester; 305 E Boscawen St **GS**: Yes **SP**: No spouse info **VI**: No further data **P**: No **BLW**: No **PH**: No **BS**: A rec 2163; B pgs 78-80 **BS**: 86 pg 51

*BARLEY, Jacob; b 24 Dec 1797, d 11 Jan 1879 **RU**: Private, 58th or 116th VMR, Capt Robert Erwin's Co, Frederick Co, attached to 2d VMR (Ballowe) **CEM**: Barley family; Frederick; 1392 Martz Rd, 100 yds uphill **GS**: Yes **SP**: Mar (10 Mar 1820) Phoebe Hurton **VI**: Enlisted Harrisonburg, 28 Jul 1814 serving 6 mos. Recd pen 6 Jun 1872, and BLW **P**: Yes **BLW**: Yes **PH**: No **BS**: A rec 2164; B pgs 181; M vol 1; pgs 21; 22 **BS**: 245

*BARNES, Henry; b unk; d 5 Jul 1840 **RU**: Lieutenant, 82nd VMR, Capt Joseph Brock, Troop of Cavalry, Madison Co, attached to 1st VMR (Clarke) **CEM**: Piedmont Episcopal; Madison; 214 Church St **GS**: Yes **SP**: Mar 1) (__) Gibbs, 2) (21 Sep 1833) Letitia Ann Rapley (__-7 May 1889) LNR Slate Mills, Rappahannock Co, 1789 **VI**: Captain rank on gravestone, He was one of the first members and vestryman at the Protestant Episcopal Church **P**: Widow **BLW**: No **PH**: No **SS**: A rec 349; B pg 126; BD pg102; M pg 23 **BS**: 30 Church Cem

*BARNETT, Lawson; b 15 Feb 1791; d 12 Jul 1864 **RU**: Corporal, 82nd VMR, Madison Co **CEM**: Barnett Family; Madison; 4.4 miles NW of Wolftown on Rt 662 **GS**: Yes **SP**: Catherine D (__), (19 Sep 1794-1 May 1877) **VI**: Son of A.G. and Ann (__) Barnett **P**: No **BLW**: No **PH**: No **SS**: A rec 843 **BS**: 30; Family Cem

*BARNETT, Robert; b 30 May 1796, d 1 Jul 1871 **RU**: Private 75th VMR, Capt James Hoge's Co, Montgomery Co, attached to 4th VMR,(Huston-Wooding) **CEM**: Piedmont; Montgomery; vic jct Rts 637 and 653, Otey, Piedmont **GS**: Unk **SP**: Mar 1) Nancy Willis 2) (Christiansburg, 19 Mar 1834) Elizabeth Jewell (30 May 1809- 25 Dec 1898), daug of Thomas Jewell (1764-1853) and Elizabeth Graham (1775-__) **VI**: No further data **P**: Both **BLW**: Yes **PH**: No **SS**: A rec 863; B pg 138; M pg 22 **BS**: 245

*BARNHOUSE, Richard; b 1789, St. Mary's Co., MD, d 24 Dec 1845 **RU**: Private, Captain Duvall's Company, Maryland Militia **CEM**: Bethel United Methodist Church; Loudoun; Stumptown **GS**: Unk **SP**: Mar (16 Jan 1816) Margaret Jane White (1755-1817) **VI**: Son of George Barnhouse, discharged (Georgetown, DC, 9 Nov 1814) **P**: No **BLW**: No **PH**: No **SS**: A rec 1077 **BS**: 245

*BARRET, William; b 29 Nov 1786, d 20 Jan 1871 **RU**: Major, 40th VMR, Louisa Co, Staff officer commissioned 1 Jan 1814 **CEM**: Hollywood; Richmond City; 14 So Cherry St **GS**: Yes **SP**: No spousal info **VI**: Son of John Barret (19 Mar 1748-9 Jun 1830). Operated large tobacco Co in Richmond **P**: No **BLW**: No **PH**: No **SS**: B pg 123 **BS**: 245

*BASS, Thomas William; b 1790, d Nov 1865 **RU**: Private, 23rd VMR, Capt Edward Johnson's Co and Capt Benjamin Good's Co, Chesterfield County attached to 2d VMR (Amber-Brown) **CEM**: Bass Family; Petersburg City; Bass St, on Appomattox River at Exeter Mills **GS**: No **SP**: Mary Ann Cosby (1827-1902) **VI**: Son of William Bass (10 May 1763-10 May 1839) and Sarah Judith Shackleford (1774-__) **P**: No **BLW**: No **PH**: No **SS**: A rec 1323; 1327; B pg 61; L pg 369 **BS**: 245

BASSETT, Alexander Hunter; b 22 Nov 1795; d 8 Oct 1880 **RU: Private, 64th VMR, Henry Co **CEM**: Bassett Family; Henry; Rt 683, Bassett **GS**: Unk **SP**: 1) Mary Koger (1763-1863) 2) (3 Dec 1872) Ann R Hardy (__-1880). LNR Spencer's Store, Henry Co **VI**: Son of Burwell Bassett (1768-1816) & Mary Hunter (1777-1866) **P**: Both **BLW**: Yes **PH**: No **SS**: A rec 1415; M pg 24; BD pg 117 **BS**: 245

Key: *Additional veteran entry   **Corrected veteran entry   ***Deleted veteran entry
RU=Rank/Unit   CEM=Cemetery   GS=Gravestone   SP=Spousal Information   VI=Other Veteran Info   P=Pension
BLW=Bounty/Land Warrant   PH=Photo   SS=Service Source   BS=Burial Source   VMR= VA Military Regiment
LNR= Last Known Residence

*** **BATES**, Fleming: deleted as service not adequate

***BATIS**, Charles; b 1794, d 13 Oct 1875 **RU**: Private, 32nd VMR. Capt Alexander Given's Co, Augusta County, attached to 5th VMR (McDowell's) **CEM**: Oak Lawn; Augusta; Rt 617, New Hope **GS**: Yes **SP**: Sallie Nutty (1797-14 Mar 1862) **VI**: LNR New Hope 1871, widower. Drew pension and BLW of 40, 1850 and 120 acres, Act of 1855 **P**: Yes **BLW**: Yes **PH**: No **SS**: M pg 25; O- BLW Fold3 **BS**: 245

***BEAHM**, Martin R; b 10 Jan 1794, Shenandoah Co, d 4 Mar 1880 **RU**: Private, 51st VMR. Capt James Sower's Co, Frederick Co, attached to 4th VMR & McDowell's Flying Camp **CEM**: Greenmount; Rockingham; Rt 772 **GS**: Yes **SP**: Mar (New Market, 9 Jun 1823) Christiana Neff or Baughman (23 Dec 1807-13 Feb 1872) **VI**: Son of Jacob Beam & Catherine (__). Recd pension and BLW **P**: Yes **BLW**: Yes **PH**: No **SS**: A rec 1889; B pg 80; M pg 26; O- BLW & pen files **BS**: 245

***BEALE**, William Churchill; b 1791, d 22 Apr 1850 1875 **RU**: Private 45th VMR. Capt Lewis Alexander's Co **CEM**: Fredericksburg City; Fredericksburg City; entrance Washington Ave **GS**: Yes **SP**: 1) Susan Vowles, 2) Jane Briggs Howison (1815-17 Jan 1882) **VI**: Was issued a BLW. Home was in Falmouth where he owned a mill **P**: Spouse Jane **BLW**: Yes **PH**: No **SS**: B pg 190; L pg 82; M pg 26; **BS**: 245

*****BEADLES**, John: Deleted as service identified belonged to his son

***BEATTY**, Henry; b 1759, d 1840 **RU**: Lt Colonel; 4th VMR, as commander, with service at Norfolk & Craney Island & 31st VMR Commander, Frederick Co **CEM**: Mt Hebron; Winchester; 305 E Boscawen St **GS**: Yes **SP**: Sarah Henning (__-1824) **VI**: He commanded the Regiment in the battle at Craney Island in June 1813 and for his accomplishment was presented a sword by U.S. Congress. He owned a saddle shop in Winchester after the war and was an elder in the Presbyterian Church, Was a Revolutionary War soldier **P**: No **BLW**: No **PH**: No **SS**: B pg 78, 226; **BS**: 245

***BELL**, Buchanan; b 1794, d 29 May 1882 **RU**: Private, Capt John Scurdy's Co of Mounted Gunman, Col Williamson' Regt serv in battles at Pensacola and New Orleans **CEM**: Bell Family; Craig; Rt 632 Upper John Creek's Rd, just N of jct with Rt 633 **GS**: Yes wooden cross **SP**: 1) Sarah Snodgrass (1793-20 Aug 1879), 2) Nancy Carr (1806-1890) **VI**: Son of Jeremiah Bell and Elizabeth Sarver. Recd pen and BLW dated Jul 1851 **P**: Yes **BLW**: Yes **PH**: No **SS**: M pg 29; O- BLW file **BS**: 245

***BELL**, David; b 1794, d 3 Jul 1848 **RU**: Private, 4th VMR, Capt John Bonnett's Co or Lt Peregrine's Co, Ohio County (now WVA) attached to 6th VMR (Dickinson, Coleman, Scott) at Norfolk **CEM**: Cedar Grove; Norfolk City; Cnr Salter St & E Princess Anne Rd **GS**: Yes **SP**: Mar (Norfolk, 28 Oct 1833) Catherine Farber (1795-3 Jun 1848) VI: No further data **P**: No **BLW**: No **PH**: No **SS**: A rec 349; B pg 155 **BS**: 245

****BELL**, James; b 4 Jun 1773; d 16 Jan 1856 **RU**: 2nd Lieutenant, 5th VMR (McDowell) **CEM**: Old Stone Presbyterian; Augusta; Rt 11, Fort Defiance **GS**: Yes **SP**: Mar 1) Sarah Allen, 2) Sarah Crawford, 3) Margaret Craig (25 Dec 1788-27 Feb 1856) **VI**: James Bell, Esquire **P**: No **BLW**: **PH**: N **SS**: B pg 78 **BS**: 1 Pt 2 pg 5

***BELL**, John Jr; b 21 Feb 1773, d 10 Mar 1838 **RU**: Private 45th VMR, Capt Lewis Alexander's Co, Stafford Co Militia **CEM**: Mount Hebron; Westminster City; 305 E Boscawen St **GS**: Unk **SP**: No spousal info however first child b 1810 **VI**: Was on the pension roll of 1835 **P**: Yes **BLW**: No **PH**: No **SS**: B pg 190; M pg 82 **BS**: 245

***BELL**, Joseph, Jr; b 25 May 1776, d 18 Apr 1855 **RU**: Sergeant, 8th VMR, Capt Archibald or Capt Isiah McBride's Co, Rockbridge County attached to 5th VMR (McDowell) **CEM**: Bell Family; Rockbridge; nr Cameron Hall across from Iron Bridge on left end of rd, Goshen **GS**: Unk **SP**: Mar (Staunton) Mary Ann Nelson (14 Apr 1785-25 Apr 1841), daug of Alexander Nelson and Ann Mathews **VI**: Son of Joseph Bell, Sr (25 May 1742-18 Apr 1823) and Elizabeth Henderson (14 Jun 1746-13 Sep 1833) **P**: No **BLW**: No **PH**: No **SS**: A rec 696; B pg 189 **BS**: 245

****BELL**, William; b 11 Feb 1772, Northumberland Co, d 25 Feb 1851 **RU**: Private, 85th VMR, Fauquier Co **CEM**: Oak Springs; Fauquier; 770 Fletcher Dr, Delaplane **GS**: Yes **SP**: No spouse info **VI**: No further data **P**: No **BLW**: No **PH**: No **SS**: A rec 924 **BS**: 3 pg 10

Key: *Additional veteran entry   **Corrected veteran entry   ***Deleted veteran entry
RU=Rank/Unit   CEM=Cemetery   GS=Gravestone   SP=Spousal Information   VI=Other Veteran Info   P=Pension
BLW=Bounty/Land Warrant   PH=Photo   SS=Service Source   BS=Burial Source   VMR= VA Military Regiment
LNR= Last Known Residence

9

*BENNETT, Bartlett; b 10 Jul 1787, d 28 Mar 1859 **RU:** Private, 101st VMR, Capt Edward Carter's Trp of Cav, Pittsylvania Co, attached to 1st VMR (Holcombe) **CEM:** Via, Patrick; N side Rt 57 vic jct Rt 635 **GS:** Unk **SP:** Mar (Pittsylvania Co, 3 Jun 1817) Mary Brown (__-c1883), LNR Buffalo Ridge, Patrick Co 1878 **VI:** No further info **P:** Widow **BLW:** Yes **PH:** No **SS:** B pg 161; M pg 30; O- pen File **BS:** 245

*BENNETT, John; b bef 24 Apr 1763, Fauquier Co, d May 1845 **RU:** Major, 101st VMR, Staff Officer, commissioned 3 Feb 1812, Pittsylvania Co **CEM:** Bennett-Lewis; Pittsylvania; Off Rt 715 toward Grassy Branch, Keeling **GS:** No **SP:** Mar (Pittsylvania Co, 12 Mar 1795) Mary Lewis (__-21 Apr 1851) **VI:** Son of John Bennett (_-1786-1787) and Elizabeth Dodson (1732, North Farnham, Richmond Co-__). Was a private in Rev War **P:** No **BLW:** No **PH:** No **SS:** B pg 161; **BS:** 245

*BENSON, John B; b 1789, Spotsylvania Co, d 28 Feb 1832 **RU:** Private, 16th VMR, Capt Anthony R. Thornton's Co, Spotsylvania County **CEM:** Maplewood; Charlottesville City; Cnr Lexington Ave and Maple St **GS:** Unk **SP:** No spouse info **VI:** No further data **P:** No **BLW:** No **PH:** No **SS:** B pg 189; L pg 774 **BS:** 245

*BERKELEY, Robert; b 21 Apr 1776, Airwell, Hanover Co, d 01 May 1818 **RU:** Private, 31st VMR, Capt William Morris, Artillery, Frederick Co, attached to Battalion of Artillery, Captain Arthur Emmerson's Company **CEM:** Mt Hebron; Winchester; 305 E Boscawen St **GS:** Yes **SP:** Julia Carter (18 Apr 1783, Nominy, Westmoreland Co-25 Aug 1855), daug of Robert Carter **VI:** Son of Nelson Berkeley and Elizabeth Wormley, daug of Landon Carter and Elizabeth (__) of Sabine Hall **P:** No **BLW:** No **PH:** Yes **SS:** A rec 2054; B pgs 79; 235 **BS:** 245

*BERRY, Henry, Jr; b 1782, NJ, d 29 Aug 1864 **RU:** Private 13th or 97th VMR, Capt Daniel Strickler's Company, Shenandoah Co, attached to 6th VMR (Coleman) **CEM:** Van Lear; Augusta; Rt 613 S 1/3 mi fr jct with Rts 742 on farm 300 yds fr road **GS:** Yes **SP:** Mar (Rockingham Co, 25 Jul 1825) Catherine Wise who recd pen and also resided Nelson Co **VI:** Son of Henry Berry (1750-1811) **P:** Widow **BLW:** No **PH:** No **SS:** A rec 1274;; B pg 185; BD pg 156 **BS:** 245

*BERRY, John; b 2 Sep 1791, d 15 Oct **RU:** Private, 5th VMR (McDowell's) **CEM:** Tinkling Spring Presbyterian Church; Augusta; Fisherville **GS:** Yes **SP:** Susan Henderson McCombs (17 Feb 1807-11 Dec 1863), daug of James & Susannah Henderson Mc Combs **VI:** No further data **P:** No **BLW:** No **PH:** No **SS:** A rec 559, B pg 39-40 **BS:** 245

*BERRY, John S; b 1789, d 7 Sep 1874 **RU:** Private, 116th VMR, Daniel Matthew's Co, Rockingham Co, attached to 4th VMR, (McDowell's Flying Camp) & (McDowell, Koontz, Chilton) **CEM:** Keezletown; Rockingham; Keezletown **GS:** Yes **SP:** Harriett Bolton (1809-1872) **VI:** Son of Benjamin Berry (1765-1834) & Johanna Berry (1765-1810); enlisted (29 Sep 1813), discharged (10 Jan 1814) **P:** Yes **BLW:** Yes **PH:** No **SS:** BD pg157 (SC-3286); B pg 182; M pg 32 **BS:** 245

BERRY, Richard J; b c1785; d aft 1850 **RU: Private, 62th VMR, Capt Daniel's Co **CEM:** Berry Family #2; Stafford; jct Rts 625 & 626 **GS:** No **SP:** Perhaps Jemina (__) listed in 1850 census **VI:** Son of Oscar Berry and Lucy (__) **P:** No **BLW:** No **PH:** No **SS:** L pg 4324 **BS:** 26 pg 149

*BEST, John Henry; b c1776, d 02 Jan 1852 **RU:** Private, 62nd VMR, Capt Daniel Eppes Co, Prince George County **CEM:** Atwood Farm; Prince George; .5 mi back of home in field, Disputanta **GS:** No **SP:** Mar (26 Jul 1821) Elizabeth D Simmons, LNR P.O. Disputanta, Prince George Co, 1878. She recd pen #20.288, Jun 1879 **VI:** Recd BLW 160 acres **P:** Widow **BLW:** Yes **PH:** No **SS:** A rec 1966; B pg 169; K pg 212; M pg 32; O- pen files, Fold3 **BS:** 245

*BIBLE, John; 1 Oct 1790, d 29 Dec 1853 **RU:** Private, 116th VMR, Captain Thomas Hopkin's Company, Rockingham Co., attached to McDowell's Flying Camp **CEM:** Radar Lutheran Church; Rockingham; Timberville **GS:** Yes **SP:** Mar 1) Sarah Branner (1795-1842) 2) 29 Aug 1829, Polly Tafflinger **VI:** Son of Johann Adam Bible (1759-1826) and Magdalena Shoemaker (1764-1840) **P:** Yes **BLW:** No **PH:** No **SS:** K pg 19; M pg 32 **BS:** 31; 245

Key: *Additional veteran entry   **Corrected veteran entry   ***Deleted veteran entry
RU=Rank/Unit   CEM=Cemetery   GS=Gravestone   SP=Spousal Information   VI=Other Veteran Info   P=Pension
BLW=Bounty/Land Warrant   PH=Photo   SS=Service Source   BS=Burial Source   VMR= VA Military Regiment
LNR= Last Known Residence

*BICKLE, Adam; b 20 Jun 1793, d 11 Nov 1859 **RU:** Private, 32nd VMR, Captain John Sower's Artillery Co, Staunton Artillery, Augusta Co, attached to Battalion of Artillery and Corporal in 32 VMR, Augusta Co. attached to 5th VMR **CEM:** Thornrose; Staunton City; 1041 W Beverly St **GS:** Yes **SP:** Margaret Gold (1798-1860) **VI:** No further data **P:** No **BLW:** No **PH:** No **SS:** A rec 213 **BS:** 245

*BICKLEY, John; b 4 Nov 1790, Castlewood, Russell Co, d 30 Aug 1864 **RU:** Ensign, 72nd VMR, Captain Andrew Caldwell, Russell Co, attached to 4th VMR (Mc Dowell, Koontz, Chilton) **CEM:** Bickley; Russell; Castlewood **GS:** Y **SP:** Elizabeth Brown (1795-13 Mar 1844) **VI:** Son of Charles Bickley (1753-1839) **P:** No **BLW:** No **PH:** No **SS:** A rec 232; B pg 183 **BS:** 245

*BIGGER, Thomas Bibb; b 22 Feb 1795, Prince Edward Co, d 05 May 1880; **RU:** Private, Capt Richard McRae's Co, Petersburg Volunteers **CEM:** Shockoe Hill, Richmond City; 100 Hospital St **GS:** Yes **SP:** Mar (14 Mar 1824) Elizabeth Meredith Bigger (4 Sep 1806-6 Mar 1875), daug of Armistead Russell and Sally Meredith **VI:** Richmond Postmaster Mar 15 1845 for 20 yrs; recd BLW and pension **P:** Yes **BLW:** Yes **PH:** No **SS:** AK pg 138; M pg 33; B pg 28 **BS:** 245

*BILLER, Christian C; b 17 Dec 1792, d 15 Aug 1873 **RU:** Private 13th VMR, Captain Reuben Morris's Co, Shenandoah County, attached to 6th VMR (Coleman) **CEM:** Tomahawk Pond (AKA Barb Schoolhouse & Biller-Estep); Shenandoah; Rt 610; Orkney Springs; **GS:** Yes, broken **SP:** Mar (Rockingham Co, 28 Nov 1815) Hannah Catherine Price (1799-1867) **VI:** Recd pen Act of Feb 1871 # 21.421 and BLW 40 acres on 27 Mar 1855. LNR P.O. Mt Clifton, Shenandoah Co **P:** Y **BLW:** Yes **PH:** No **SS:** A rec 891; M pg 33; O- pen & BLW files **BS:** 245

*BISHOP, Elijah; b 5 Sep 1793, Washington Co, d 13 Apr 1862 **RU:** Private, Captain Jillian Hamilton's Co, 2d Regt (Lillard's) East TN Militia. Enl 12 Oct 1813, discharged 8 Feb 1814 **CEM:** Bishop Family; Lee; Rt 665 behind Mt Moriah Ch, Jonesville **GS:** Yes **SP:** Mar (3 Oct 1811) Lavinia Clark (30 Mar 1795-1Aug 1886), daug of Robert Clark (1770-1857) and Judith Weaver (1768-__). Pen WO 10155 & WO-31040 **VI:** Recd BLW #36411-80 acres and #6912 80 acres **P:** Widow **BLW:** Y **PH:** No **SS:** A rec 2069; M pg 34; O BLW and Pen files **BS:** 245

*BISHOP, James Jr; b 1790, d 1 Nov 1855 **RU:** Private, 83rd VMR, Capt Theodoric Walker's Co, Dinwiddie Co **CEM:** Historic Bishop Family; Dinwiddie; jct Sapony Rd and Mealy Branch **GS:** Yes **SP:** No spouse info **VI:** Son of James Bishop, Sr **P:** Yes **BLW:** N **PH:** No **SS:** A rec 2159; B pg 46; L pg 803; O- pen files **BS:** 245

*BISHOP, John; b 1767, d 1822 **RU:** Sergeant, 98th VMR, served in a company of Mecklenburg County, under the command of LTC Grief Green **CEM:** Bishop Family; Brunswick; Rt 644 GPS: 36,62640,-77.95580 **GS:** Unk **SP:** Elizabeth Keene Jones (__-1850, AR) **VI:** Son of Mathew Bishop (1747-1810) and Martha Evelyne Mc Cullar (1750-1828) **P:** No **BLW:** No **PH:** No **SS:** A rec 2205; B pgs 130; 131 **BS:** 245

*BLACKBURN, Thomas Rolander; b 23 Dec 1794, Jefferson Co, WV, d 28 Aug 1867 **RU:** Sergeant, 16th VMR, Capt George Hamilton's Co, Spotsylvania County, attached to Major Stapleton Crutchfield's Detachment, Aug- Dec 1814 **CEM:** Thornrose; Staunton City; 1041 West Beverly St **GS:** Yes **SP:** Mar (Spotsylvania Co, 6 Apr 1817) Mary Ann H Wright (5 Mar 1796-14 Feb 1879), Recd widow pen **VI:** Son of William Blackburn and Mary Adham. Was architect under tutelage of Thomas Jefferson and designed Augusta Co C.H. in 1835, Western State Hospital and many homes in Staunton. He recd BLW **P:** Widow **BLW:** Yes **PH:** No **SS:** B pg 188; O- BLW and pen files Fold3 **BS:** 245

*BLACKWELL, Joseph; b 1795, d Dec 1875 **RU:** Private 70th and 105th VMR, served in a company of Washington County attached to the 5th VMR (Mason-Preston) **CEM:** Blackwell Chapel United Methodist Church; Washington; Vic jct Rts 700 and 746; At 8329 Blackwell Chapel Rd, Meadowview **GS:** Unk **SP:** No spousal info **VI:** Son of William B Blackwell and Mary Polly Smith **P:** No **BLW:** No **PH:** No **SS:** A rec 1152; B pgs 198; 199 **BS:** 245

Key: *Additional veteran entry   **Corrected veteran entry   ***Deleted veteran entry
RU=Rank/Unit   CEM=Cemetery   GS=Gravestone   SP=Spousal Information   VI=Other Veteran Info   P=Pension
BLW=Bounty/Land Warrant   PH=Photo   SS=Service Source   BS=Burial Source   VMR= VA Military Regiment
LNR= Last Known Residence

*BLACKWELL, Moses, Jr; b 1 Mar 1794, Bedford Co, d14 Mar 1863 **RU:** Private, 10th or 91st VMR, drafted Apr 1813, Bedford Co in company of one of the VMRs and was used for hauling salt from Washington Co to Norfolk (Widow pen applic) **CEM:** Red Oak Grove; Floyd; Rt 684 to top of hill past Dobbin's farm another 1/2 mi **GS:** Yes **SP:** Mar 1) (26 Nov 1812) Elizabeth Williamson (1790-1835), 2) (18 Aug 1836) Rebecca Walter (1813-1885), daug of Martin Walter and Christina Clem **VI:** Son of Moses Blackwell, Sr and Susan Wall **P:** Widow Rebecca **BLW:** Yes **PH:** No **SS:** B pgs 42, 43; M pg 35; O- pen Applic, Fold3 **BS:** 245

*BLACKWELL, Samuel B; b 27 Apr 1785, d 23 May 1837 **RU:** Lieutenant, 37th VMR, Captain Samuel Downing's Co, Northumberland County **CEM:** Roseland; Northumberland; Reedville **GS:** Yes **SP:** Mar (Northumberland Co, Nov 1818) Ann (__) (__-bef May 1837) **VI:** Obtained title of Colonel; He and his family were first buried at his home site in Burgess and which later were removed to the Roseland cemetery. One of the graves moved was that of William Blackwell that may have been his Revolutionary War father **P:** No **BLW:** No **PH:** No **SS:** A rec 1174; B pg 152 **BS:** 271 pg 197

*BLAIR, John; b 5 Jul 1772, Somerset Co, PA, d 12 Jul 1852 **RU:** Private, Montgomery's Regt, PA Militia **CEM:** Blair Family; Carroll; Sylvia Drive off Rt 721 **GS:** Yes **SP:** Charity Bourne (1 Nov 1776-19 Jul 1860), daug of William Bourne and Rosamond Jones **VI:** Son of Thomas Blair and Mary Ann Jones **P:** No **BLW:** No **PH:** No **SS:** A rec 1425 **BS:** 245

*BLAIR, Samuel Brittain; b 11 Apr 1789, d 22 Mar 1870 **RU:** Private, 101st VMR, Captain Edward Carter's Troop of Calvary, Pittsylvania Co attached to VMR of Calvary (Holcombe) **CEM:** Samuel Blair Family; Pittsylvania; Rondo **GS:** Yes **SP:** Mar 1) (26 Jul 1810) Mary (Polly) S Reynolds (1791-1833), 2) 22 Feb 1834, Clarissa W, Fuller (1802-1890), LNR P.O. Callards, Pittsylvania Co **VI:** Son of (__) Blair and Sarah Suter (1760-1807) **P:** Widow Clarissa **BLW:** Yes **PH:** No **SS:** B pg 161; L pg 200; O- pen applic file; Fold3 **BS:** 245

*BLAIR, Thomas; b 1789, d 20 Sep 1876 **RU:** Private, 78th VMR, Captain James Anderson's Co, Grayson County **CEM:** Blair Family; Carroll; Cliffview **GS:** No **SP:** Sallie Patton **VI:** Son of John Blair (1772-1852) and Charity Bourne (1776-1860) **P:** No **BLW:** No **PH:** No **SS:** B pg 86 **BS:** 245

*BLAIR, Walter Dabne; b 9 Jul 1797, d 3 Apr 1878 **RU:** Private, 4th VMR **CEM:** Shockoe Hill; Richmond City; 100 Hospital St **GS:** Yes **SP:** Louisa Edmonna Willis (1 Oct 1804-14 Feb 1886) **VI:** Son of John Durbarrow Blair (1759-1823) and Mary Winston (1763-1831) **P:** No **BLW:** No **PH:** Yes **SS:** A rec 1520 **BS:** 245

*BLAKEMORE, Joseph; b 29 Mar 1793, d 27 Jul 1868 **RU:** Private, 13th or 97th VMR, Captain Joshua Ruffner's Co, Shenandoah County, attached to 6th VMR (Coleman) **CEM:** Prospect Hill; Warren; 200 West Prospect St, Front Royal **GS:** Yes **SP:** Mar (18 May 1814) Polly Connell (__-aft 1871), LNR Linden, Warren Co 187, recd pen Act of 14 Feb 1871 # L718 **VI:** Recd BLW **P:** Widow **BLW:** Yes **PH:** No **SS:** A rec 1972; B pg 184,185; O- pen files Fold3 **BS:** 245

*BLAND. Edward; b1774, Fergusonville, Nottoway, d 24 Feb 1850 Centreville, Nottoway Co **RU:** Private, 46th VMR, Captain Jesse Heinkle's Co, Pendleton Co, WV. **CEM:** Bland Family; Nottoway; Centreville **GS:** No **SP:** Mar (Amelia Co, 18 Sep 1814) Mary P Perkinson, LNR Petersburg **VI:** Son of Edward Bland (16 Dec 1746-1795) **P:** Widow # 18.129 **BLW:** Yes **PH:** No **SS:** B pg 261; M pg 37 **BS:** 245

*BLAND. Richard; 1792, d 29 Jun 1864 **RU:** Corporal 62nd VMR, Captain Edward Mark's Troop of Cavalry, Prince George Co. **CEM:** Hollywood; Richmond; 100 Hospital St **GS:** Yes **SP:** Mar 1) Adeline Manton (__-1827 or 1828), 2) (Prince George Co, 13 Mar 1829) Martha Elizabeth Ledbetter (1812- 25 Jan 1889) **VI:** Son of Richard Bland (1762-1806) and Susannah Poythress (1769-1839) **P:** Widow # 47.666 **BLW:** Yes **PH:** No **SS:** A rec 2417; B pg 170; L pg 562; M pg 37 **BS:** 245

Key: *Additional veteran entry   **Corrected veteran entry   ***Deleted veteran entry

RU=Rank/Unit	CEM=Cemetery	GS=Gravestone	SP=Spousal Information	VI=Other Veteran Info P=Pension
BLW=Bounty/Land Warrant PH=Photo		SS=Service Source	BS=Burial Source	VMR= VA Military Regiment
LNR= Last Known Residence				

*BLEDSOE, Isaac R.; b 1795, NC, d 1883 **RU**: Private, 94th VMR, Capt Jeremiah Skelton's Co, Lee County, attached to Bradley's Regt **CEM**: Bledsoe family; Scott; Rt 600 bef Neeley Dr, end of road to right **GS**: Yes **SP**: Amelia Emily Wallen **VI**: Memorialized in cem as exact burial unk; two BLW one for 80 acres, one for 120 acres; pen later rejected as did not have enough service before end of war **P**: Yes **BLW**: Yes **PH**: Yes **SS**: B pg 118; M pg 38; O- BLW file Fold3 **BS**: 245

*BOARD, Henry; b 26 May 1794, Bedford Co, d 17 Jul 1880 **RU**: Private, 43rd VMR, Capt Cassimer Cabiness Co, Franklin Co **CEM**: Board Family; Franklin; Off unnamed road .25 mi from connecting to Horse Hollow Rd at jct with Rt 40 and Rt 703 **GS**: No **SP**: Nancy Majors (1794-1889), LNR 1880 Taylor's Store, Franklin Co **VI**: Recd BLW of 40 acres and 120 acres and pen **P**: Both **BLW**: Yes **PH**: Yes **SS**: B pg 76; M pg 39; O BLW file; Fold 3 **BS**: 245

*BOARD, John, III; b 27 Aug 1789, MD, d 29 Jan 1877 **RU**: Corporal, 10th VMR, Capt Joshua Early & Lt Nicodemus Leftwich, Bedford County, attached to 3rd VMR (Dickinson) **CEM**: Board Family, Bedford, Rt 655, Moneta **GS**: Unk **SP**: Mar 1) Elizabeth McCabe, 2) (Bedford Co, 23 Sep 1834) Cleopatra A. McDaniel **VI**: No further data **P**: Both **BLW**: Yes **PH**: No **SS**: A rec 1607; B pg 42; M pg 39; BD pg 190 **BS**: 245

*BOATRIGHT, Meador; b 19 Jul 1796, Cumberland Co, d 7 Dec 1876, Fort Blackmore, Scott Co **RU**: Private, 17th VMR, Capt Benjamin Allen's Co, Cumberland Co, attached to 1st VMR (Trueheart) **CEM**: Green Family; Scott; So of Lick Creek and just E of Rt 777, Duncan Mill **GS**: Yes **SP**: Mar (17 Sep 1828) Mary Maud Pendleton (Scott Co, May 1808-Fort Blackmore, 26 Feb 1903) She recd pen **VI**: Son of Valentine Boatwright (Hanover Co 1770- 28 May 1849) and Sarah Meador. He recd pen and BLW 1850-40 acres, 1855-120 acres **P**: Both **BLW**: Yes **PH**: No **SS**: B pg 64; M pg 39; O fold3 pen rec **BS**: 245

*BOAZ, Robert; b c1797, d Aug 1873 **RU**: Private, 24th or 100th VMR, Capt William Jones's Co, Franklin County, attached to 5th VMR (Mason & Preston) **CEM**: Boaz Family; Henry; Ridgeway **GS**: Yes (Govt) **SP**: Mar (Patrick Co, 17 Feb 1818) Martha W Sandifer **VI**: LNR P.O. Ridgeway, Henry Co, 1871. Recd pen #21739. Recd BLW, 40 acres #135816, Act of 1855 at age 58 **P**: Yes **BLW**: Yes **PH**: No **SS**: B pg 77; M pg 39; O- pen file, Fold3 **BS**: 245

BOGGS, James; b 26 Jul 1787; d 27 Mar 1855 **RU: Private, 2nd VMR (Bayley), Accomack Co **CEM**: Rodgers-Boggs; Accomack; located one mi north of Evans Wharf on east side county road 638 **GS**: Yes **SP**: Elizabeth P (__) (6 Oct 1791-26 Apr 1856) **VI**: Son of John & Mary (__) Boggs **P**: No **BLW**: No **PH**: No **SS**: A rec 2374 **BS**: 178; 129

***BOLES, William: Deleted as service or burial not satisfactory

BOOKER, George; b 14 Nov 1785; d 3 Mar 1848 **RU: Captain, 100th VMR, Company Commander, Buckingham Co, attached to 5th VMR (Mason & Preston) **CEM**: Booker / Main; Buckingham; Montrose, Rt. 653 **GS**: Yes **SP**: Unk wife is buried here **VI**: Son of Samuel Booker **P**: No **BLW**: No **PH**: No **SS**: B pg 50 **BS**: 66 pg 66

*BOOTH, George H; b 1788 Montgomery Co, d 14 Dec 1872 **RU**: Private, 48th or 121st VMR, a company of Botetourt Co, attached to 5th VMR **CEM**: Reed; Floyd; GPS: 36.94919,-80.38864; South across Stream fr Spirit Wind Industries, Rt 705 **GS**: Yes **SP**: Lucy Reed (1795-20Jun 1856), daug of Peter Reed and Lucy S (__) **VI**: Son of Abijah Booth and Rhoda Ellen Howard **P**: No **BLW**: No **PH**: No **SS**: A rec 1721; B pgs 45, 46 **BS**: 245

*BOOTH, William; b 30 Nov 1798, d 31 Jun 1866 **RU**: Private, 19th VMR, attached to 2nd VMR (Ballowe's) **CEM**: Shockoe Hill; Richmond City; 100 Hospital St **GS**: Yes **SP**: Miranda Wright (6 Oct 1802-15 Apr 1883) **VI**: No further info **P**: No **BLW**: No **PH**: Yes **SS**: A rec 1827 **BS**: 31

*BOOTWRIGHT (BOATWRIGHT), William; b 1792, d 17 Dec 1869 **RU**: Private, 19th VMR, Capt Robert Gamble's Troop of Cavalry, Richmond City **CEM**: Shockoe Hill; Richmond City; 100 Hospital St **GS**: Yes **SP**: No info **VI**: Gov't Gr Stone had incorrect service but is being replaced with corrected serv. Recd BLW #55 80 2200 **P**: No **BLW**: Yes **PH**: Yes **SS**: A rec 1990; G; O Serv Rec Fold3 **BS**: 31

Key: *Additional veteran entry   **Corrected veteran entry   ***Deleted veteran entry
RU=Rank/Unit   CEM=Cemetery   GS=Gravestone   SP=Spousal Information   VI=Other Veteran Info   P=Pension
BLW=Bounty/Land Warrant   PH=Photo   SS=Service Source   BS=Burial Source   VMR= VA Military Regiment
LNR= Last Known Residence

13

*BOSHER, William; b 9 Oct 1789, d 1 Jul 1884 **RU:** Private, 19th VMR, Capt Andrew Stevenson's Co, Richmond City, attached to 2nd VMR (Ballowe's) **CEM:** Mount Columbia; King William; off Rt 649 in Manquin **GS:** Unk **SP:** Mar (King William Co, 24 Dec 1811) Gabrilla H Lipscombe **VI:** Recd pen #28753 and BLW **P:** Yes **BLW:** Yes **PH:** No **SS:** B pg 175; M pg 43; O- pen file Fold3 **BS:** 245

*BOSSERMAN, Frederick; b 22 May 1786, d 04 Apr 1877 **RU:** Private, 48th VMR, Capt David Rowland's Co, Berkeley Co, attached to 4th VMR (Boyd's) **CEM:** St Johns Reformed United Church of Christ; Augusta; 1515 Arbor Hill Rd, Middlebrook **GS:** Yes **SP:** Mar (1811) Margaret Burbank **VI:** LNR P.O. Box Mint Springs, Augusta Co. Recd pen #12723, 1871, age 83 and BLW in 1856 **P:** Yes **BLW:** Yes **PH:** No **SS:** B pg 46; M pg 43; O- pen files Fold3 **BS:** 245

*BOSSERMAN, George, b 11 Jan 1793, Shenandoah Co, d 26 Nov 1875 **RU:** Private, 32nd or 93rd VMR, Capt Samuel Doka's Co, Augusta County, attached to 5th VMR (McDowell) **CEM:** St Johns Reformed United Church of Christ; Augusta; 1515 Arbor Hill Rd, Middlebrook **GS:** Yes **SP:** Mar (1 Oct 1816) Catherine Brubeck (1797-1881), LNR P,O, Box, Middlebrook, Augusta Co. Recd widow pen Act of 1878 **VI:** Recd pen and BLW of 40 acres **P:** Both **BLW:** Yes **PH:** No **SS:** B pg 39; M pg 43; O- pen BLW files Fold3 **BS:** 245

*BOULDIN, Robert E; b 15 Jun 1795, d 25 Mar 1881 **RU:** Private, 26th VMR, Capt George Hannah's Troop of Cavalry, Charlotte Co **CEM:** Mount Carmel United Methodist Church; Charlotte; Brookneal **GS:** Yes **SP:** Mar (Halifax Co, 20 Jan 1850) Sarah B Britton (__-1884), LNR Charlotte Co, recd widow pen # W31856 **VI:** Was Doctor, recd pen #22297 $8.00 per mo 25 Mar 1881 and BLW 40 acres # 135594 **P:** Both **BLW:** Yes **PH:** No **SS:** A rec 1194; B pg 57; M pg 44; O- pen File Fold3 **BS:** 245

BOULDIN, Thomas Tyler; b nr Charlotte Co Court House, d 11 Feb 1834, Washington DC **RU: Private, 26th VMR, Capt Henry A Watkin's Co, Charlotte County, attached to 5th VMR (Mason and Preston) **CEM:** Golden Hills Estate; Charlotte; Drakes Branch **GS:** Yes **SP:** No spouse info **VI:** Son of Wood Bouldin & Joanna Tyler. Judge of Circuit Court, US Congress 1829-1833. Died in Congress while giving a eulogy to John Randolph. A cenotaph to his memory, Congressional Cemetery in Washington DC **P:** No **BLW:** No **PH:** No **SS:** A rec 1197 **BS:** 49; 245

*BOULDIN, Thomas, Jr; b 1769, d 1845 (court rec) **RU:** Corporal 64th VMR, Capt Thomas Grave's Co, Henry County **CEM:** Bouldin Family; Henry; Grassy Creek Rd (Rt 829) loc behind Jehovah Witness's Church **GS:** Yes **SP:** Armine Estes Cox (1770-__) **VI:** Son of Thomas Bouldin (1738-1827) and Martha Moseley (1742-___) Govt Gr St gives incorrect death date of 1740 or 1749 **P:** No **BLW:** No **PH:** Yes **SS:** B pg 101 G **BS:** 49; 245

*BOURN(E), Stephen Grey; b 26 Feb 1779, d 29 Apr 1849 **RU:** Private 32nd or 93rd VMR, Capt John Trimble's Co, Augusta Co, attached to 7th VMR (Saunders) **CEM:** Stephen G Bourne; Grayson; Rt 777 Liberty Hill Rd in pasture field **GS:** Yes **SP:** Mar 1) (20 Jan 1820) Milla Martin, 2) Martha (Patsy) Mayes (23 Feb 1778-29 Apr 1849) **VI:** Son of William Bourne (1743-1836) and Rosamond Jones (1750-1821). Recd pen # SO-25187 and SO-15958 and two BLW of 80 acres ea #24754 and #45282 **P:** Yes **BLW:** Yes **PH:** No **SS:** A rec 1440; B pg 40; M pg 44; O- pen file Fold3 **BS:** 245

*BOWEN, Arthur; b 1772, d 1845 **RU:** Private, 112th VMR, Capt William Gillespie's Co, Tazewell County, attached to Major Bradley's Command **CEM:** Henry Bowen's Family Farm; Tazewell; Maiden Spring **GS:** Unk **SP:** No spousal info **VI:** Son of Rees Tate Bowen (1729-1780 killed at Kings Mtn battle) and Margaret Louisa Smith (1741-1834) **P:** No **BLW:** No **PH:** No **SS:** A pg 1794; B pg 196 **BS:** 245

*BOWEN, Henry; b 18 Mar 1770, Maiden Spring, Tazewell Co, d 18 Apr 1850 **RU:** Captain, Probably commanded a company in Tazewell Co **CEM:** Bowen Family; Tazewell; Nr VA Historical road sign "Maiden Spring Fort, Cove Creek **GS:** Unk **SP:** No spousal info **VI:** Son of Le Rees Bowen (killed at Kings Mountain battle, (1729-1780) and Margaret Louisa Smith (1741-1834). Represented Tazewell in House of Delegates, Justice in Co Court, titled Colonel after war period **P:** No **BLW:** No **PH:** No **SS:** Z **BS:** 245

Key: *Additional veteran entry   **Corrected veteran entry   ***Deleted veteran entry
RU=Rank/Unit   CEM=Cemetery   GS=Gravestone   SP=Spousal Information   VI=Other Veteran Info   P=Pension
BLW=Bounty/Land Warrant   PH=Photo   SS=Service Source   BS=Burial Source   VMR= VA Military Regiment
LNR= Last Known Residence

14

BOWERS, John; b 17 Dec 1794, Forestville, Shenandoah Co, d 1 Jun 1848 **RU**: Private, 6th VMR (Coleman) **CEM**: Solomon Church; Shenandoah; Rt 727, 9 mi SW of Mt Jackson **GS**: No **SP**: 1) Hannah Fultz, 2) (abt 1815) Anna Marie (__) **VI**: Son of Ludwig Bowers, (1767-1814) and Elizabeth Fultz (1768-1847) **P**: No **BLW**: No **PH**: No **SS**: A rec 2226 **BS**: 217

*BOWERS, Philip; b 1777, d 1867 **RU**: Private, 31st VMR, Capt Thomas Robert's Co, Frederick County, attached to 4th VMR (Beatty) **CEM**: Solomon Lutheran; Shenandoah; 9 mi SW of Mt Jackson on Rt 727 **GS**: Yes **SP**: No spouse info **VI**: Son of Christian Bowers **P**: No **BLW**: No **PH**: No **SS**: A rec 2261; S pg 106 **BS**: 245

*BOWMAN, Archelaus; b 1793, d 3 Jan 1884, **RU**: Private. 18th VMR, Capt John A Conn's Co, Patrick County **CEM**: Borman Family; Patrick; on hill N side Rt 631 at Squirel Creek **SP**: Mar c1811, Patrick Co, Elizabeth (__) (1797-1877) **VI**: Son of (__) Bowman and Becky (__). Recd Pen $8.00 mo, BLW 180 acres; LNR San River, Patrick Co **P**: Yes **BLW**: Yes **PH**: No **SS**: B pg 157; M pg 46; O- pen files **BS**: 245

*BOWMAN, George; b 27 Jan 1780, d 29 Apr 1850 **RU**: Private, 13th VMR, Captain George Shrum's Company, Shenandoah Co., attached to Major Perkins Garrison **CEM**: Radar Lutheran Church; Rockingham; Timberville **GS**: Yes **SP**: Margaret Miller (23 Dec 1783-30 Sep 1824) daug of Matthias Mueller/Miller **VI**: Son of John Bowman (1756- 1816) and Mary Magdalena Zervus (1755-1835) **P**: No **BLW**: **PH**: No **SS**: B pg 185; K pg 227 **BS**: 31; 245

*BOYD, James; b 17 Jan 1779, d 29 Jun 1848 **RU**: Sergeant, 93rd VMR, Captain Samuel Steele's Co, Augusta County, attached to Armistead's Bn and Cocke's Detachment **CEM**: Pine Creek Primitive Baptist Ch; Floyd; Spangler Mill Rd **GS**: Yes **SP**: No spouse info **VI**: No further data **P**: No **BLW**: No **PH**: No **SS**: B pg 40; K pg 198 **BS**: 245

*BOYD, John; b Aug 1778, d 14 Oct 1833 **RU**: Private, 51th VMR, Capt John Gilkerson's Co. Frederick County, attached to (McDowell, Koontz, Chilton} **CEM**: Boyd Family; Warren; Brownstown **GS**: Yes **SP**: No spouse info **VI**: Recd Pen **P**: Yes **BLW**: No **PH**: No **SS**: B pg 79; O- pen files Fold3 **BS**: 245

*BOYER, John; 26 Jan 1777, d Feb 1821 **RU**: Ensign, 78th VMR a company of Grayson Co, attached to the 4th VMR (Boyd) **CEM**: Churchwell-Boyer Family; Grayson; GPS: 36.685,-81.175 loc 1.2 mi N jct Rt 611 and US 21, on pvt rd ¾ mi, left in field **GS**: Yes **VI**: Polly Long (1763-1864), daug of William Long, Jr (1777-1861) and Catherine "Caty" (__-1828) **VI**: Son of William Henry Boyer (1763-Mar 1821) and Elizabeth (__) **P**: No **BLW**: No **PH**: No **SS**: A rec 1533; B pg 86 **BS**: 245

*BOYER, b 1787, d Feb 1871 **RU**: Private, 78th VMR, Capt Timothy Dalton or Capt Lewis Hail's(Hale's) Co, Grayson Co attached to 4th VMR **CEM**: Long-Phipps Family; Grayson; GPS: 36,57641,-81.20306; West of Rt US 21 and S of Peach Bottom Creek, Long Gap **GS**: Unk **SP**: Polly Long (1787-1864), daug of William Long, Jr (1777-1861) and Catherine "Caty" (__-1828) **VI**: No further info **P**: No **BLW**: No **PH**: No **SS**: A rec 1534; B pg 86 **BS**: 245

*BOYER, Peter; b 4 Apr 1771, d 19 May 1845 **RU**: Private 13th or 97th VMR, Capt John W Bayliss's Artillery Co, Shenandoah County, attached to a Bn of Artillery **CEM**: Peter Boyer Family; Shenandoah; Oranda **GS**: Unk **SP**: Mar (7 Apr 1771) Elizabeth Keller (2 May 1765- 9 Dec 1843), daug of Hans George Keller (1731-1788) and Barbara Zimmerman (1740-1798) **VI**: No further data **P**: No **BLW**: No **PH**: No **SS**: A rec 1556; B pg 184 **BS**: 245

BOYLE, David; b 1771; d 18 Jun 1818 **RU: Paymaster, 36th VMR (Reno), Prince William Co **CEM**: Dumfries; Prince William; off Cameron St, SW of Dumfries Elementary School **GS**: Yes **SP**: Jane (__) (c1791-10 Apr 1852) **VI**: No further data **P**: No **BLW**: No **PH**: No **SS**: L pg 16, 17 **BS**: 11 pg 21; 245

*BRADLEY, Philip; b 21 May 1792, d 21 Nov 1826 **RU**: Private, 43rd VMR, Capt Joel Estis's Co, Franklin Co, attached to 4th VMR (McDowell, Koontz, Chilton) **CEM**: Bradley/Bird Family; Franklin; Scruggs **GS**: Unk, Spouse, Yes **SP**: Mar (Franklin Co, 18 Oct 1817) Elizabeth Forbes (25 Aug 1797-5 Jul 1884), daug of John Forbes (1765-1826) and Sarah R Bernard, recd pen **VI**: Pen file indicates he also served in Capt Estridge Pepper's Co but no listed source B. **P**: Spouse **BLW**: Both **PH**: No **SS**: A rec 914; B pg 76; M pg 48; O- pen files Fold3 **BS**: 245

Key: *Additional veteran entry   **Corrected veteran entry   ***Deleted veteran entry
RU=Rank/Unit   CEM=Cemetery   GS=Gravestone   SP=Spousal Information   VI=Other Veteran Info   P=Pension
BLW=Bounty/Land Warrant   PH=Photo   SS=Service Source   BS=Burial Source   VMR= VA Military Regiment
LNR= Last Known Residence

*BRADLEY, Solomon C; b 1790, Buchanan Co, d 9 Mar 1860 **RU:** Private, A company of Infantry, U.S. Army **CEM:** George W Cleek; Bath; GPS: 38.19310,-79.73220, Rt US 220 at Mill Run, Warm Springs **GS:** yes **SP:** Harriet (__) (1808-1899) **VI:** Son of John Bradley and Martha (__) **SS:** O- pen files Numerical Index; Fold3 **BS:** 245

*BRAGG, John; b 1 Dec 1787, d 24 Sep 1858 **RU:** Surgeons Mate, 1st VMR, on staff of Regt commanded by Lt Col James Byrne, Petersburg City **CEM:** Blandford; Petersburg City; 111 Rochelle Ln **GS:** Yes **SP:** Mar (Petersburg, 24 Apr 1839) Ann Maria Hill (1809-1892), LNR Raleigh, NC 1878, recd pen **VI:** A Dr after war, recd BLW **P:** Widow **BLW:** Yes **PH:** No **SS:** A rec 1555; B pg 159; L pg 1; M pg 48 **BS:** 245

*BRAXTON, Charles Hill; b 13 Jun 1782, d 02 Mar 1845 **RU:** Captain, 87th VMR Troop Commander of 4th Cavalry, King William Co. attached to 1st VMR (Holcombe's) **CEM:** Oak Spring Farm; King William; (see property rec for location) **GS:** Unk **SP:** Mar (20 Apr 1812) Elizabeth Pope Grimes (25 Nov 1794-18 Dec 1831) **VI:** Son of George Braxton **P:** No **BLW:** No **PH:** No **SS:** B pg 115 **BS:** 245

*BRENT, George; b 1787, Alexandria, d 16 Dec 1869 **RU:** Second Lieutenant, 22nd U.S. Army Inf, promoted to his rank, 7 Sep 1814, discharged Carlisle Barracks, PA, 26 Jun 1815 **CEM:** Mt Hebron; Frederick; 305 E Boscawen St, Winchester **GS:** Yes **SP:** Mar (10 Dec 1812) Susannah Anderson **VI:** Interred 10 Dec 1870 **P:** No **BLW:** No **PH:** No **SS:** C pg 20; **BS:** 81; 245

*BRIGGS, David, b Unk, d 3 Dec 1815 **RU:** Private, 45th VMR, Capt Alexander's Co, Stafford County **CEM:** Briggs Family, AKA Stoney Hill; Stafford; Stoney Hill Rd adjacent to Curtis Memorial Park **GS:** Yes **SP:** Mary Frazier Vowells (Vowles), daug of Henry Vowells (Vowles) and Mary Frazier **VI:** Perhaps a son of Jane (__) Briggs (c1760-6 Jun 1810) who is buried next to him. A Fredericksburg attorney **P:** No **BLW:** No **PH:** No **SS:** L pg 82 **BS:** 130

BROCKENBROUGH, Austin; b 9 Oct 1782; d 31 Dec 1858 **RU: Surgeon, 6th VMR, Essex Co **CEM:** Blake/Brockenbrough; Essex; Walter Ln by Museum Tappahannock **GS:** Yes **SP:** Frances Blake (1809-1867) **VI:** Son of Dr John & Sarah (Roane) Brockenbrough; Physician; Magistrate of Essex Co **P:** No **BLW:** No **PH:** No **SS:** A rec 1989 **BS:** 291 pg 122

*BROCKENBROUGH, John; b 8 May 1773, Essex County, d 3 Jul 1852 **RU:** Private, 19th VMR, Richmond City **CEM:** Warm Springs; Bath; 12 mi S of Warm Springs **GS:** Yes **SP:** Gabriella Harvie Randolph (1772-1853) **VI:** Doctor, graduating from the University of Edinburgh in 1795; President Bank of Virginia **P:** No **BLW:** No **PH:** No **SS:** A rec 1990 **BS:** 245

*BRONAUGH, Thomas; b 1790, Spotsylvania Co, d 5 Jul 1866 **RU:** Ensign,40 VMR, Capt Reuben Chewning's Co, Louisa Co, attached to 7th VMR (Gray) **CEM:** Oak Grove; Louisa; town or village of Oak Grove (in 1871) **GS:** Y -Govt **SP:** Mar (25 Feb 1813) Judith Hart (25 Dec 1795-27 Nov 1872); LNR Oak Grove, 1871 Louisa Co, Recd pen and two BLW **VI:** Plaque in cem lists burials **P:** Widow **BLW:** Widow **PH:** No **SS:** B pg 123; K pg 326; M pg 54; O- pen files **BS:** 245

*BROOKS, Alexander; b 1789, d 22 Dec1862 **RU:** Private, 1st VMR Cav, Capt Henry Heth's Troop of Cavalry, serving as a sub for Zachary Brooks, Chesterfield Co **CEM:** Hollywood; Richmond; 412 S Cherry St **GS:** Yes **SP:** Mar (26 Aug 1830) Mary McRae (1809-1899),recd pen #19886 **VI:** Recd BLW #68.228 **P:** Widow **BLW:** Yes **PH:** No **SS:** A rec 263; B pg 60; M pg 54 **BS:** 245

*BROOKS, John; b 2 Sep 1791, d 15 Oct 1863 **RU:** Private, 5th VMR (McDowell's) **CEM:** Tinkling Springs Presbyterian Church; Augusta; Fisherville **GS:** Yes **SP:** Susan Henderson McCombs (17 Feb 1807-11 Dec 1863), daug of James McComb (1765-1846) and Susannah Henderson (1769-1848) **VI:** No further info **P:** No **BLW:** No **PH:** No **SS:** A rec 546,569, B pg 39-40 **BS:** 245

*BROOKS, Lewis Durham; b 1793, d 3 Jul 1876 **RU:** Private, 6th VMR, Capt Samuel Muse's Co, Essex Co, transferred to 111th VMR (Capt William Waring's Co) **CEM:** Brooks Family; Essex; N side Rt 689, Howertons Rd **GS:** No **SP:** 1) Maria Alexander (__-1821) 2) Fanny Griggs (__-1826) 3) (8 Nov 1827) Sarah Roy (__-3 Jul 1885), recd pen **VI:** Son of Lewis Brooks and Rebecca Durham. Recd BLW and Pen, buried on family farm **P:** Both **BLW:** Yes **PH:** No **SS:** B pg 69, 70, 202; M pg 54; O- pen files; Fold 3 **BS:** 245

Key: *Additional veteran entry   **Corrected veteran entry   ***Deleted veteran entry
RU= Rank/Unit   CEM=Cemetery   GS=Gravestone   SP=Spousal Information   VI=Other Veteran Info   P=Pension
BLW=Bounty/Land Warrant   PH=Photo   SS=Service Source   BS=Burial Source   VMR= VA Military Regiment
LNR= Last Known Residence

*BROOKS, William; b 25 Jul 1789, d 11 Nov 1857 **RU**: Private 93rd VMR Capt Archibald Stuart's Co, Augusta County, attached to McDowell's Flying Camp **CEM**: Old Presbyterian Church; Waynesboro City; 203 New Hope Rd **GS**: Yes **SP**: Mar (Waynesboro, 29 Apr 1824) Elvira A Dold (20 Nov 1804-28 Apr 1890), daug of William Dold and Sarah Brent, LNR Staunton City; recd pen # 6973 Mar 1878 **VI**: Son of Samuel Brooks and Margaret (__); recd BLW **P**: Widow **BLW**: Yes **PH**: No **SS**: B pg 40; M pg 55; O- pen files, Fold3 **BS**: 245

*BROWN, Christopher Strophel, Jr; b 7 Nov 1774, Hanover, York Co, PA, d 20 Sep 1850 **RU**: Captain, 35th VMR, commanded a company, Wythe Co Militia **CEM**: St John's Lutheran Church; Wythe; GPS 36.95500,-81.10110; N of jct Rt 52 and Holston Rd, Sauers, Wytheville **GS**: Yes **SP**: Mar (Wytheville, 15 Nov 1809) Anna Marie Rader (14 Jul 1778- 19 Dec 1849), daug William Rader and Regina Gerhardt **VI**: Son of Christopher Brown (1750-1816) and Anna Maria Mason (1754-1822) **P**: No **BLW**: No **PH**: No **SS**: A rec 1749; B pg 204 **BS**: 245

*BROWN, Ellis; b 14 Mar 1793, Hanover Co, d 22 Jul 1870 **RU**: Private 19th VMR Capt Samuel Jones's Co, Richmond City, attached to 2d VMR (Ballowe) **CEM**: Hollywood, Richmond City; 412 So Cherry St **GS**: Unk, bur Sec H, lot 37 **SP**: 1) Louisa Bosher (__-1829), daug of John Bosher, 2) (17 Jan 1833) Virginia Hughes (__-aft 23 Dec 1878), daug of Achelous (sp) Hughes, recd pen $8.00 mo commencing 23 Dec 1878 **VI**: Recd BLW 80 acres, Jun 1853 **P**: Widow **BLW**: Yes **PH**: No **SS**: B pg 175; M pg 55, 56; O- pen and BLW files, Fold 3 **BS**: 245

*BROWN, John; b 1768, d 1840 **RU**: Private, US Army 12th Regt **CEM**: Goose Creek Burying Ground; Loudoun; Rt 722 Lincoln **GS**: Yes **SP**: Ann Hirst (c1781-d 1834) **VI**: Enlisted 19 May 1814, discharged Pittsburgh, PA **P**: No **BLW**: No **PH**: No **SS**: C pg 22 **BS**: 245

*BROWN, Peter; b 23 Aug 1795, d 1 Jan 1858 **RU**: Private, 58th or 116th VMR, Capt Robert Erwin's Co, Rockingham County, attached to Lt Col Thomas Ballowe's 2d VMR **CEM**: Friedens's United Church of Christ; Rockingham; 3960 Friedens Church Rd **GS**: Yes **SP**: Mar (Rockingham Co, 16 Aug 1821) Elizabeth Huffman (__-1880), recd pen **VI**: Recd BLW **P**: Widow **BLW**: Yes **PH**: No **SS**: B pg 181; M pg 57; O- pen files, Fold3 **BS**: 245

*BROWN, Samuel: b Unk, d 5 Oct 1847 **RU**: Captain, 33 VMR, Company Commander, Henrico Co **CEM**: Cauthorn Cemetery; Henrico; Short Pump **GS**: Yes **SP**: No spouse info **VI**: No further data **P**: No **BLW**: No **PH**: No **SS**: K pg 117; B pg 99 **BS**: 245

*BROWN, Samuel; b 1783, d 10 Aug 1852 **RU**: Private, 31st, 51st & 122nd VMR, Frederick Co, attached to 4th VMR (Boyd's) **CEM**: Brown; Winchester City; check city property records for location **GS**: Unk **SP**: No spouse info **VI**: No further data **P**: No **BLW**: No **PH**: No **SS**: A rec 1658; B pgs 78-80 **BS**: 245

*BROWN, William; b 17 Jul 1782, d 4 May 1849 **RU**: Private, 8th VMR, Capt Henry McClung's Artillery Co, Rockbridge County, attached to Battalion of Artillery **CEM**: Walkerland; Rockbridge; Vic jct Rts 602 & 724 **GS**: Yes **SP**: No spouse info **VI**: No further data **P**: Yes **BLW**: No **PH**: No **SS**: B pg 180; BD pg 268 **BS**: 245

*BROWN, William Cox; b 30 May 1788, d 7 Sep 1851 **RU**: Private, 56th VMR. Loudoun Co company **CEM**: Goose Creek; Loudoun; Rt 722 Lincoln **GS**: Yes **SP**: Sarah Piggott (1798-1877) **VI**: No further data **P**: No **BLW**: No **PH**: No **SS**: A rec 2375 **BS**: 245

*BROWN, William H; b 4 Oct 1779, d 2 Nov 1865 **RU**: Private, 31st or 51st VMR, Frederick Co company, attached to 4th VMR (Beatty or Boyd) or 5th VMR McDowell's Flying Camp **CEM**: Upper Ridge; Frederick; Apple Pie Ridge Rd, Rt 739, Nain **GS**: Yes **SP**: No spouse info **VI**: No further data **P**: No **BLW**: No **PH**: No **SS**: A recs 2295; 2267 2271; B pgs 78-80 **BS**: 245

BROWN, William H; b 2 Feb 1793; d 4 Apr 1879 **RU: Private, 2nd VMR (Ballowe) **CEM**: Wright Family; Giles; Rt 42, 4 mi W of Poplar Hill **GS**: Yes **SP**: Margaret (__) (__-1891) **VI**: No further data **P**: No **BLW**: No **PH**: No **SS**: A rec 2396 **BS**: 14 pg 167; 245

Key: *Additional veteran entry   **Corrected veteran entry   ***Deleted veteran entry
RU=Rank/Unit   CEM=Cemetery   GS=Gravestone   SP=Spousal Information   VI=Other Veteran Info   P=Pension
BLW=Bounty/Land Warrant   PH=Photo   SS=Service Source   BS=Burial Source   VMR= VA Military Regiment
LNR= Last Known Residence

*BROWNE, William; b 1787, d 5 Dec 1856 **RU:** Surgeon, 16th VMR, Spotsylvania Co **CEM:** Fredericksburg City; Fredericksburg; GPS: 38.30112, -77.46628 **GS:** No **SP:** Margaret Emily (__) (__-3 Sep 1857) **VI:** Name listed Dr William Browne on gr st **P:** No **BLW:** No **PH:** No **SS:** A rec 2569; B pg 188 **BS:** 245

*BROWNLEE, John; b 1 Apr 1790, d 10 Mar 1877, **RU:** Private, Capt Jesse Dold's Co, Troop of Cavalry, Augusta Co, attached to Maj Woodford's Squadron **CEM:** Bethel Presbyterian Church; Augusta; Middlebrook **GS:** Yes **SP:** Mar (28 Oct 1819) Nancy Bell (1792-7 Dec 1886), Recd pen, Act of 1878 **VI:** Son of John Brownlee, Jr and Mary Moffett. Recd pen and BLW of 120 acres **P:** Both **BLW:** Yes **PH:** No **SS:** B pg 39; M pg 58; O- pen and BLW files; Fold3 **BS:** 245

BRUCE, Charles; b 18 Jan 1768, King George County; d 18 Sep 1845 **RU: Private, 2nd Corps De Elite(Green) **CEM:** William's Family #1; Stafford; vic Bethlehem Baptist Church, Rt 602, on dirt road 0.5 miles past church **GS:** Yes **SP:** Sarah Jane (__) (1787-11 Jan 1839) **VI:** Son of William Bruce and Elizabeth Grant **P:** No **BLW:** No **PH:** No **SS:** A rec 59 **BS:** 26 pg 397

*BRYAN, Edward Jr; b 22 Mar 1797, d 12 Feb 1838 **RU:** Private, 8th VMR, Captain James Paxton's Co, Rockbridge County, attached to 2nd Corps D'Elite (Green) **CEM:** Old Ebenezer Church; Rockbridge; Rockbridge Baths **GS:** Yes **SP:** Mar (Rockbridge, 20 Jul 1819) Mary "Polly" Shaw **VI:** Son of Edward Bryan Sr. and Polly Parker **P:** No **BLW:** No **PH:** No **SS:** B pg 180 **BS:** 245

*BUCHANAN, John; b 1784, d 14 Sep 1858 **RU:** Private,105th VMR, Capt William Smith's Co of Artillery, Washington County attached to Battalion of Artillery **CEM:** Buchanan; Washington; GPS 36.84163,-81.75199, Clark Farm Rd **GS:** Yes **SP:** Mar (Smyth Co, 19 Feb 1807) Mary Ryburn (2 Apr 1785-5 May 1858), daug of William Ryburn & Mary (__) **VI:** Son of John Buchanan (1745-1824) & Anne Ryburn **P:** No **BLW:** No **PH:** No **SS:** A rec 1646; B pg 199 **BS:** 245

*BUCKNER, William Aylett; b 13 Feb 1766, d 2 Jan 1830 **RU:** Sergeant, 30th VMR, in a company of Caroline County, attached to the 41st VMR (Branham) **CEM:** Greenlawn; Caroline; Bowling Green **GS:** Yes **SP:** No spouse info **VI:** No further data **P:** No **BLW:** No **PH:** No **SS:** A rec 2702; B pg 56 **BS:** 245

*BURCH, Samuel L; b Unk, d 11 Jun 1853 **RU:** Private, 26th VMR Capt John Pollock's Co, Charlotte County, attached to 7th VMR (Gray) **CEM:** Presbyterian Church; Lynchburg; 220 Grace St **GS:** Unk, bur lot 6, row 5 **SP:** Mar (23 Nov 1823) Mary Puryear **VI:** Recd pen and two BLW of 80 acres each **P:** Yes **BLW:** Yes **PH:** No **SS:** B pg 56; K pg 349; M pg 62; O BLW file Fold3 **BS:** 245

*BURK, Edmond; b Culpeper Co, d 28 Jul 1813 **RU:** Private, 5th VMR, Capt Jesse Nalle, Culpeper Co, attached to 5th VMR (Mason – Preston) **CEM:** Trinity Baptist Church; Louisa; 135 Mansfield Rd, Mineral **GS:** No **SP:** Mar (Culpeper Co, Mar 1797) Frances Weaver **VI:** Died (29 Jul 1813) on travel home from service in Norfolk; they resided in Slate Mills in now Rappahannock Co **P:** No **BLW:** No **PH:** No **SS:** A rec 584; B pg 63 **BS:** 31

*BURK, James; b 1795, d 8 Dec 1879 **RU:** Private, 56th VMR, Capt Edward Grady's Co, Loudoun Co, under command of Colonel Minor in Baltimore defense **CEM:** Cedar Hill; Covington City; 1521 So Carpenter Dr **GS:** Yes **SP:** 1) Sophia Landrum (1794-1851), 2) (Covington City, 10 Mar 1852) Jane (__) Clark (1802-1881). Widow Jane drew pen **VI:** Loudoun Co 1850 was a hotel keeper. Recd pen and two BLW of 80 acres each **P:** Both **BLW:** Yes **PH:** No **SS:** B pg 120; M pg 63; O BLW files Fold3 **BS:** 245

*BURNER, John Rhodes; b 30 Aug 1788, d 4 Oct 1860 **RU:** Fifer, 13th or 97th VMR, Capt Joshua Ruffner's Co, Shenandoah County, attached to 6th VMR (Coleman) **CEM:** Evergreen Memorial Gardens; Page; Massanutten Ave, Luray **GS:** Yes **SP:** 1) Elizabeth Stricker (1793-1838), 2) Susannah (__) Hershberger, LNR Page Co 1878. Widow drew pen **VI:** Son of Joseph Burner (1759-1821) and Mary Ann Rhodes (1766-1812), Veteran drew BLW of 40 acres and 120 acres **P:** Widow **BLW:** Yes **PH:** No **SS:** B pg 185; M pg 64; O-BLW files Fold3 **BS:** 245

***BURNHAM, James D: Deleted as too young, thus did not serve

---

Key: *Additional veteran entry   **Corrected veteran entry   ***Deleted veteran entry
RU=Rank/Unit   CEM=Cemetery   GS=Gravestone   SP=Spousal Information   VI=Other Veteran Info   P=Pension
BLW=Bounty/Land Warrant   PH=Photo   SS=Service Source   BS=Burial Source   VMR= VA Military Regiment
LNR= Last Known Residence

BURWELL, Armistead; b 26 Jun 1770; d 28 Jan 1820 **RU:** Major, 3rd VMR **CEM:** Burwell (AKA Hamlin; Mecklenburg; Mecklenburg **GS:** Yes **SP:** Mar (Mecklenburg Co, 14 Nov 1791) (bond, Robert Crawley surety) Lucy Crawley (30 Dec 1775-14 Nov 1825) **VI:** Son of Colonel Lewis Burwell & Ann Spotswood. Had been Captain in 83rd VMR in Dinwiddie Co. Promoted to Major of the 3rd VMR on 20 Mar 1813 **P:** No **BLW:** No **PH:** N **SS:** A rec 568; C pg 74 **BS:** 24 pg 174

BURWELL, Nathaniel; b 1785; d 21 Jul 1866 **RU:** Lieutenant, 48th VMR, Capt Andrew Hamilton's, Troop of Cavalry, Botetourt Co **CEM:** Burwell Family (moved); Roanoke; East Hill Cem; Salem **GS:** Unk **SP:** Lucy Carter (__-1845) **VI:** Son of Captain Nathaniel Burwell & Martha Digges **P:** No **BLW:** No **PH:** No **SS:** L pg 391 **BS:** 157 pg 33

*BURWELL, Nathaniel; b 16 Feb 1779, Carter's Grove, James City Co, d 1 Nov 1849 at "Saratoga", Frederick Co **RU:** Lieutenant, in Frederick or Gloucester Co, in company attached to 5th VMR **CEM:** Old Chapel; Clarke; Millwood **GS:** Yes **SP:** Elizabeth Nelson, daug of Nathaniel Nelson of York Co **VI:** Son of Colonel Nathaniel Burwell (1750-1814) and Susanna Grymes **P:** No **BLW:** No **PH:** No **SS:** A rec 591 **BS:** 85 pg 37

*BURWELL, Robert Carter; b 24 Jul 1785, Carter Grove, James City Co, d 22 Aug 1813, New Market **RU:** Captain of Frederick Co company, attached to McDowell's Flying Camp **CEM:** Old Chapel; Clarke; Millwood **GS:** No **SP:** No spouse info **VI:** Son of Nathaniel Burwell and (__) Wormley of Isle of Wight Co (or Susanna Grymes) **P:** No **BLW:** No **PH:** No **SS:** A rec 595; B pg 78; **BS:** 245

*BURWELL, William Nelson; b 23 Apr 1791, Carter's Grove, d 1822, Glen Owen **RU:** Private, in Frederick County Co attached to 4th VMR **CEM:** Old Chapel; Clarke; Millwood **GS:** Yes **SP:** Mary Marshall Brooke of Fauquier Co **VI:** Son of Nathaniel Burwell (1750-1814) and Lucy Page **P:** No **BLW:** No **PH:** No **SS:** A rec 604; B pg 78; **BS:** 245

BUSH, Andrew; b c1763; d 1839 (Inv) **RU:** Corporal, 31st VMR, Captain Thomas Roberts, Frederick Co, attached to 4th VMR (Boyd) **CEM:** Mt Hebron; Winchester; 305 E Boscawen St **GS:** Unk **SP:** Mar (Frederick Co, 13 Sep 1815) Mary Currell by George M Frye **VI:** No further data **P:** No **BLW:** No **PH:** No **SS:** A rec 713; B pgs 79-80 **BS:** 68 pg 55

*BUSHONG, Phillip, b 3 May 1790, d 21 Oct 1873, Shenandoah Co **RU:** Private, 13th VMR, Captain William Newell's Co, Shenandoah County, attached to 2nd VMR (Ambler/ Brown) **CEM:** Radar Lutheran Church; Rockingham; Timberville **GS:** Yes **SP:** Mary Knop (15 Jun 1791-6 Jan 1878) **VI:** No further data **P:** No **BLW:** No **PH:** No **SS:** B pg 40 **BS:** 31; 245

BUSSELL, Charles; b 20 May 1799; d 16 or 18 Nov 1875 **RU:** Private, 36th US Infantry, Capt Randolph **CEM:** Wine Family; Stafford; Mountain View Rd (Rt 627) past jct Kellogg Mill Rd & Rt. 651, in woods past fifth house **GS:** Yes **SP:** 1) Mary Black, 2) (Staunton, 18 Jan 1867) Lucy Ann Wine (__-8 Nov 1903); LNR Mountain View (Stafford Co, 1878) **VI:** Son of Randall Bussell and Frances Black. Enlisted at Aquia or Fredericksburg into the regular army (13 Jun 1814) discharged, (Washington, DC, 13 Mar 1815). The gravestone indicates he was also in the Civil War **P:** Widow **BLW:** Yes **PH:** No **SS:** C pg 26; BD pg 306 **BS:** 26 pg 399

*BUTCHER, John Humphrey; b 17 Aug 1788, Bloomfield, Loudoun Co, d 1861 **RU:** Sergeant, Served in the 56th VMR in an undetermined company, Loudoun Co **CEM:** Ebenezer Baptist Church; Loudoun; 20421 Airmont Rd **GS:** Unk **SP:** 1) Nancy Ann Overfield (5 Oct 1794-7 Aug 1824), daug of Martin Overfield (__-1814) and Elizabeth Botts (__-1818), 2) Mary Glasscock (11 Mar 1798-3 Aug 1856) **VI:** Son of Samuel Butcher (28 May 1756-2 May 1847) and Hannah Drake (16 Aug 1761-2 Feb 1844) **P:** No **BLW:** No **PH:** No **SS:** A rec 1270; B pgs 119-121 **BS:** 245

*BUTTS, Daniel Claiborne; b 1776, d 4 Jan 1850 **RU:** Captain Commanded a company 35th Regt USA **CEM:** Blandford; Petersburg; 319 So Crater Rd **GS:** Yes **SP:** 1) Elizabeth Randolph Harrison (1776-1837), daug of Gen Charles Harrison, 2) (24 Oct 1837) Mary Ann Parsons who drew widows pen **VI:** Recd BLW of 180 acres. Made rank of General after war. **P:** Widow **BLW:** Yes **PH:** No **SS:** M pg 67; O BLW files Fold3 **BS:** 245

Key: *Additional veteran entry   **Corrected veteran entry   ***Deleted veteran entry
RU=Rank/Unit   CEM=Cemetery   GS=Gravestone   SP=Spousal Information   VI=Other Veteran Info   P=Pension
BLW=Bounty/Land Warrant   PH=Photo   SS=Service Source   BS=Burial Source   VMR= VA Military Regiment
LNR= Last Known Residence

*BUTTS, William, Esq; b c1791,Shepardstown, WVA, d 01 Dec 1871 **RU:** Lieutenant, 57th VMR, Capt Van Bennett, Loudoun Co, and 67th VMR Capt Robert Wilson, Berkeley Co (now WVA), attached to 6th VMR(Reade) **CEM:** Arnold Grove Methodist Episcopal; Loudoun; jct Rts 9 & 690, Hillsboro **GS:** Yes **SP:** Mar (Apr 1816) Margaret Howsworth **P:** Both **BLW:** Yes **PH:** No **SS:** A rec 2445; B pgs 44, 119; BD pg 316; M pg 767 **BS:** 31

BYRAM, John M; b unk, d aft 27 Nov 1823 **RU: Private, 16th VMR, Capt Claiborne Wigglesworth's Co, Spotsylvania County **CEM:** Byram Family; Stafford; jct Rts 672 & 630 **GS:** Marker lists names **SP:** Mar (Spotsylvania Co, 27 Nov 1823) Frances Fell or Frances Jett **VI:** No further data **P:** No **BLW:** No **PH:** N **SS:** L pg 832 **BS:** 26 pg 161

*BYRD, Francis Otway; b 20 Aug 1790, d 2 May 1860, Baltimore **RU:** Brevet, First Lieutenant, US Army Artillery, appointed 20 Feb 1815 because of distinguished gallantry on 15 Aug 1814 **CEM:** Old Chapel; Clarke; Millwood **GS:** Yes **SP:** Mar (Philadelphia, 9 Jun 1817) Elizabeth Rhodes Pleasants (Philadelphia, 1793-1880) **VI:** Son of Captain Thomas T. Byrd (1752-1821) and Mary A Armistead (1753-1824). For his great valor in service in the capture of an Algerian frigate, he was awarded a handsome Turkish sword and a pair of Algerian pistols from Commodore Decatur. Also the VA government in 1848 awarded him a sword for his gallant service **P:** No **BLW:** No **PH:** No **SS:** AF; B pg 271 **BS:** 85 pg 42; 245

BYRNE, William; b 7 Oct 1787, Manassas, Prince William Co; d 22 Mar 1861 **RU: Corporal, 85th VMR, Fauquier Co **CEM:** Byrne Family; Fauquier; "Byrnley," Rt 704, The Plains **GS:** Yes **SP:** Mar (18 Jan 1809) Annie Turner (22 Aug 1789-24 Feb 1881) **VI:** Son of Uriah Byrne (1758-1836) and Lydia Elizabeth Brown (__-1762) **P:** No **BLW:** No **PH:** No **SS:** A rec 2932 **BS:** 3 pg 5

CALDWELL, John; b 1769; d 29 Oct 1823 **RU: Private, 93rd VMR, in company, Augusta Co, attached to 5th VMR **CEM:** Tinkling Spring; Augusta; 11 mi NE of Staunton **GS:** Unk **SP:** No spouse info **VI:** Died age 54 **P:** No **BLW:** No **PH:** No **SS:** A rec 1011; B pg 39; L pg 282 **BS:** 183

*CAMPBELL, Alexander; b 1782, d 5 Jun 1822 **RU:** Captain, 8th VMR, commanded a company in Rockbridge Co, attached to 2d Corps De Elite (Green) **CEM:** Timber Ridge Presbyterian Churchyard; Rockbridge; Vic Rt 716 next to jct with Rt 758 **GS:** Unk **SP:** No spousal info **VI:** Son of Duncan Campbell (1757-1812) and Margaret Newell **P:** No **BLW:** No **PH:** No **SS:** B pg 179 **BS:** 245

*CAMPBELL, David; b 12 Mar 1784, Scotland, d 10 Jul 1855 **RU:** Private 53rd VMR Capt James Dunningham's Artillery Co, Campbell County, attached to Cocke's Detachment **CEM** Hodges Family; Henry; Rt 627, 4.5 mi W of Fieldale **GS:** Yes **SP:** 1) Milly Dempsey (__-24 Jun 1824), 2) (5 Jul 1831) Mary C Dillon (1799-1891) as widow recd pen and BLW 80 acres **VI:** Was a tailor by trade. Recd BLW 1855 **P:** Widow **BLW:** Both **PH:** No **SS:** B pg 53; M pg 70; O- BLW and Pen files Fold3 **BS:** 245

CAMPBELL, Henry; b 1766; d 1844 **RU: Private, 91st VMR, Capt John Gray's Company, Russell Co, attached to 2nd VMR (Ambler/Brown) **CEM:** Campbell Family; Russell; Mountain Rd **GS:** Yes **SP:** Sarah (__) **VI:** Son of Abraham Campbell (1736-1805) and Dorcas (__) (1739-1805) **P:** Spouse appl **BLW:** No **PH:** No **SS:** A rec 422; B pg 42; BD pg 326; M pg 70 **BS:** 245

*CAMPBELL, James; Jr: b 1786, Cecil County, MD, d 10 Mar 1856 **RU:** Private, Captain Burke's Co, MD 6th Regt **CEM:** Campbell; Lee; SW of Rose Hill **GS:** No **SP:** Mar (Westmoreland County, 7 Dec 1808) Eliza Ferguson Murphy **VI:** Son of James Campbell & Mary Gibbs **P:** No **BLW:** No **PH:** No **SS:** V pg 238 **BS:** 245

*CAMPBELL, John II; b 1768, d 1843 **RU:** Private, 48th VMR Capt James Cartmill's Co, Botetourt County, attached to Flying Camp (McDowell) **CEM:** Rich Valley Presbyterian Church; Smyth; Rt 610, I mi E of jct with Long Hollow Rd **GS:** Yes **SP:** Dorcas Tate (1782-1817) **VI:** No further data **P:** No **BLW:** No **PH:** No **SS:** B pg 244; K pg 7 **BS:** 245

Key: *Additional veteran entry **Corrected veteran entry ***Deleted veteran entry
RU=Rank/Unit   CEM=Cemetery   GS=Gravestone   SP=Spousal Information   VI=Other Veteran Info   P=Pension
BLW=Bounty/Land Warrant   PH=Photo   SS=Service Source   BS=Burial Source   VMR= VA Military Regiment
LNR= Last Known Residence

20

*CAMPBELL, Robert; b 5 Sept 1781, d 11 Feb 1853 **RU:** Private 32nd VMR, Capt Alexander Given's Company, Augusta Co attached to 5th VMR (McDowell) **CEM:** Thomas Campbell Family; Bedford; Irving **GS:** Unk **SP:** No spouse info **VI:** Son of Thomas Campbell (c1749-27 Jun 1827, age 78 years, 7 months, 9 days) and Mary (__) (__-4 Aug 1826, age 71 years and 14 days). **P:** No **BLW:** Yes **PH:** No **SS:** B pg 39; BD pg 328 **BS:** 245

*CARPENTER, John; b 1785; d 8 Sep 1824 **RU:** Private, 6th VMR **CEM:** Carpenter and Lohr; Madison; Rt 616, Oak Park Rd about one mi from Rt 636 **GS:** Yes **SP:** Mildred Blankenbaker (c1790-10 Jun 1866) **VI:** Son of John Carpenter (__-1804) in Madison Co **P:** No **BLW:** No **PH:** No **SS:** A rec 1839 **BS:** 30; Family Cem

*CARR, Dabney B; b 4 Mar 1791, Chesterfield Co, d 3 Mar 1858 **RU:** Regular Army Artillery **CEM:** Portlock; Portsmouth; part of Oak Grove Cem; jct Peninsula Ave & London St **GS:** Yes **SP:** No spouse info **VI:** Enlisted (Richmond, Dec 1813), discharged (18 Sep 1815) **P:** No **BLW:** No **PH:** No **SS:** C pg 29 **BS:** 245

*CARR, David; b 1795, d 9 May 1884, Hamilton **RU:** Private, 57th VMR, Captain George Washington Ball's Troop of Cavalry, Loudoun Co, attached to Green's Mounted Infantry **CEM:** Union; Loudoun; Leesburg **GS:** Yes **SP:** Mar (Frederick Co, MD, 1819) Susan Brown (1801-1881) **VI:** LNR Hamilton, (Loudoun Co, 1879) **P:** Yes **BLW:** Yes **PH:** No **SS:** A rec 54; B pg 119; BD pg 344; M pg 74 **BS:** 245

*CARR, John: b 1777; d 12 Nov 1861 **RU:** Sergeant, 68th or 116th VMR, company of Rockingham Co, attached to 6th VMR, (Coleman) **CEM:** Linville Creek Church of the Brethren; Rockingham; Broadway **GS:** Yes **SP:** Margaret Holsinger (1779-1850) **VI:** No further data **P:** No **BLW:** No **PH:** No **SS:** A rec 200; B pgs 181-2 **BS:** 245

*CARR, William Crow; b 14 Aug 1791, Montgomery Co, d 14 Feb 1858 **RU:** Private, 86th VMR, Lt Ralph Lucas Co, Giles County attached to 4th VMR (Koontz-Boyd's) **CEM:** Shannon-King; Giles Co; GPS: 37.2181,-80.7417 **GS:** Yes **SP:** Elizabeth Bane (1797-1883) **VI:** Son of John Carr and Hannah Crow **P:** No **BLW:** No **PH:** No **SS:** B pg 81; pgs 226; 227 **BS:** 245

*CARRELL, James S (P); b 13 Feb 1787, d 28 Oct 1854 **RU:** Private, 35th VMR, Lt Col Kent, Wythe Co **CEM:** North Church St (AKA Carrell-Price); Russell; Lebanon **GS:** Unk **SP:** Martha G. (__) (1786-7 Apr 1872) **VI:** Son of Charles Carrell and Agnes Perry **P:** No **BLW:** No **PH:** No **SS:** A rec 507; B pg 204 **BS:** 245

*CARTER, Bane (Baynes)(Bains); b 1795, Henry Co, d 29 Nov 1865 **RU:** Private, 64th VMR, Capt Thomas Graves Co, Henry County **CEM:** Carter; Murphy; Tatum Families; Patrick; No specific directions, check property rec for loc **GS:** No **SP:** Mar (26 Apr 1817) Julia B Philpott (1800-9 Jul 1884); recd pen (9 Mar 1878) **VI:** Son of Jesse Carter (1775-1855) and Elizabeth Philpott (1780-1829). Recd BLW of 120 acres **P:** Widow **BLW:** Yes **PH:** No **SS:** B pg 101; M pg 75; O- pen files Fold3 **BS:** 245

*CARTER, Cary; b unk, d 27 Oct 1877 **RU:** Corporal, 64th VMR, Capt John Dillard's Co of Artillery, Henry County, attached to Battalion of Artillery **CEM:** Dillon (AKA Daniel-Dodson-Shelton Families); Henry; nr Mount Herman Church on Philpott Dam Rd **GS:** Yes **SP:** Mar (Henry Co, 15 Jan 1812) Mahala Lewis (__-9 Jul 1884) Note: a person this name, perhaps son or nephew, mar (bond) (31 Dec 1840) Elizabeth Dillon **VI:** Recd pen and two BLW of 80 acres each acres **P:** Yes **BLW:** Yes **PH:** No **SS:** B pg 101; M pg 75; O- pen files Fold3 **BS:** 245

*CARTER, Charles Warner Lewis; b Apr 1792, Blenheim, Fairfax Co, d 7 Nov 1867 **RU:** Surgeon, 2d Corp de Elite, VMR commanded by Lt Col Moses Green and as Surgeons Mate in 92nd VMR Lancaster Co commanded by Lt Col John Chowning **CEM:** Maplewood; Charlottesville City; cnr Lexington Ave and Maple St **GS:** Yes **SP:** Chastain Cocke (Oct 1796-5 Mar 1888). She recd pen **VI:** Son of Edward Hill Carter and Mary Randolph Lewis. In1851 representative Albemarle Co House of Delegates. Surgeon Confederate Army. Recd BLW **P:** Widow **BLW:** Yes **PH:** No **SS:** B pg 116; 243; M pg 75; O- pen & BLW files Fold3 **BS:** 245

Key: *Additional veteran entry   **Corrected veteran entry   ***Deleted veteran entry
RU=Rank/Unit   CEM=Cemetery   GS=Gravestone   SP=Spousal Information   VI=Other Veteran Info   P=Pension
BLW=Bounty/Land Warrant   PH=Photo   SS=Service Source   BS=Burial Source   VMR= VA Military Regiment
LNR= Last Known Residence

21

*CARTER, John; b 1774, d 1836 **RU:** Third Lieutenant, East TN Regiment (Brown) **CEM:** Taylor Family; Scott; Rye Cove **GS:** No **SP:** 1) Mary Magdalene, 2) Sarah Margaret (__) **VI:** No further data **P:** No **BLW:** No **PH:** No **SS:** A rec 1583 **BS:** 245

CARTER, John; b 4 Apr 1787, d 15 Oct 1842 **RU: Private, 31st VMR, Capt Thomas Robert's Company, Frederick Co, attached to 1st VMR (Taylor) **CEM:** Mt Hebron; Winchester; 305 E Boscawen St **GS:** Yes **SP:** Mar (Shenandoah Co, 16 Feb 1818) Susan Pitman (1795-1866), daug of Lawrence and Mary Catherine (__) Pitman **VI:** No further data **P:** No **BLW:** No **PH:** No **SS:** A rec 1618 **BS:** 93

*CARTER, John, Jr; b 1792, d 1872 **RU:** Private, Rockbridge Co company attached to 5th or 6th VMR (Coleman) **CEM:** Carter Family; Rockbridge; Kerr's Creek **GS:** Yes **SP:** Margaret Replogle **VI:** No further data **P:** No **BLW:** No **PH:** No **SS:** A recs 1638; 1639; 1640; or 1642 **BS:** 245

*CARTER, John R, Jr; b 10 Jun 1796, d 9 Dec 1869 **RU:** Private, 94th VMR, Capt Jeremiah H. Neill's Company, Lee County, attached to 7th VMR (Saunders) **CEM:** Carter Family; Scott; Hunter Valley Rd **GS:** Yes **SP:** Jemina W (__) (29 Jul 1797-11 Feb 1862) **VI:** No further data **P:** No **BLW:** No **PH:** No **SS:** A rec 1647; B pg 118 **BS:** 245

*CARTER, Littleberry (Littleburry); b 1795, died 4 Jun 1870 **RU:** Private 35th Regt U.S. Army and 90th VMR Capt Cornelius Sales Co, Amherst County, attached to 8th VMR (Walls) **CEM:** Carter Family; Bedford; Sheep Creek Rd, nr Prospect Baptist Church on farm of Eugene Boyer **GS:** Yes **SP:** 1) Nancy Stinnett, 2) (Bedford Co, 15 Mar 1848) Elizabeth F Parker (1830-3 Mar 1897). Widow recd pen **VI:** Recd BLW of 160 acres **P:** Widow **BLW:** Yes **PH:** No **SS:** B pg 38; M pg 76; O- Serv Index Card; Pen & BLW files Fold3 **BS:** 245

CARTER, Sanford, Jr; b 1785 Overwharton Parish; d Oct 1873 **RU: Private, 45th VMR (Peyton), Capt John Edrington's Company, Stafford Co **CEM:** Carter Family; Stafford; Ruby, formerly Tackett's Mill, Locust Hills **GS:** Yes **SP:** Hannah Read (__-bef Apr 1871) before soldier received pension **VI:** Son of James and Millender Carter. LNR P.O. Stafford Church, 1871. Recd pen $8.00 per mo commencing 14 Feb 1871 **P:** Yes **BLW:** Yes, Recd 80 acres in 1850 and another 80 acres in 1855 for 210 days of total service **PH:** No **SS:** A rec 1909; BD pg 353; B pg 190 **BS:** 49

*CARTER, Thomas, Jr; b 15 Feb 1792, d 16 Jun 1863 (Pen Rec) 11 Dec 1868 (Gr St) **RU:** Private 28th VMR, Capt William Scott's Co, Nelson County, attached to 8th VMR (Walls) **CEM:** Wilkinson-Carter Family; Nelson; off Rt 721 Norwood Rd, vic Carter's Creek **GS:** Yes **SP:** 1) Pollie Scruggs, 2) (20 Apr 1838) Elizabeth A Moore (1820-17 Feb 1895) who recd pen **VI:** Recd BLW **P:** Widow **BLW:** Yes **PH:** No **SS:** B pg 142; M pg 76; O- Serv Index Card; Pen & BLW files Fold3 **BS:** 245

*CARTER, Thomas, Sr; b 8 Feb 1767, d 14 Feb 1830 **RU:** Corporal, 8th VMR (Wall), Halifax Co **CEM:** Bennett Family; Halifax; Rt 49, 10 mi SE of South Boston **GS:** Unk **SP:** No spouse info **VI:** Son of Theodrick Carter and Judith Cunningham **P:** No **BLW:** No **PH:** No **SS:** A rec 1970 **BS:** 245

CARTER, William Fitzhugh; b 16 Aug 1782, Sudley, Prince William Co, d 1855 (Inv) **RU: Private, 16th VMR, Capt James H Fox's Company, Spotsylvania Co **CEM:** City Cemetery; Fredericksburg; William St & Washington Ave **GS:** Yes **SP:** Elizabeth Lucy Ball (16 Aug 1791-Jan 1855), daug of Capt Spencer Mottrom Ball and Elizabeth Landon **VI:** Son of John K Carter of Sudley **P:** No **BLW:** No **PH:** No **SS:** L pg 335 **BS:** 18 pg 6; 89 v4 CN-46; 245

*CASSIN, Stephen; b 6 Feb 1783, d 29 Aug 1857 **RU:** Master Commandant, U.S. schooner *Ticonderoga,* during the Battle of Plattsburg, NY on Lake Champlain, 11 Sep 1814 **CEM:** Arlington National; Arlington; off George Washington Parkway **GS:** Unk, buried in section 1, lot 299 **SP:** No spouse info **VI:** Obtained rank of Midshipman, 21 Feb, 1800 and Lieutenant, 12 Feb, 1807. After war he obtained rank of Captain, 3 March, 1825 with service at Norfolk. He was on the reserve list, 13 September, 1855 **P:** No **BLW:** No **PH:** No **SS:** AL **BS:** 53 pg19

Key: *Additional veteran entry   **Corrected veteran entry   ***Deleted veteran entry
RU=Rank/Unit   CEM=Cemetery   GS=Gravestone   SP=Spousal Information   VI=Other Veteran Info   P=Pension
BLW=Bounty/Land Warrant   PH=Photo   SS=Service Source   BS=Burial Source   VMR= VA Military Regiment
LNR= Last Known Residence

*CASTLEMAN, Alfred; b 22 Dec 1795, d 22 Apr 1880 **RU:** Private 31st VMR, Capt Eben (Elen)Taylor's Troop of Calvary, Frederick Co, attached to 1st VMR (Griffin Taylor) **CEM:** Runnymede Farm; Clarke; 3838 Shepherds Mill Rd, Berryville **GS:** Yes **SP:** Margaret Lucinda Milton (23 Dec 1808-10 Nov 1864) **VI:** Son of William Castleman (1762-1832) and Massey Osborn (__-1833); recd BLW of 40 and 120 acres and pen #31456 and 25069 **P:** Yes **BLW:** Yes **PH:** No **SS:** B pgs 80; 247; M pg 78; O- Serv Index Card; Pen & BLW files Fold3 **BS:** 245

*CATHER, James; b 15 Nov 1795, d 1 Sep 1875 **RU:** Sergeant, 122nd VMR, Lt James Ship's Detachment of Infantry, Frederick Co **CEM:** Back Creek Quaker; Frederick; Gainesboro **GS:** Yes **SP:** Nancy Ann Howard (1798-1878) **VI:** Son of Jasper Cather (1740-1812) and Sarah Moore (1732-1810); Applied for BLW 1855 and recd 160 acres **P:** No **BLW:** Yes **PH:** No **SS:** B pg 81; Fold 3 BLW application files **BS:** 245

*CHAMBERLAYNE, Lewis Webb; b 9 Jan 1798, King William Co, d 28 Jan 1854 **RU:** Corporal, 87th VMR, Capt James Ruffin's Artillery Co, King William County **CEM:** Brook Hill; Henrico; Montrose, Brook Hill **GS:** Yes **SP:** Mar (New Kent Co, 11 Apr 1820) Martha Burwell Dabney (15 Sep 1802-16 Mar 1888); recd pen **VI:** A founder of the Medical College of VA. Recd BLW of 80 acres **P:** Widow **BLW:** Yes **PH:** No **SS:** B pg 115; M pg 80; O- Pen files Fold3 **BS:** 245

*CHANDLER, Robert Bradenham; b 27 Nov 1791, d 19 Jul 1864 **RU:** Corporal, 52d VMR, Capt Robert Perkin's Co, New Kent County **CEM:** Chandler Family; New Kent; Laurel Springs **GS:** Unk **SP:** No spousal info **VI:** No further data **P:** No **BLW:** No **PH:** No **SS:** B pg 144; K pg 131 **BS:** 245

*CHANDLER, William; see page 104

*CHAPPELL, James; b c1790, d 1871 **RU:** Sergeant, 15th VMR, Capt William P Wyche's Troop of Cav, Sussex Co **CEM:** Chappell family; Sussex; Fowlkes Bridge Rd, betw Burtons Bridge and Rt 644 **GS:** Unk **SP:** Louisa Marshall Seay (1808-__), daug of Austin Seay **VI:** Son of John Chappell (__-1826) and Elizabeth (__) **P:** No **BLW:** No **PH:** No **SS:** B pg 195; L pg 853 **BS:** 245

*CHAPPELL, John; Sr; b Unk, d 1826 **RU:** Private, Capt William Ross's Troop of Cavalry, Dinwiddie Co **CEM:** Chappell Family; Amelia; Rt 644 toward Burton's bridge fr Paineville **GS:** Unk **SP:** Elizabeth (__) **VI:** Will dated (Amelia Co, 25 Oct 1825) **P:** No **BLW:** No **PH:** No **SS:** A rec 829; B pg 66 **BS:** 245

*CHEVALLIE, Peter Joseph; b 15 May 1791, NY, d 21 Feb 1837 **RU:** Corporal; 19th VMR, Capt Robert Gamble's Co, Richmond City **CEM:** Shockoe Hill; Richmond City; 100 Hospital St; **GS:** Yes **SP:** Elizabeth Gilliam (21 Oct 1796-26 Jul 1865) **VI:** Son of Jean Auguste Marie Chevallie and Sarah McGee; flower manufacturer **P:** No **BLW:** Applied for but rejected **PH:** Yes **SS:** A rec 2653; B pg 174; L pg 344 **BS:** 31; 245

**CHEWNING, Reuben; see page 104

*CHRISTIAN, James H; b 22 Jan 1794, d 11 Mar 1873 **RU:** Corporal, 52nd VMR Capt Jones Rivers Christian's Troop of Cavalry **CEM:** Manoah; Charles City; Rt 615, the Glebe Lane, Ruthville **GS:** Yes **SP:** 1) Susan Hill (__-1856), 2) (15 Jun 1859) Bettie Bowry (__-aft 1878) LNR Wilcox's Wharf, Charles City Co. Recd pen **VI:** Recd BLW 40 acres **P:** Widow **BLW:** Yes **PH:** No **SS:** B pg 143; M pg 84; O- pen files Fold3 **BS:** 245

*CLAIBORNE, Ferdinand Leigh; b 9 May 1772, Sussex Co, d 18 Mar 1815, Adams Co, MI **RU:** Brigadier General, Brigade Commander of Volunteers from Mississippi and Louisiana at Fort Stoddard 30 Jul 1813 **CEM:** Claiborne Family; Franklin; Claibrook Plantation, Rocky Mount **GS:** Yes **SP:** Mary Magdalene Hutchins, daug of Col Anthony Hutchins **VI:** Son of William Claiborne and Mary Leigh. Originally buried in MI but reinterred in family cemetery in Rocky Mount 1862; He became second Governor of the Mississippi Territory **P:** No **BLW:** No **PH:** No **SS:** A rec 2325; T pg 13 **BS:** 245

CLARK, James; b 30 Dec 1789; d 30 Dec 1847 **RU: Private, Lt Col Abraham Bradley's Regiment, 17th Brigade **CEM:** Glade Spring Presbyterian; Washington; 33234 Lee St, Glade Springs **GS:** Yes **SP:** Eleanor Porterfield (__-28 Mar 1846, age 56 yrs, 11 mos, 26 days) **VI:** No further data **P:** No **BLW:** No **PH:** No **SS:** A rec 787 **BS:** 116 pg 177

Key: *Additional veteran entry   **Corrected veteran entry   ***Deleted veteran entry
RU=Rank/Unit   CEM=Cemetery   GS=Gravestone   SP=Spousal Information   VI=Other Veteran Info   P=Pension
BLW=Bounty/Land Warrant   PH=Photo   SS=Service Source   BS=Burial Source   VMR= VA Military Regiment
LNR= Last Known Residence

CLARK, James, Sr; b 21 Mar 1789, d 6 Aug 1871 **RU: Private, 5th VMR **CEM:** Glade Spring Presbyterian; Washington; 33234 Lee St, Glade Springs **GS:** Yes **SP:** Ann Ryburn (4 Oct 1792-21 Dec 1869) **VI:** Son of James Clark, Sr (1754-1818) and Isabella Mary Breckenridge (1764-1848) **P:** No **BLW:** No **PH:** N **SS:** A rec 862 **BS:** 116 pg 177

*CLARK, John; b 1 Mar 1782, KY, d 26 Jul 1865 **RU:** Corporal 35th VMR, Capt Samuel Graham's Co, Wythe County, attached to 4th VMR (McDowell, Koontz, Chilton) **CEM:** McMillan; Wythe; Clarkes Summit **GS:** Unk **SP:** 1) Mary Rebecca Seybert, (24 Jun 1787-11 Dec 1832), 2) Mary Crawford (1811-1876) **VI:** Recd BLW 80 acres in 1850 **P:** No **BLW:** Yes **PH:** No **SS:** B pgs 204; 216 **BS:** 245 cites Kegley, Mary, *Early Adventures on the Western Waters,* vol II, part 2

*CLARK, Nelson H; b 1793 (pen rec) Bedford Co, d 7 Aug 1871 (GS) 19 Sep 1871(BLW Rec) Lexington City **RU:** Private, 90th VMR, Capt Samuel B Jeter's Co of Artillery, Amherst County, attached to 7th VMR (Gray) **CEM:** Clark Family; Rockbridge; no specific directions. Findagrave shows photo of large cemetery in wooded area on a knoll **GS:** Unk **SP:** No spouse listed however 2 children at death **VI:** Recd pen $8.00, 1871; BLW 80 acres 1850 and 80 acres 1855 **P:** Yes **BLW:** Yes **PH:** No **SS:** B pgs 38; M pg 85; O- BLW files Fold3 **BS:** 245

*CLARK, Robert; b 12 Jan 1796, Rockbridge Co, d 06 Aug 1873, **RU:** Private, 8th VMR, Capt James Elliott's Co, Rockbridge County **CEM:** Falling Spring Presbyterian Church; Rockbridge; Hickory Hill **GS:** Yes **SP:** Mar (14 Apr 1825) Sarah Jane Wilson **VI:** Son of Robert Clark and Phebe Beach **P:** Self and widow **BLW:** Yes **PH:** No **SS:** B pg 179; BD pg 402 **BS:** 245

*CLARKE, Samuel W; b 1795 CT, d 15 Mar 1851, **RU:** Private, 8th CT Militia (Belchers) **CEM:** Western State Hospital; Staunton City; Village Drive **GS:** No **SP:** No further info **VI:** Was a Tinnier in Buckingham Co. Death fr cerebral excitement **P:** No **BLW:** No

*CLARKSON, Richard; b 4 Sep 1778, d 21 Dec 1824 **RU:** Private; 6th VMR, Capt Arthur S Brockenbrough' s Co, Essex County, attached to 1st VMR (Clarke) and 2nd VMR (Sharp) **CEM:** Clarkson Family; Essex; GPS: 37.93851, -76.95794; on Rt 627, .25 mi fr jct with Rt 717 **GS:** Unk **SP:** Mar (Essex Co, 18 Nov 1813) Susan Lorinda Chittenden (3 May 1795-15 Aug 1872), daug of Lemuel Crittenden and Susannah Fisher **VI:** Son of James Clarkson (16 Mar 1749-24 Sep1872) and Mary Adams (__-17 Aug 1830) **P:** No **BLW:** No **PH:** No **SS:** B pg 69; K pg 477 **BS:** 245

*CLAYTOR, Robert Mitchell; 25 Dec 1792, d 03 Jul 1865 **RU:** Private, 1st Cav, Captain James Leftwich' s, Troop of Cavalry, Bedford Co, attached to Maj Woodford's Squadron **CEM:** Longwood; Bedford; jct Oakwood & Rt 221 **GS:** Yes **SP:** Mar (Bedford Co, 30 Jan 1817) Julia Ann Graham (1796-1880), LNR Liberty, Bedford, 1878 **VI:** No further info **P:** Spouse **BLW:** Yes **PH:** No **SS:** B pg 43; BD pg 408 M pg 87 **BS:** 245

CLEEK, John; b 1777, Rockbridge Co; d 16 or 26 Apr 1848 **RU: Private, 32nd VMR Captain James Todd's Co Augusta County, attached to Flying Camp (Mc Dowell's), **CEM:** Cleek Family; Bath; Rt 220 12.5 mi N of Warm Springs **GS:** Yes **SP:** Jane Gwin (1 Jul 178x-16 Oct 1856). She was born & died Bath Co **VI:** War of 1812 service inscribed on stone **P:** No **BLW:** No **PH:** No **SS:** B pg 48 **BS:** 212

CLOYD, Thomas; b 21 Aug 1774; d 27 Jul 1849 **RU: Private, 1st VMR (Connell), Leftwich' s Brigade **CEM:** New Dublin Presbyterian; Pulaski; 5331 New Dublin Church Rd, Dublin **GS:** Yes **SP:** McGavock (31 Aug 1788-15 Feb 1866), daug of Hugh McGavock and Nancy Kent **VI:** Service is indexed as Lloyd, Son of Joseph Cloyd (1742-1833) and Mary Gordon (1750-__) **P:** No **BLW:** No **PH:** No **SS:** A rec 22261 **BS:** 254 pg 196; 245

*COCKE, John; b 24 Feb 1798, d 27 Sep 1879 **RU:** Private 38th VMR, Capt Charles Hopkin's Co, Goochland County, attached to 1st VMR (Clarke) Also was a drummer in Capt John E Cocke's Co **CEM:** Cedar Grove; Portsmouth; Fort Lane **GS:** Yes **SP:** Mar (Great Bridge, Chesapeake) Ann Bressie Webb (Sep 1795-25 Feb 1868) **VI:** No further data **P:** No **BLW:** No **PH:** No **SS:** B pgs 82, 223; K pg 485 **BS:** 245

Key: *Additional veteran entry   **Corrected veteran entry   ***Deleted veteran entry
RU=Rank/Unit   CEM=Cemetery   GS=Gravestone   SP=Spousal Information   VI=Other Veteran Info   P=Pension
BLW=Bounty/Land Warrant   PH=Photo   SS=Service Source   BS=Burial Source   VMR= VA Military Regiment
LNR= Last Known Residence

*COCKE, John E or F; b 19 Jan 1777, d 10 Jun 1841 **RU:** Captain, 102 VMR, Commanded company in Powhatan Co, attached to 1st Corps D 'Elite (Randolph) **CEM:** Wagner Community; Pittsylvania; Rt 790 Piney Rd **GS:** Yes **SP:** Mar (Powhatan Co, Harriet West) **VI:** No further data **P:** No **BLW:** No **PH:** No **SS:** A rec 1255; B pg 163; L pg 231 **BS:** 245

*COCKE, William Irby; b 2 Nov 1790, Surry Co, d 2 May 1853 **RU:** Sergeant, 71st VMR (Allen), Surry Co **CEM:** Portlock; Portsmouth; a portion of Oak Grove cem, jct Peninsula Ave & London St **GS:** Yes buried in plot 9, # 429 **SP:** No spouse info **VI:** A Doctor **P:** No **BLW:** No **PH:** No **SS:** A rec 1294; B pg 192 **BS:** 245

*COCKRELL, Samuel; b unk, d 13 Apr 1855 **RU:** Private, 56th VMR, company in Loudoun Co. **CEM:** Saint Pauls Lutheran Church; Loudoun; Neersville **GS:** Yes **SP:** Elizabeth (__) (1791-1863) **VI:** No further data **P:** No **BLW:** No **PH:** No **SS:** A rec 484 **BS:** 245

*COFFMAN, Daniel; b 8 Feb 1794, d 21 Jan 1880 **RU:** Private, 13th or 97th VMR, Capt Samuel Colville's Co, Shenandoah County, attached to 6th VMR (Coleman) **CEM:** Zion Lutheran Church; Shenandoah; Rt 707; Hamburg **GS:** No **SP:** Rebecca White (1794-1861) **VI:** Son of Daniel Kauffman (1764-1807) and Barbara Houtz (1762-__). Recd pen and BLW of 40 and 120 acres **P:** Yes **BLW:** Yes **PH:** No **SS:** B pg 184; M pg 90; O-Pen files Fold3 **BS:** 245 cites Obituary Shenandoah Herald; Woodstock, (25 Feb 1880) providing age at death, 1812 service and place of death

*COINER, Michael; b 9 Oct 1790, d 20 Jun 1864 **RU:** Sergeant, 2nd Corps D 'Elite commanded by Lt. Colonel Moses Green **CEM:** Trinity Lutheran; Augusta; River Rd (Rt 12), Crimora **GS:** Yes **SP:** Catharine Koiner (16 Dec 1797- 23 Jul 1866), daug of George & Susan Hawpe Koiner **VI:** Son of Kasper & Anna Margaret Barger Koiner **P:** No **BLW:** No **SS:** K pg 220 **BS:** 245

*COLE, John Gatewood; b 30 May 1792, d Nov 1864 **RU:** Sergeant, 30th VMR, Caroline Co **CEM:** Old Salem Baptist Churchyard; Caroline; Nr Alps **GS:** Unk **SP:** Nancy Robinson Broadus (12 Nov 1794-4 Mar 1863) **VI:** Son of Susannah (__) **P:** No **BLW:** No **PH:** No **SS:** A rec 508; B pg 56; L pg 445 **BS:** 245

*COLEMAN, Thomas; b 1788, d 21 Jun 1834 **RU:** Captain, 68th VMR, Company Commander, James City Co. **CEM:** Bruton Parish Episcopal Church; Williamsburg; 331 W Duke of Gloucester St **GS:** Yes **SP:** Francis Catherine Hill, daug of Baylor Hill **VI:** Son of William Coleman and Elizabeth Holt **P:** No **BLW:** No **PH:** No **SS:** B pg 105; L pg 242 **BS:** 245

*COLEMAN, William; b 1776, d 8 Dec 1828 **RU:** Private, 56th or 77th VMR, Loudoun Co in company attached to 98th VMR (Green) **CEM:** Leesburg Presbyterian Church; Loudoun; Leesburg **GS:** Unk **SP:** No spouse Info **VI:** No further data **P:** No **BLW:** No **PH:** No **SS:** A rec 1305 **BS:** 245

*COLEMAN, William Harris; b 10 Jan 1787, d 7 Sep 1842 **RU:** Corporal, 88th VMR, Capt Samuel Carr's Company, Albemarle Co, attached to 1st Corps D 'Elite (Randolph) **CEM:** Coleman Family; Nelson; Wintergreen **GS:** Unk **SP:** No spouse Info **VI:** No further data **P:** No **BLW:** No **PH:** No **SS:** B pg 35 **BS:** 245

COLLINS, Charles; b unk; d 18 Jun 1853, Mount View **RU: Private; 30th VMR, Capt William F. Gray's Company, Caroline Co, attached to 9th VMR (Boyd) **CEM:** Collins Family; Caroline; Rt 638 **GS:** Yes **SP:** Mar (Caroline Co, 21 Feb 1822) Catherine Jesse (1810-May 1887), daug of John Jesse & Mary Todd **VI:** Son of James & Clarissa (Todd) Collins **P:** Spouse **BLW:** Yes **PH:** No **SS:** G; L pg 32; M pg 92; BD pg 435; B pg 55 **BS:** 10 pg 44

*COMPHER, Peter; b 21 May 1793, d 3 Mar 1886 **RU:** Private 57th VMR Capt George Huff's Co, Loudoun County **CEM:** New Jerusalem Lutheran Church; Loudoun; 12942 Lutheran Church Rd, Lovettsville **GS:** Yes **SP:** Margaret Spring (13 Jan-11 Sep 1867) **VI:** Recd pen $8.00 per mo and BLW 120 acres **P:** Yes **BLW:** Yes **PH:** No **SS:** B pg 120; M pg 94; O-Pen files Fold3 **BS:** 245

Key: *Additional veteran entry   **Corrected veteran entry   ***Deleted veteran entry
RU=Rank/Unit   CEM=Cemetery   GS=Gravestone   SP=Spousal Information   VI=Other Veteran Info   P=Pension
BLW=Bounty/Land Warrant PH=Photo   SS=Service Source BS=Burial Source   VMR= VA Military Regiment
LNR= Last Known Residence

25

*COMPHER, Peter; b 24 Aug 1776, d 5 Nov 1858 **RU:** Private 61st VMR, Capt Gabriel Hughes's Co, Mathews County **CEM:** New Jerusalem Lutheran Church; Loudoun; 12942 Lutheran Church Rd, Lovettsville **GS:** Yes **SP:** Mar 1) (16 Mar 1816) Susanna (__) (12 Mar 1779-5 Apr 1815), 2) Anna Catherine Kadel (Cordell) (28 May 1785-9 Oct 1860), daug of Johann Adam Cordell and Elizabeth A Prill **VI:** Recd pen, resided after war in Loudoun Co **P:** Yes **BLW:** No **PH:** No **SS:** B pg 198; BD pg 442 **BS:** 245

*CONRAD, Jacob; b 15 Sep 1787, d 17 Sep 1841 **RU:** Private, 5th and 116th VMR, Rockingham Co, company attached to 5th or 6th VMR **CEM:** Elk Run; Rockingham; Rockingham and Spotswood Aves, Elkton **GS:** Yes **SP:** No spouse info **VI:** No further data **P:** No **BLW:** No **PH:** No **SS:** A rec 116; 177 **BS:** 245

*CONRAD, John; b 18 Oct 1788, d 23 Jun 1853 **RU:** Private, 15th VMR, Hardy Co (now WVA) **CEM:** East Point; Rockingham; Elkton **GS:** Yes **SP:** Anna Maria Nicholas (1788-1830) **VI:** No further data **P:** No **BLW:** No **PH:** No **SS:** A rec 136; B pg 96 **BS:** 245

*CONWAY, John F; b 29 Feb 1782, d 10 Nov 1856 **RU:** Private, 82nd VMR, Capt George C. Allen's Company, Madison Co, attached to 1st VMR (Clarke) **CEM:** John F. Conway; Madison; Rt 230 before the Conway River bridge **GS:** Yes **SP:** Mary F (__) (21 Oct 1793-3 Mar 1865) **VI:** No further data **P:** No **BLW:** No **PH:** No **SS:** A rec 384; B pg 126 **BS:** 30; Family Cem

CONWAY, Thomas Barrett; b 16 Sep 1779; d 1 Dec 1825 **RU: Ensign, 19th VMR, Capt Wilson Bryan's Co, Lt Col Ambler's Regiment, Richmond City **CEM:** Miller Family; Stafford; On a hill off Rt 692; Quarry Rd **GS:** Yes **SP:** No spouse info **VI:** Was residing in Falmouth, Stafford Co according to 1810 and 1820 US federal census, obituary Richmond newspapers Enquirer and Whig **P:** No **BLW:** No **PH:** No **SS:** K pg 359 **BS:** 26 pg 276

COOK, Jacob; b c1790; d c1869 **RU: Sergeant, VMR, Battalion of Artillery **CEM:** Sangersville; Augusta; nr Sangersville **GS:** Yes **SP:** Mar (Rockbridge Co, 11 Apr 1811) Mary Elizabeth Clylee **VI:** Died age 79 yrs, 1 mo, 8 days, thus assuming age 21 at time of marriage, (c1790-c1869) **P:** No **BLW:** No **PH:** No **SS:** A rec 67 **BS:** 183

*COOKE, Warner T; b Unk, d aft 1814 **RU:** Private, 87th VMR, December 1814, King William Co **CEM:** Ware Episcopal Church; Gloucester; 7825 John Clayton Memorial Rd **GS:** Footstone only **SP:** Ellen C Booth (1777-15 Apr 1820), daug of Dr Mordecai & Eliza M (__) Booth **VI:** No further data **P:** No **BLW:** No **PH:** No **SS:** A rec box 45 mfilm M602 **BS:** Church website burial list

*COOKE, William, Jr; see page 104

*COOPER, George; b 1771, d 7 Mar 1837 **RU:** Private, 13th or 97th VMR, a company of Shenandoah Co attached to 4th VMR **CEM:** Benjamin Bowman; Shenandoah; Narrow Passage **GS:** Unk **SP:** Mary Bowman (1777-1849) **VI:** No further data **P:** No **BLW:** No **PH:** No **SS:** A rec 173; B pgs 184-5 **BS:** 245

*COOPER, George; b 14 Feb 1770, d 18 Aug 1846 **RU:** Private, 56th VMR, probably Captain Thomas Gregg's Co, Loudoun County, attached to 4th VMR (Beatty) **CEM:** New Jerusalem Lutheran Church; Loudoun; Lovettsville **GS:** Yes **SP:** Mary (__) (1774-1858) **VI:** Son of Michael Cooper (1742-1815); need to obtain service record to verify service unit **P:** No **BLW:** No **PH:** No **SS:** A rec 173; B pgs 119-121 **BS:** 245

*COOPER, Henry; b 30 Jun 1794, d 7 May 1869 **RU:** Private, 31st, 51st or 122nd VMR, Frederick Co company attached to 4th VMR (specific company may be obtained from NARA) **CEM:** Richards Family; Frederick; Mountain Falls **GS:** Yes **SP:** Magdalene Eshelman (1801-1881) **VI:** Son of Frederick Cooper (1764-1815) & Hannah Richards (1770-1857) **P:** No **BLW:** No **PH:** No **SS:** A rec 202; B pgs 78-80 **BS:** 245

COOPER, Jacob; b 1782, Chester Co, PA, d 25 Aug 1855, **RU: Private, 47th VMR, Capt Robert McCulloch's Company, Albemarle Co, attached to 7th VMR (Gray) **CEM:** Cooper; Kent; Heslep; Roanoke; W of Salem **GS:** Yes **SP:** Rebecca Barnett (1786-1850) **VI:** Son of John Cooper and Barbara Traut **P:** No **BLW:** No **PH:** No **SS:** K pg 343; A rec 233 **BS:** 157 pg 98; 245

Key: *Additional veteran entry   **Corrected veteran entry   ***Deleted veteran entry
RU=Rank/Unit   CEM=Cemetery   GS=Gravestone   SP=Spousal Information   VI=Other Veteran Info   P=Pension
BLW=Bounty/Land Warrant   PH=Photo   SS=Service Source   BS=Burial Source   VMR= VA Military Regiment
LNR= Last Known Residence

*COOPER, John; b 1 Aug 1797, d 12 Dec 1863 **RU:** Private, 56th VMR, Loudoun Co **CEM:** New Jerusalem Lutheran Church; Loudoun; Lovettsville **GS:** Yes **SP:** No spouse info **VI:** Son of George Cooper (1770-1846) and Mary (__) (1774-1858) **P:** No **BLW:** No **PH:** No **SS:** A rec 418 **BS:** 245

*COOPER, John, Sr; b 1794, d 1833 **RU:** Private, 56th VMR, Loudoun Co **CEM:** New Jerusalem Lutheran Church; Loudoun; Lovettsville **GS:** Yes **SP:** No spouse info **VI:** No further data **P:** No **BLW:** No **PH:** No **SS:** A rec 417 **BS:** 245

*COOPER, John, Sr; b 29 Feb 1782 d 10 Nov 1856 **RU:** Private, 52st VMR, Captain James H. Sower's Co, Frederick County, attached to Flying Camp (McDowell's) **CEM:** New Jerusalem Lutheran Church; Loudoun; Lovettsville **GS:** Yes **SP:** No spouse info **VI:** No further data **P:** No **BLW:** No **PH:** No **SS:** B pg 80; K pg 33 **BS:** 245

*COOPER, Joseph; b 9 Dec 1783; d 5 Apr 1859 **RU:** Private, 122nd VMR, Frederick Co attached to 1st VMR (Taylor) **CEM:** St Pauls Lutheran Church; Shenandoah; Strasburg **GS:** Yes **SP:** Mar (Shenandoah Co, 17 Jun 1820) Rebecca Ann Hoffman (__-1839) **VI:** No further data **P:** No **BLW:** No **PH:** No **SS:** A rec 463 **BS:** 245

*COOPER, Samuel; b 13 Feb 1790, d 28 Aug 1829 **RU:** Corporal, 13th VMR or 97th VMR, Captain Peter Hay's Company, Shenandoah Co, attached to 1st VMR(Yancey) **CEM:** Saint Pauls Lutheran Church; Shenandoah; Strasburg **GS:** Yes **SP:** No spouse info **VI:** No further data **P:** No **BLW:** No **PH:** No **SS:** B pg 242 **BS:** 245

COOPER, Samuel; b 12 Jun 1798; d 03 Dec 1876 **RU: Private, Major Phillip Alexander's Independent Battalion of VA Volunteers **CEM:** Christ Church Episcopal; Alexandria; Wilkes & Hamilton **GS:** Yes **SP:** Sarah Maria Mason (1800-1890) **VI:** After war in 1815 graduated from US Military Academy, commissioned LT of Artillery, in 1837 appointed Chief Clerk, US War Department, then served in Seminole War, promoted to Colonel, then in Mexican War appointed Adjutant General in 1852. Resigned Commission 1861 and in Confederate Army promoted to 4 Star General. In this capacity, after the war turned over Confederate Army records to US Government **P:** No **BLW:** No **PH:** No **SS:** A rec 585 **BS:** 34 pg 92; 245

*COOTES, Samuel; b 1792, Berks Co, PA, d Mar 1882 **RU:** Private, 19th VMR, Capt William H. Richardson's Co Richmond City, attached to 1st Corps D 'Elite' (Randolph) **CEM:** Woodbine; Harrisonburg City; Cnr E. Market & Ott St **GS:** Yes **SP:** 1) Emily Graham, 2) Paulina Neff (__-1864) **VI:** Magistrate and elected as Rockingham Co's representative to VA Legislature; LNR Coote's Store, Oakland, Rockingham Co,1878 **P:** Yes **BLW:** Yes **PH:** No **SS:** B pg 175, 240; M pg 97; O- pen files Fold3 **BS:** 245

COPENHAVER, George; b 6 May 1792; d 14 Mar 1822 **RU: Private, 4th VMR (Boyd) **CEM:** Mt Hebron; Winchester; 305 E Boscawen St **GS:** Yes **SP:** Rebecca (__) (__-18 Jul 1824 age 32) or/and Susan Crebs (1794-1853) **VI:** No further data **P:** No **BLW:** No **PH:** No **SS:** A rec 1008; 1025 **BS:** 87 pg 46; 245

*COPENHAVER, Michael; b 1759, d 18 Nov 1834 **RU:** Private; 31st VMR, Capt Michael Coyle's Co, Frederick County, attached to 1st VMR (Taylors) **CEM:** Mt Hebron; Winchester; 305 E Boscawen St **GS:** Yes **SP:** Mar (Frederick County, 9 Mar 1786) Margaret Price **VI:** No further data **P:** No **BLW:** No **PH:** Yes **SS:** A rec 1015; S pg 106 **BS:** 86 pg 43

*CORNETT, Reuben; b 7 Dec 1790, d 14 Feb 1875 **RU:** Private, 78th VMR, Capt Lewis Hail's Co, Grayson County, attached to 4th VMR (Koontz) **CEM:** Reuben Cornett Family; Grayson; Corners Rock Rd proceed1.6 mi to private rd on left, thence is on left 400 yds in field at top of knoll **GS:** Yes **SP:** Mar (Grayson Co, 19 Oct 1816) Celia Pennington **VI:** Son of David Canute Cornett and Aldine Sarah Platt. Recd pen $8.00 mo commencing (16 Sep 1872), and recd BLW 80 acres. LNR P.O. Bridle Creek, Grayson Co **P:** Yes **BLW:** Yes **PH:** No **SS:** B pg 86; M pg 99; O- pen files Fold3 **BS:** 245

Key: *Additional veteran entry   **Corrected veteran entry   ***Deleted veteran entry
RU=Rank/Unit   CEM=Cemetery   GS=Gravestone   SP=Spousal Information   VI=Other Veteran Info   P=Pension
BLW=Bounty/Land Warrant   PH=Photo   SS=Service Source   BS=Burial Source   VMR= VA Military Regiment
LNR= Last Known Residence

*CORNICK, Lemuel, III; b 24 May 1777, d 16 Mar 1836 **RU:** Captain, 20th VMR, commanded a company, Princess Anne County **CEM:** Atwood-Cornick Family; Virginia Beach; Broad Bay Farm, Great Neck Rd **GS:** Yes **SP:** Mary (__) (12 Apr 1783-13 Jun 1839) **VI:** Son of Lemuel Cornick, II (24 May 1749-16 Mar 1804) **P:** No **BLW:** No **PH:** Yes **SS:** B pg 164-5 **BS:** 245

*COTTRELL, Samuel Smith Sr; b 27 Jan 1782, Henrico Co, d 14 Jul 1855 **RU:** Private, 33rd VMR, Captain Samuel Brown's Co, Henrico Co. **CEM:** Hollywood; Richmond City; 412 S Cherry St **GS:** Unk **SP:** Mar (1816) Sarah Warrock Shephers (3 Jul 1797-23 Apr 1855), daug of Reuben Shepherd and Sarah Cocke **VI:** Son of Charles Waddell and Mary (__) **P:** No **BLW:** No **PH:** No **SS:** L pg 177 **BS:** 245

*COTTRELL, William; b 13 Jun 1785, d 7 Jan 1838 **RU:** Private, 33rd VMR, Captain Samuel Brown's Company, Henrico Co. **CEM:** Hollywood; Richmond City; 412 S Cherry St **GS:** No **SP:** No spouse info **VI:** No further data **P:** No **BLW:** No **PH:** No **SS:** L pg 177 **BS:** 245

*COULLING (COULING), James Mathias; b 25 Mar 1786, Baltimore, MD, d 24 Dec 1863 **RU:** Sergeant, 19th VMR, Capt Anthony Turner's Co, attached to 6th VMR (Coleman) **CEM:** Shockoe Hill; Richmond City; 100 Hospital St **GS:** No **SP:** Mar (19 Aug1812) Mary Warrock DuVal (1787-1852) **VI:** Son of James Coulling and Elizabeth Sullers **P:** No **BLW:** No **PH:** No **SS:** A rec 917; B pg 175; L pg 788 **BS:** 245

*COUNCIL, James; b Unk, d 8 Aug 1871 **RU:** Fifer, 65th VMR, Capt Rice B Pierce Co, Southampton Co, attached to 1st VMR (Allen) **CEM:** Council Family; Southampton; Franklin City **GS:** Unk **SP:** Mar 1) Parthenia (__), 2) (Halifax Co, NC, 15 Mar 1858) Sarah M. Slade who as widow remarried (4 Jan 1883) (__) Turner. Her LNR Vicksville, Southampton 1885; recd pension **VI:** Recd BLW of 160 acres **P:** Widow **BLW:** Yes **PH:** No **SS:** B pg 187; M pg 100; O- pen files Fold3 **BS:** 245

COX, Berryman; b c1787; d 4 Jan 1866/67 **RU: Private, 45th VMR, Capt Thomas Alexander's Company, Stafford Co **CEM:** Bloxton Family #3; Stafford; 122 Newton Road **GS:** Yes **SP:** Mar (Falmouth, Jan 1818) Delila Payne, LNR (Stafford Co, 1878) **VI:** No further data **P:** Widow pension commencing 9 Mar 1878 of $8.00 per mo **BLW:** Yes, recd 40 acres of bounty land on 23 Jan 1852 **PH:** No **SS:** BD pg 482; B pg 190; K pg 124 **BS:** 26 pg A17; 49

*COX. David "River Dave"; b 28 Nov 1790, d 14 Sep 1878 **RU:** Private, 78th VMR, Grayson Co **CEM:** Cox Chapel; Grayson; Independence **GS:** Yes **SP:** Irene Jane Doughton (27 Aug 1793-9 Mar 1883), daug of Joseph Bain Doughton and Mary Ann "Polly" Reeves **VI:** Son of Joshua McGowen Cox (1766-1820) and Ruth Osborne (1770-1851) **P:** No **BLW:** No **PH:** No **SS:** A rec 409; B pg 86 **BS:** 245

*COX, David Leach; see page 104

*COX, James Robertson; b 25 Jun 1797, Wythe Co, d 1850 **RU:** Private, 78th VMR, a company in Grayson Co, attached to 4th VMR (McDowell, Koontz, Chilton) **CEM:** James Cox Family; Grayson; Bridle Creek **GS:** No **SP:** Mar 1) possibly Sarah (__), 2) Elizabeth Belcher with whom he moved to KY **VI:** Son of James Cox (1763-1842) and Elizabeth (__) (24 Feb 1753-__), widow of Timothy Terrill **P:** No **BLW:** No **PH:** No **SS:** A rec 282; B pg 86 **BS:** 245

*COX, John; b 4 Apr 1771, d 23 Jun 1856 **RU:** Sergeant, 78th VMR, Captain Timothy Dalton or Captain Lewis Haiz (Hale) Co, Grayson County, attached to 4th VMR (McDowell, Koontz, Chilton) **CEM:** Potato Creek; Grayson; mouth of Wilson Creek **GS:** Yes **SP:** Lucy Terrill (1780-1852) **VI:** Son of David Cox (1737-1818) and Margaret Ann McGowan. **P:** No **BLW:** No **PH:** No **SS:** A rec 380; B pg 86 **BS:** 245

*COX, John; b 1788, d 1850 **RU:** Private, 78th VMR, a company in Grayson Co **CEM:** LT David Cox; Grayson; Baywood **GS:** No **SP:** Mar (13 Sep 1821) Eleanor "Nellie" Ward **VI:** Family monument in row 2 of Bridle Creek Methodist Cemetery states burial in this family cem; son of Joshua McGowan Cox and Ruth Osborne **P:** No **BLW:** No **PH:** No **SS:** A rec 409; B pg 86 **BS:** 245

*COX: William: b 1783, Fauquier Co, d 1835 **RU:** Private; 44th VMR, Captain Enoch Jeffries' Co, Fauquier County, attached to 1st VMR (Clarke) **CEM:** Cox Homestead; Warren; Howellsville **GS:** Unk **SP:** Mar (1820) Nancy McDonald (11 Sep 1793-__) **VI:** Son of Samuel Cox (1755, Ireland-12 Jul 1828) and Elizabeth (__) (1765-aft 1820) **P:** No **BLW:** No **PH:** No **SS:** A rec 646; B pg 74 **BS:** 245

Key: *Additional veteran entry   **Corrected veteran entry   ***Deleted veteran entry
RU=Rank/Unit   CEM=Cemetery   GS=Gravestone   SP=Spousal Information   VI=Other Veteran Info   P=Pension
BLW=Bounty/Land Warrant   PH=Photo   SS=Service Source   BS=Burial Source   VMR= VA Military Regiment
LNR= Last Known Residence

*COX, William; b 19 Sep 1765, d 1849 **RU**: Private, 35th VMR, Lt Braxton Davenport, Wythe County, attached to 4th VMR (Boyd) **CEM**: Cox Family; Smyth; vic jct rts 610 & 622 **GS**: Unk **SP**: No spouse info **VI**: No further data **P**: No **BLW**: No **PH**: No **SS**: A box 48, mfilm M602; B pg 227 **BS**: 245

*COYLE, Michael; b unk, d aft 1820 **RU**: Captain 31st VMR, company commander, Frederick Co, attached to 1st VMR (Taylor) **CEM**: Mt Hebron; Winchester; 305 E Boscawen St **GS**: Yes **SP**: No spouse info **VI**: Listed as head of household in the 1810 and 1820 US Census in Winchester. A person this name was a private in the Revolutionary War in MD **P**: No **BLW**: No **PH**: No **SS**: B pg 79 **BS**: Named on DAR plaque in cemetery

*CRALLE, Darius Griffin; b bef 1799, d 7 Jul 1848 **RU**: Private, 37th VMR, Captain William Henderson's Co, Northumberland County **CEM**: Gibeon Baptist Church; Northumberland; 48 John Deere Rd, Gibeon **GS**: Yes, plat of the cem lists his plat # as 191 **SP**: Mariah Gatewood Gordon (1805-1885) **VI**: One of the founders of the church where he is buried. Gravestone gives erroneous birth date of 1801 **P**: No **BLW**: No **PH**: No **SS**: B pg 153; K pg 391 **BS**: 245

*CRAWFORD, James; b 16 Feb 1794, d 6 Jul 1874 **RU**: Private, 116th VMR, Capt James Mallory's Co, Rockingham County, attached to 5th VMR (McDowell) **RU**: Augusta Stone Presbyterian Church; Augusta; Fort Defiance **GS**: Yes **SP**: Cynthia A McClung (1807-1898) **VI**: Styled "Major" on GS, rank obtained after war period **P**: No **BLW**: No **PH**: No **SS**: A rec 185; B pg 182; K pg 176 **BS**: 245

*CRAWFORD, Robert, b 1791, d 13 Feb 1852 **RU**: Corporal, 8th VMR, served in either Captain Archibald's or Captain Isiah Mc Bride's Co, Rockbridge County, attached to 5th VMR (McDowell) **CEM**: Walkerland; Rockbridge; Rt 602; Walkers Creek Rd, top of hill, behind Maxwelton Cabins bef jct with Rt 724 **GS**: Yes **SP**: No spousal info **VI**: No further data **P**: No **BLW**: No **PH**: No **SS**: A rec 345; B pg 179; **BS**: 245

*CREWS, Isaac Seat; b 1788, d 1883 **RU**: Private, 69th or 84th VMR, Capt William Bailey's Co, Halifax County, attached to 4th VMR (Greenhill) **CEM**: Crews Family, AKA William R Carr Farm; Halifax; one mi W of Ellis Creek Church **GS**: No **SP**: Mar (Halifax Co, 1 Jun 1809) Susan L. Sutherlin **VI**: Memorialized on family Gr St at the Ellis Creek Bapt Ch. LNR Martin's Store, Halifax Co, 1878. Recd BLW of 40 acres and 120 acres **P**: Yes **BLW**: Yes **PH**: No **SS**: B pg 89; M pg 104; O- pen & BLW files Fold3 **BS**: 245

CRIDLER, John; b Jan 1789, d 16 Nov 1854 **RU: Private, 57th VMR, Capt Thomas Wilkinson's Artillery Co, Loudoun County, attached to 6th VMR (Read) **CEM**: Old Stone Methodist; Loudoun; 110 Cornwall St, Leesburg **GS**: Yes **SP**: 1) Elizabeth (__) (__-14 Dec 1852) John Crider (not Cridler), 2) Elizabeth Taylor (Loudoun Co, 19 Nov 1812) **VI**: No further data **P**: No **BLW**: No **PH**: No **SS**: A rec 1448; B pg 122 **BS**: 73 pg 70; 245

*CRIGLER, Lewis; b 10 Jun 1799, d 15 Jul 1855 **RU**: Corporal, 1st VMR (Taylor) **CEM**: Lewis; Madison; Crigler, Rt 636, 0.2 miles E of intersection with Rt 637 **GS**: Yes **SP**: No spouse info **VI**: No further data **P**: No **BLW**: No **PH**: No **SS**: A rec 1466 **BS**: 30; Family Cem

*CROUCH, John Thomas; b1793, d 17 Jan 1878 **RU**: Private, 10th & 91st VMR, Capt William Green's Co, Bedford Co, attached to 5th VMR (Hairston) **CEM**: Overstreet-Crowder- Foster; Bedford; Rt 24 at jct with Crowder Pt Rd, Chestnut Fork **GS**: Unk **SP**: Mar 1) (6 Mar 1815) Barsheba Overstreet (1794-1858), 2) (17 Jan 1859) Mary Ann Crowder (__-12 Feb 1899) who drew pen $12 mo, and LNR P.O. Thaxton, Bedford Co, 1878 **VI**: No further data **P**: Both **BLW**: Yes **PH**: No **SS**: B pg 42; M pg 105; O-pen & BLW files Fold3 **BS**: 245

*CROUCH, Thomas H; b 14 Aug 1788, d 31 Oct 1856 **RU**: Private 38th VMR, Capt William Bolling's Calvary Co, Goochland County, attached to a Detachment of Calvary at Norfolk **CEM**: Hollywood; Richmond City; 412 S Cherry St **GS**: Yes **SP**: Mar (Ampel Hill, Chesterfield Co, 25 Nov 1815) Mollie Brooks Temple (1798-1879). LNR P.O. Henrico Co, 1878. She drew pen 1878 **VI**: Recd BLW (30 Sep 1851, age 62) **P**: Widow **BLW**: Yes **PH**: No **SS**: B pgs 85; 235; M pg 105; O-Pen & BLW files Fold3 **BS**: 245

Key: *Additional veteran entry   **Corrected veteran entry   ***Deleted veteran entry

RU=Rank/Unit	CEM=Cemetery	GS=Gravestone	SP=Spousal Information	VI=Other Veteran Info	P=Pension
BLW=Bounty/Land Warrant	PH=Photo	SS=Service Source	BS=Burial Source	VMR= VA Military Regiment	
LNR= Last Known Residence					

*CRUISE, John Mortimer, Sr; b 5 Oct 1794, Halifax Co, d 21 Mar 1868 **RU:** Private, 18th VMR, Lt John Corn's Co, Patrick Co **CEM:** Cruise Family; Patrick; Vesta **GS:** No **SP:** 1) Mazy Minnie Martin (1790-7 Aug 1859, 2) (27 Jun 1861) Hannah Cockram, who recd pen. LNR P.O. Snake Creek, Carroll Co, 1878 **VI:** Recd two BLW # 9029, 40629. Burial loc is probable **P:** Widow **BLW:** Yes **PH:** No **SS:** B pg 157; M pg 106; O- pen & BLW files Fold3 **BS:** 245

CRUMP, John; b unk, d 6 Aug 1847 (Admin) **RU: Private, 16th VMR (Waller), Spotsylvania Co **CEM:** City Cemetery; Fredericksburg; William St & Washington Ave **GS:** Yes **SP:** Susanna Hart (1785-1860) **VI:** Enumerated in 1840 Census in Fredericksburg **P:** No **BLW:** No **PH:** No **SS:** A rec 1717 **BS:** 18 pg 8; 245

*CRUMP, Joshua; b 1784, d 18 Apr 1851 **RU:** Sergeant, 19th VMR, in 1813, Capt William McCabe's Co; in 1814 Capt John McPherson's Co, Richmond City **CEM:** Shockoe; Richmond City; 100 Hospital St **GS:** Unk **SP:** No further info **VI:** No further data **P:** No **BLW:** No **PH:** No **SS:** B pg 175; O- Serv Index Cards, Fold3 **BS:** 245

*CULLERS, Henry Samuel; b 17 Jun 1792, d 22 Feb 1865 **RU:** Corporal, 13th or 97th VMR, Capt Jesse Carter's Co, Shenandoah County attached to 4th VMR (Wooding) **CEM:** Henry Culler Family; Shenandoah; off Rt 678 to East, N of Seven Fountains **GS:** Yes **SP:** Mar (6 Jul 1817) Sarah Keyser (8 Jun 1800-9 Nov 1880), daug of Andrew Keyser (1758-1833) and Sarah Margaret Rinehart. She recd pen. LNR Powell's Fort, Shenandoah Co, 1878 **VI:** Son of John Kullers (1747, Germany-1796) and Anna Marie Muller (1750-1832). Recd BLW 80 acres **P:** Widow **BLW:** Yes **PH:** No **SS:** B pg 184; M pg 107; O- Serv Index Card and Pen & BLW files, Fold3 **BS:** 245

*CUNNINGHAM, James B: b 1771, d 18 Aug 1823 **RU:** Sergeant, 54th VMR, Captains Julian Magagnos or James R Nimmos, Norfolk Borough, attached to 5th VMR **CEM:** Cedar Grove; Norfolk; 238 E Princess Anne Rd **GS:** Yes **SP:** No spouse info **VI:** No further data **P:** No **BLW:** No **PH:** No **SS:** A rec 853; B pg 145 **BS:** 245

*CUNNINGHAM, William; b 1795, Rockingham Co, d 1881 **RU:** Private, 8th VMR A company of Rockbridge Co, attached to 6th VMR (Dickinson, Scott, Coleman) **CEM:** Egypt-Cunningham; Rockbridge; past end of Rt 629, Waterloo; Kerrs Creek **GS:** Unk **SP:** No further data **VI:** Obituary indicates he served in War of 1812 **P:** No **BLW:** No **PH:** No **SS:** B pg 179; O-Serv Index Card, Fold3 **BS:** 245

*CUPP, Frederick; b 17 Mar 1784, d 14 Jan 1867, Rockingham Co **RU:** Private, 58th or 116th VMR, Capt Robert M McGill's Co, Rockingham County, attached to Flying Camp (McDowell's) **CEM:** Emmanuel Church; Augusta; Mount Solon **GS:** Yes **SP:** 1) Elizabeth Baker, 2) (Rockingham Co, 26 May 1851) Elizabeth Shaver (__-aft 1879), widow of Isaac Carson. LNR P.O. Mt Solon, Augusta Co. Recd Pen **VI:** Recd BLW 40 acres **P:** Widow **BLW:** Yes **PH:** No **SS:** B pg 182; M pg 108; O- Serv Index Card and Pen & BLW files Fold3 **BS:** 245

CURTIS, George, Jr; b abt 1767; d abt 1845 **RU: Sergeant, 45th VMR Capt Levi Swelman's Co, Stafford County **CEM:** Curtis Family #3; Stafford; Stefania Rd (Rt 648) **GS:** Unk **SP:** Prob Mar 1) (abt 1792) Mary McIlhaney, 2) (Fauquier Co, 12 Mar 1804) Jemina Payne, daug of Francis Payne and Susannah Jett **VI:** Son of George Curtis, Sr, (1730-1806) and Elizabeth Jett who was head of household in Aquia in Stafford County in the 1810, 1820 and 1840 US Federal Census, between ages of 70 and 79 in the latter **P:** No **BLW:** No **PH:** No **SS:** A rec 2003 **BS:** 26 pg 187

CUSTIS, George Washington Parke; b 30 Apr 1781, Mt Airy, MD, d 10 Oct 1857, Arlington House **RU: Colonel; Aide de Camp of General Charles Pickney and as a citizen served in the defense of DC during the war **CEM:** Arlington National; Arlington; Off George Washington Parkway **GS:** Unk **SP:** Mar (Alexandria, 14 Jul 1804) Mary Lee Fitzhugh (1788, MD-__) **VI:** Son of John Parke Custis (1754-1781) and Eleanor (Calvert) Stuart (1754-1811), and grandson of Martha Washington by her first marriage. His father dying, soon after his birth, he was adopted by General and Mrs. Washington and was raised at Mount Vernon. He was enrolled at St John's College and at Princeton. He built the Arlington House now located in the cemetery. Before the war he was a Cornet on 8 Jan 1799 and promoted to 2/LT, (15 Mar 1799) discharged (15 Jun 1800) **P:** No **BLW:** No **PH:** No **SS:** A rec 2005; D **BS:** 96 pg 71

Key: *Additional veteran entry **Corrected veteran entry ***Deleted veteran entry
RU=Rank/Unit   CEM=Cemetery   GS=Gravestone   SP=Spousal Information   VI=Other Veteran Info   P=Pension
BLW=Bounty/Land Warrant PH=Photo   SS=Service Source BS=Burial Source   VMR= VA Military Regiment
LNR= Last Known Residence

30

*DABNEY, Robert Kelso; b 1787, Prince Edward Co, d 1867 **RU:** Orderly, Field Staff, 8th VMR, attached to Colonel John H Cocke **CEM:** Dabney; Powhatan; Montpelier **GS:** Y **SP:** Lucy Ann Pope (1793-21 Mar 1834), daug of William Pope (1762-1852) and Ann Woodson (1774-1823) **VI:** Son of John Dabney and Ann Harris **P:** No **BLW:** No **PH:** No **SS:** K pg 81 **BS:** 245

DAFFIN, William; b abt 1780; d 2 Apr 1855 **RU: Private, 16th VMR (Waller), in a company in Spotsylvania Co **CEM:** Daffan Family; Stafford; Rt 625, south of railroad jct **GS:** Yes **SP:** Nancy Davis (1795), daug of G Davis and Patsy (__) **VI:** Son of Vincent and Betty M (__) Daffin, enumerated in 1850 Census, Eastern District Stafford County at age 70 **P:** No **BLW:** No **PH:** No **SS:** A rec 804 **BS:** 26 pg 188

*DALTON, Timothy; b 1783, Grayson Co, d 15 Dec 1872 **RU:** Captain 78th VMR, Company commander, Grayson Co, attached to 4th VMR (McDowell, Koontz, (Chilton) **CEM:** William Dalton Family; Carroll; Dugspur **GS:** Yes **SP:** Elizabeth Phillips (1783-1850), daug of Tobias Phillips and Margaret Peggy Jennings **VI:** Son of William Dalton (1740-1811) and Elizabeth Struman **P:** No **BLW:** No **PH:** No **SS:** B pg 86 **BS:** 245

*DANIEL, John; b 1790, Campbell Co, d 12 Oct 1862 **RU:** Private. 66th or 96th VMR Captain Baker Pegram's Cavalry Co, Brunswick Co, attached to 1st VMR (Byrne) **CEM:** Daniel-Turpin; Rockbridge; vic Rappa Mill **GS:** Unk **SP:** 1) (12 May 1814) Nancy Newcombe (__-22 Oct 1840) 2) (So Buffalo, Rockbridge Co, 11 Feb 1841) Cassandra F.S. Layne (__-8 Dec 1888). She recd pen. LNR Rappa Mill, Rockbridge Co, 1878 **VI:** Son of Leonard Daniel and Mary Spears. He recd BLW of 80 acres **P:** Widow **BLW:** Yes **PH:** No **SS:** B pg 23; K pg 153; M pg 111; O- pen file of widow, Fold3 **BS:** 245

DANIEL, William, Sr; b 1770, Cumberland Co, d 20 Nov 1839, Lynchburg **RU: Lt Colonel, 17th VMR, Cumberland Co **CEM:** Old City Cemetery; Lynchburg; 401 Taylor St **GS:** Yes **SP:** Margaret Baldwin (1786-1824) **VI:** Member of Virginia House of Delegates, was circuit judge and ex-officio member of General Court of Virginia; plaque on gravestone **P:** No **BLW:** No **PH:** No **SS:** K pg 336 **BS:** 87 pg 43; 245

*DARNE, Robert; b 1778, d unk **RU:** Private, Capt George Graham, Troop of Calvary, Fairfax Co, attached to 1st Corps DeElite (Randolph) **CEM:** Wren Darne; Fairfax; Hillsman and Mahalia Lane, Falls Church **GS:** No **SP:** Verline Wren **VI:** Son of Henry and Penelope (Minor) Darne. Obtained rank of captain, probably after the war **P:** No **BLW:** No **PH:** No **SS:** A rec 406; B pg 71 **BS:** 31

*DASHIELL, Thomas Bennett, b 2 May 1787, probably Alexandria, d 30 Aug 1859 at Parish Rectory **RU:** Private 1st Regt DC Militia in Capt John Davidson's Co, Col George McGruder's Command **CEM:** Yeocomico Episcopal Church; Westmoreland Co; 1233 Old Yeocomico Rd **GS:** Yes **SP:** Mar (Alexandria, 29 Mar 1825) Mary Ann (__), who mar 2) Weston McCobb (Richmond City, __-12 Jan 1885). She recd pen. LNR Richmond City, 1878 **VI:** Recd BLW of 40 acres and 120 acres **P:** Widow **BLW:** Yes **PH:** No **SS:** B pg 23; K pg 153; M pg 111; O- pen file of widow, Fold3 **BS:** 245

*DAVENPORT, John; b 1778, d 17 Feb 1882 **RU:** Fifer, 16th VMR, Capt Gullielmus Smith's Co, Spotsylvania County, attached to 1st VMR (Crutchfield) **CEM:** Davenport Family; Spotsylvania; Post Oak **GS:** Unk **SP:** Mar 1) Ellen Branchett Luck (1780-bef 1843), 2) (19 Mar 1843) Harriet A Hart (1809-c1900). She recd pen. LNR P.O. Post Oak, Spotsylvania Co, 1882 **VI:** Recd pen and two BLW of 80 acres each **P:** Both **BLW:** Yes **PH:** No **SS:** B pg 189; M pg 112; O- pen file of widow, Fold3 **BS:** 245

*DAVENPORT, Martin W; b c1795, d 16 Jan 1872 **RU:** Sergeant 26th VMR, Capt John P Richardson's Artillery Co, Charlotte County, attached to 4th VMR (Greenhill) **CEM:** Presbyterian; Lynchburg City; 2020 Grace St **GS:** Yes **SP:** Mar (26 Jun 1838) Ann Elizabeth Thompson (__-1883), who recd pen at Princeton Mercer Co, NJ, 1878 **VI:** Recd BLW 40 acres **P:** Widow **BLW:** Yes **PH:** No **SS:** B pg 57; M pg 112; O- pen files, Fold3 **BS:** 245

Key: *Additional veteran entry   **Corrected veteran entry   ***Deleted veteran entry
RU=Rank/Unit   CEM=Cemetery   GS=Gravestone   SP=Spousal Information   VI=Other Veteran Info   P=Pension
BLW=Bounty/Land Warrant   PH=Photo   SS=Service Source   BS=Burial Source   VMR= VA Military Regiment
LNR= Last Known Residence

31

*DAVIDSON, Robert; b 5 Jan 1797, d 16 Dec 1867 **RU:** Private 48th and 121st VMR, served in either Capt Thomas N Burwell's Co or Captain David Rowland's Co, Botetourt County, attached to 5th VMR (McDowell) **CEM:** Rich Valley Presbyterian Church; Smyth; Valley Rd Rt 610, Saltville **GS:** Unk **SP:** Nancy Rebecca Pennington (1796-17 Nov 1887) **VI:** Son of Benjamin Davidson (1765-1810) and Rebekah Newman (1770-1844) **P:** No **BLW:** No **PH:** No **SS:** A rec 1650; B pgs 45, 46 **BS:** 245

*DAVIDSON, William M: b 1768, d 14 May 1809 **RU:** Private, 52nd VMR, Capt John Armistead, New Kent & Charles City Co, attached to 2nd VMR (Ambler-Brown) **CEM:** St John's Episcopal Church; Richmond; 24th & Broad, Church Hill **GS:** Yes **SP:** No spouse info **VI:** No further data **P:** No **BLW:** No **PH:** No **SS:** A rec 1732; B pg 143 **BS:** 245

*DAVIS, Henry; b 1778, d 11 Dec 1863 **RU:** Corporal 38th VMR, Capt William Sales Troop of Calvary, Amherst Co, attached to Sales Battalion **CEM:** Presbyterian; Lynchburg City; 2020 Grace St **GS:** Yes **SP:** No spousal info **VI:** No further data **P:** No **BLW:** No **PH:** No **SS:** B pg 38; O- Serv Index Card, Fold3 **BS:** 245

*DAVIS, John B; b bef 1800, d 1848 Williamsburg **RU:** Private, 47th or 88th VMR, Enlisted 1814, Charlottesville, Capt John Rothwell's Co, Albemarle County, attached to 7th VMR (Gray) **CEM:** University of Virginia and Columbarium; Charlottesville City; Alderman and Cemetery Rd **GS:** Yes **SP:** Mar (21 Dec 1837) Jiney D McCauley (c1807-2 Jun 1895}. Recd two BLW, one of 80 acres Act of 1750, and one Act of 1855 and pension. LNR P.O. Free Union, Brown's Cove, Albemarle Co, **VI:** Died in asylum in Williamsburg 1848 **P:** Widow **BLW:** Widow **PH:** No **SS:** A rec 814; B pg 36; M pg 113; O- pen file of widow, Fold3 **BS:** 245

*DAVIS, Levi; b 1772, d 1847 **RU:** Private, 78th VMR, Captain James Anderson's Company, Grayson Co. **CEM:** Hebrew; Richmond City; Hospital St **GS:** No **SP:** No spouse info **VI:** No further data **P:** No **BLW:** No **PH:** No **SS:** A rec 126 **BS:** 245

*DAVIS, Robert; b 8 Oct 1794, d 19 Apr 1883 **RU:** Sergeant, 58th or 116th VMR, a company from Rockingham Co attached to 6th VMR (Coleman) **CEM:** Keezletown; Rockingham; Rt 925 at jct with Miss Elton Ln **GS:** Yes **SP:** Mar (15 Mar 1815) Lucinda Shifflett Grafton (1 Jul 1792-23 Mar 1870) **VI:** Local newspaper indicates died nr Montevideo, Rockingham Co. Recd pen and BLW **P:** Yes **BLW:** Yes **PH:** No **SS:** B pg 181-2; O- Serv Index Card, Fold3 **BS:** 245

DAVIS, Samuel; b 1776; d 7 Apr 1819 **RU: Lieutenant, 2nd VMR (Sharp) **CEM:** Cedar Grove; Portsmouth; Effington St & Fort Ln **GS:** Yes **SP:** Lydia Gorman (25 Aug 1778-3 Oct 1823) and/or Caroline (__) **VI:** No further data **P:** No **BLW:** No **PH:** No **SS:** A rec 1619 **BS:** 65 pg 108; 245

*DAVIS, Thomas: b 1775, d 2 Feb 1845 **RU:** Private, 101st VMR, Captain Edward Carter's Troop of Cavalry, Pittsylvania County, attached to Regiment of Calvary (Holcombe) **CEM:** Davis Family; Pittsylvania; Cherrystone Plantation **GS:** Yes **SP:** Sarah Meadows (1775-1849) **VI:** Son of William Davis (1729-1791) & Susannah Wells (__-1789) **P:** No **BLW:** No **PH:** No **SS:** B pg 161; L pg 200 **BS:** 245

*DAVIS, Thomas; b 1795; d Unk **RU:** Private, 75th VMR, company from Montgomery Co, attached to 4th VMR at Norfolk **CEM:** Piedmont; Montgomery; Otey, Piedmont **GS:** Yes **SP:** Martha Jewell (1804-3 Jan 1822), daug of Thomas Jewell (1764-1853) and Elizabeth Graham (1775-__) **VI:** No further data **P:** No **BLW:** No **PH:** No **SS:** A rec1780; B pg 138 **BS:** 245

*DAVIS, William; b 1770, d 27 May 1848 **RU:** Private, 54th VMR, Captain John Ott's Flying Artillery Co, Norfolk Borough, attached to 6th VMR (Reade) **CEM:** Basilica of Saint Mary Churchyard; Norfolk City; 232 Chapel St **GS:** Yes **SP:** Margaret (__), (1770-1 Jan 1848) **VI:** No further data **P:** No **BLW:** No **PH:** No **SS:** A rec 2149 **BS:** 245

*DAVIS, William Jr; b 1770, d 19 Mar 1853 **RU:** Private, 10th VMR, Captain Walter Otey's Troop of Calvary, Bedford Co., attached to Major Woodland's Squadron **CEM:** South River Meeting House; Lynchburg City; 5810 Fort Ave **GS:** Yes **SP:** Zalinda Lynch (6 Feb 1772-9 May 1839), daug of John Lynch and Mary Bowles **VI:** No further data **P:** No **BLW:** No **PH:** No **SS:** A rec 2226 **BS:** 245

Key: *Additional veteran entry   **Corrected veteran entry   ***Deleted veteran entry
RU=Rank/Unit   CEM=Cemetery   GS=Gravestone   SP=Spousal Information   VI=Other Veteran Info   P=Pension
BLW=Bounty/Land Warrant   PH=Photo   SS=Service Source   BS=Burial Source   VMR= VA Military Regiment
LNR= Last Known Residence

*DEAN, William; b 5 Feb 1796, d 31 Jan 1857, Page Co **RU:** Private, 58th or 116th VMR, a company of Rockingham Co, attached to 6th VMR (Coleman) **CEM:** Saint Peters; Rockingham; nr Massanattan United Methodist Ch, off Rt E Point Rd, Elkton **GS:** Yes **SP:** Mary Dean (15 Dec 1794-27 Jan 1879). Pensioned in Orange Co. LNR P.O. Roadside, Rockingham, 1878 **VI:** Recd BLW **P:** Widow **BLW:** Yes **PH:** No **SS:** B pg 181-2; M pg 117; O-Serv Index Card, and pen files Fold3 **BS:** 245

DEARING, Alfred; b 22 Mar 1791, d 18 Mar 1856 **RU: Private, 34th VMR, Capt Charles Shackleford, Culpeper Co, attached to 1st VMR (Crutchfield) **CEM:** Dearing, Caledonia Farm; Rappahannock; 47 Dearing Rd, Flint Hill, Rt 630 **GS:** Yes **SP:** Mar (Culpeper Co, 17 Apr 1817) Anne Jackson, LNR Flint Hill, Rappahannock Co **VI:** Son of John Dearing a captain in the Rev War and Anna "Nancy" Jett, daug of Francis. Died near Flint Hill, Rappahannock Co **P:** Widow **BLW:** Yes **PH:** No **SS:** BD pg 564; B pg 63 **BS:** 74 pg 51; 31

*DeHAVEN, John; b 31 Mar 1784, d 9 Nov 1859 **RU:** Private, Captain Charles Boeut's Co, Frederick County, attached to 4th VMR **CEM:** Gainesboro Methodist Church; Frederick; Gainesboro **GS:** Yes **SP:** Rhoda Doster (11 Nov 1787-3 Sep 1855), daug of Thomas Doster and Mary Crumley **VI:** Son of Peter DeHaven (1741-1822) and Abigail West (1746-1827) **P:** No **BLW:** No **PH:** No **SS:** A rec 588 **BS:** 245

*DEMORY, Wilham; b 1790, d 1 May 1873 **RU:** Private, 56th VMR, a company in Frederick Co. **CEM:** Old Ebenezer Methodist Episcopal Church; Loudoun; Neersville **GS:** Yes **SP:** No spouse info **VI:** No further data **P:** Applied for **BLW:** No **PH:** No **SS:** A rec 1785; M pg 118 **BS:** 245

*DENTON, William D; b 1798, d 1870, TN **RU:** Private, 105th VMR, Captain William Smith's Co of Artillery, Washington County, attached to Battalion of Artillery **CEM:** Montgomery; Washington; Abingdon **GS:** No **SP:** Rachel Gibson (1811-1889) **VI:** Memorialized in cem **P:** No **BLW:** No **PH:** No **SS:** A rec 261 **BS:** 245

*DIBRILL, Edwin; b 19 Sep 1794, d 17 Dec 1871 **RU:** Private 19th VMR, Capt George Booker's Co, Richmond City **CEM:** Hollywood; Richmond City; 412 S Cherry St **GS:** Yes **SP:** Mar (Wayne Co, KY, 31 Oct 1816) Martha Shrewsbury (1800-1866) **VI:** Son of Anthony Dibrell, Jr and Wilmuth Watson. Was 20 yrs recorder and ex-officio clerk of Mayor's Court, Nashville, TN. Conducted a Tobacco Commission house in Richmond and was clerk in Federal Treasury Dept. Recd two pensions and two BLW, one 40 acres, one 120 acres. Resided Wash DC 1851-1855, then Henrico Co 1871 **P:** Yes **BLW:** Yes **PH:** No **SS:** B pg 174; M pg 120; O- pen files Fold3 **BS:** 245

*DICKENSON, Robert; b unk, d 25 Dec 1818 **RU:** Lieutenant, 49th VMR, probably Capt Samuel Thomas Co, Nottoway Co, marched to Norfolk Aug 1813 by Lt Col Grief Green and upon arrival attached to 6th VMR (Sharps) **CEM:** Fowles Family; Nottoway; 6808 W Courthouse Rd (Rt 625), Burkeville **GS:** No **SP:** No spouse info **VI:** No further data **P:** No **BLW:** No **PH:** No **SS:** A rec 216; B pgs 154; 230 **BS:** 245

*DICKERSON, John Bethel; b 1791, Montgomery Co, d 28 Dec 1837 **RU:** 75th VMR, Capt James Hoge's Co, Montgomery Co, attached to 4th VMR (Huston-Wooding) **CEM:** Webb Family; Carroll; Snake Creek Rd, Rt 670 l mi E of jct with Rt 870, Hillsville **GS:** No **SP:** Mar (Grayson Co, 5 Aug 1814) Mary Cock (c1802-c1890), daug Andrew Cock and Penelope Ward. Recd pen 1878 **VI:** Recd BLW **P:** Widow **BLW:** Yes **PH:** No **SS:** B pg 138; M pg 121; O- pen files Fold3 **BS:** 245

DIGGES, Dudley; b 6 Aug 1766; d 4 Apr 1839 **RU: Sergeant, 9th VMR (Boyd) **CEM:** Fork Church Episcopal; Hanover; 12566 Old Ridge Rd, Doswell **GS:** Yes **SP:** Alice Grymes Page (__-1846) **VI:** Also served as a Private in the 6th VMR Artillery. Son of Dudley Power Digges (1728-1790) **P:** No **BLW:** No **PH:** No **SS:** A rec 1216, 1217 **BS:** 143 pg 570; 195

*DILL, Adolph; b 8 Jan 1792, d 13 Aug 1867 **RU:** Corporal, 19th VMR, Capt George Booker's Co, Richmond City **CEM:** Shockoe Hill; Richmond City, 4th & Hospital Sts **GS:** Yes **SP:** Mar 1) (1817) Hannah Heisler (1793-1822), 2), (Philadelphia, 2 Dec 1826) Hannah Keyser Gorgas (1804-12 Dec 1878) who drew a pen **VI:** Home was at 00 Clay St, where now stands Black History Museum. He drew BLW. Was a baker **P:** Widow **BLW:** Yes **PH:** No **SS:** B pg 174; M pg 121; O- pen files Fold3 **BS:** 245

Key: *Additional veteran entry   **Corrected veteran entry   ***Deleted veteran entry
RU=Rank/Unit   CEM=Cemetery   GS=Gravestone   SP=Spousal Information   VI=Other Veteran Info   P=Pension
BLW=Bounty/Land Warrant   PH=Photo   SS=Service Source   BS=Burial Source   VMR= VA Military Regiment
LNR= Last Known Residence

33

*DILLON, William; b 26 Aug 1774, d 20 Dec 1832 **RU:** Private, 57th VMR, Loudoun Co **CEM:** White Hall United Methodist; Frederick Co; 3265 Apple Pie Ridge, White Hall **GS:** Yes **SP:** Mar (16 May 1825) Elizabeth Haines **VI:** No further data **P:** No **BLW:** No **PH:** No **SS:** A rec 1769 **BS:** 56 pg 15

DISHMAN, John; b 1793, King George Co; d 24 Sep 1843 **RU: Sergeant, 25th VMR, Capt Caldwell Dade's Co, King George County **CEM:** Dishman Family; King George; Pine Hill Hunt Club Rd, Carruthers Corner **GS:** Yes **SP:** Anne Edmond Jones (1799-1845); wife of John Dishman (stone) **VI:** Son of Samuel Dishman, (1756-1817) & Susanna Baker (1750-1813) **P:** No **BLW:** No **PH:** N **SS:** K pg 265; A rec 2276 **BS:** 50; 80

*DIXON, John; b 1787, d 05 Sep 1830, Mount Pleasant **RU:** Quartermaster, 21st VMR Headquarters Staff Officer, Gloucester Co **CEM:** Ware Episcopal Church; Gloucester; 7825 John Clayton Memorial Rd **GS:** Yes **SP:** Sally Throckmorton of Airville **VI:** Son of John Dixon and Elizabeth Peyton; gravestone moved from Mount Pleasant in 1948; served 13 months and 15 days **P:** No **BLW:** No **PH:** Yes **SS:** L pg 11 **BS:** 82 pg 75

*DIXON, John B; b 1782, d 12 Jul 1818 **RU:** Matross, 1st District of Columbia Militia **CEM:** Trinity United Methodist; Alexandria; Cameron Mills Rd **GS:** Yes **SP:** No spouse info **VI:** No further data **P:** No **BLW:** No **PH:** No **SS:** A box 58; mfilm M602 **BS:** 245

DIXON, Thomas; b 2 May 1776; d 13 Apr 1854 **RU: Private, 5th VMR **CEM:** Old Timber Grove; Rockbridge; between Fairfield & Timber Ridge **GS:** Yes **SP:** Sarah Paxton (21 Nov 1792-16 Jul 1834), daug of Samuel Paxton Sr (1748-1807) and Sarah Coalter **VI:** No further data **P:** No **BLW:** No **PH:** No **SS:** A rec 2847 **BS:** 261 v10 pg 113; 245

*DOAKE, Robert; b 6 Jun 1775, d 13 Aug 1818 **RU:** Captain, East TN Regiment of Volunteers (Wears) **CEM:** Thompson- Buchanan Family; Tazewell; check county rec for loc **GS:** Unk **SP:** Mar (2 Nov 1801) Rachel Thompson **VI:** No further data **P:** No **BLW:** No **PH:** No **SS:** A rec 29 **BS:** 245

*DOLD, Samuel Miller; b 22 Aug 1798, d 9 Feb 1883 **RU:** Private, 63th VMR, Capt Benjamin Dyer's Detachment, Henry Co, attached to 5th VMR (Mason-Preston) **CEM:** Stonewall Jackson Memorial; Lexington City; jct White St & Rt 11 **GS:** Yes **SP:** Mar (17 Oct 1820) Elizabeth McFadden (21 Apr 1799-28 Mar 1871) **VI:** Son of William Dold (Staunton Co, 6 Nov 1771-3 Jul 1856, Waynesboro) and Sarah Brent (1773-1848). Was a banker. Recd pen and BLW of 120 acres 1871 **P:** Yes **BLW:** Yes **PH:** No **SS:** B pg 101; M pg 124; O- pen and BLW files, Fold3 **BS:** 245

*DOLD, William; b c1771, d 3 Jul 1856 **RU:** Private, 32nd or 93rd VMR. Augusta Co in company attached to 5th VMR (McDowell) **CEM:** Tinkling Spring Presbyterian Church; Augusta; Fishersville, 11 mi NE of Staunton **GS:** Yes **SP:** No spouse info **VI:** No further data **P:** No **BLW:** No **PH:** No **SS:** A rec 1108 **BS:** 245

DORNIN, Thomas Alysius; b c1800, Ireland; d 22 Apr 1874, Norfolk **RU: Midshipman, US Navy **CEM:** Cedar Grove; Norfolk City; 238 E Princess Anne Rd **GS:** Yes **SP:** Mar (Cedar Grove Church, Portsmouth, 29 Jul 1837) (__) Thornburn **VI:** Entered US Navy 2 May 1815 as a Midshipman from MD. Commissioned a Lieutenant 1825. Commanded the USS *Relief* and made a 5-year cruise with United States Exploring Expedition. Promoted in 1841 to Commander. Commissioned a Captain 1855. During Civil War was commissioned a Commodore and placed on retired list 16 July 1862 **P:** No **BLW:** No **PH:** No **SS:** AQ **BS:** 49

*DOSS, Stephen; b 1786, d 23 Feb 1866 **RU:** Private 42nd or 101st VMR, Capt Thomas H Clark's Trp of Cav, Pittsylvania Co, attached to 7th VMR (Saunders) **CEM:** Dalton-Adams-Doss; Pittsylvania; end of Georges Creek Rd; N by Creek, Gretna **GS:** Yes **SP:** Mar (Franklin Co, c1810) Judith Hodges (__-aft 1781), LNR P.O. Chalk Level; Pittsylvania Co, 1781 Recd pen **VI:** Recd two BLW of 80 acres each **P:** Widow **BLW:** Yes **PH:** No **SS:** B pg 161; M pg 125; O- pen files Fold3 **BS:** 245

Key: *Additional veteran entry **Corrected veteran entry ***Deleted veteran entry
RU=Rank/Unit   CEM=Cemetery   GS=Gravestone   SP=Spousal Information   VI=Other Veteran Info   P=Pension
BLW=Bounty/Land Warrant   PH=Photo   SS=Service Source   BS=Burial Source   VMR= VA Military Regiment
LNR= Last Known Residence

*DOVE, Elisha; b bef 1796, d 9 Apr 1814 **RU:** Private 101st VMR, Lt William Lewis Co and Lt William Wimbish's, Detachment, Pittsylvania County, attached to 6th VMR (Coleman) **CEM:** Fort Tar mass burial site; Norfolk; nr Wharf vic Fort Tar Lane & VA historical road sign this name **GS:** No **SP:** Not married **VI:** Son of Levicy Dove his mother. Serv rec indicates died in service in Norfolk **P:** No **BLW:** No **PH:** No **SS:** B pg 161; P **BS:** 49- 2017; S. Butler; *Defending the Old Dominion, VA and its Militia in the War of 1812,* 2013, Univ Press of VA, pgs 513-525

*DOVE, Reuben; b 15 Nov 1795, d 18 Feb 1890 **RU:** Private, 116th VMR, Capt Thomas Hopkin's Co, Rockingham Co, attached to 6th VMR (Coleman) **CEM:** Myers Family; Rockingham; vic Old Linville Creek **GS:** Unk **SP:** Mar (9 Jul 1818) Catherine Dove (1797-1881), daug of Jacob Dove and Sarah Wetzel **VI:** Son of George Dove (1760-Jul 1831) and Elizabeth Dean (1764-__), Recd two BLW, one 40 and one 120 acres and two pen **P:** Yes **BLW:** Yes **PH:** Yes **SS:** B pg 182; M pg 125; O- pen files Fold3 **BS:** 245

*DOVE, William: b c1784, d 20 May 1863 **RU:** Corporal, 101st VMR ,Capt William Payne's Co, Pittsylvania Co, attached to 6th VMR (Coleman) **CEM:** Old Dove Family; Pittsylvania; Chalk Level, nr Gretna **GS:** Yes **SP:** Mar (18 Apr 1811) Mary Polly Mustain, (21 Aug 1784-14 Jun 1880), daug of Avery Mustain (Rev War soldier) and Mary Barber; widow LNR P.O. Chalk Level, Pittsylvania Co, 1781 **VI:** Son of William Dove (27 Nov 1758, Charles Co, MD-20 Sep 1847). He and father operated a mill on So Fork of Stinking River **P:** Widow **BLW:** Yes **PH:** Yes **SS:** A rec 1050; B pgs 162-3; M pg 126 **BS:** 31

*DOVEL, William; b 19 Sep 1793, d 7 Dec 1860 **RU:** Private, 97th VMR Capt Walter Hambaugh's Co, Shenandoah County, attached to 6th VMR (Coleman), discharged Camp Crossroads, MD I Dec 1814 **CEM:** William Dovel; Page; off Stroll Farm Rd (Rt 613) **GS:** Family monument lists veteran **SP:** Mar (May 1816) Christina Long (1801-30 Sep 1885), she recd pen, LNR P.O. Box, East Liberty, Page Co **VI:** Son of David S Dovel (15 Apr 1768-Feb 1832) and Keziah Short (20 Aug 1774-1859) **P:** Widow **BLW:** Yes **PH:** No **SS:** B pg 184; M pg 126; O- pen files Fold3 **BS:** 245

*DOWNING, Rufus; b 26 Jul 1784 Canterbury, Windham Co, CT, d 2 Aug 1844 **RU:** Cornet, 30th VMR, Capt Armistead Hoomes's (Hoomes's) Co, Caroline County, attached to LTC Cocke's Detachment **CEM:** Greenlawn; Caroline; Rt 608, Lakewood Rd, Bowling Green **GS:** Yes **SP:** Mar (2 Sep 1813) Fanny Sales (__-aft 1871). Recd widow's pens # 4251, 1672 **VI:** Came to VA 1807, Recd BLW 80 acres in 1850 and again in 1855 **P:** Widow **BLW:** Yes **PH:** No **SS:** B pg 56; M pg 126; O- pen files Fold3 **BS:** 245

*DRAKE, Francis T; b 5 Mar 1797, Culpeper Co, d 25 Feb 1891 **RU:** Corporal, 57th VMR, Capt Benjamin Shreve's Co, Loudoun County, attached to 5th VMR (Mason) **CEM:** Union; Loudoun; Leesburg, 323 North King St **GS:** Yes, gr loc plat B, lot 368, site 9 **SP:** Maria Ann Washington (11 Sep 1799-6 Apr 1856) **VI:** Applied for pen and BLW but later rejected for insufficient service of 25 days. Worked Snowden's Iron Works; became ill and hired substitute, Thomas Rogers for remaining service needs **P:** No **BLW:** No **PH:** No **SS:** B pg 121; M pg 127; O- pen & BLW files, Fold3 **BS:** 245

*DREWEY, Henry Tandy; b 1794, King William Co, d 26 May 1866, Chesterfield Co **RU:** Private, 19th VMR, Capt George Booker's Co, Richmond City **CEM:** Hollywood; Richmond City; 412 South Cherry St **GS:** Yes **SP:** Mar (28 Sep 1830) Martha Amelia Davis (__-Centralia, Chesterfield Co, 7 Jun 1886, daug of John W Davis of Amelia Co. Recd pen of $8.00 per mo, 9 Mar 1878. LNR Proctors Creek, Chesterfield Co, 1878 **VI:** Son of John Drewry and Sarah S (__). Recd BLW 160 acres, 1855 **P:** Widow **BLW:** Yes **PH:** No **SS:** B pg 174; M pg 127; O- pen files Fold3 **BS:** 245

*DRINKARD, Beverly; b 1793, Charles City Co, d 1 Dec 1875 **RU:** Private, 62d VMR, Capt Daniel Epps Co, Prince George Co, placed under command of Capt William Allen, 35th U.S. Inf, Fort Powhatan **CEM:** Blandford; Petersburg City; 319 South Crater Rd **GS:** Yes **SP:** 1) Elizabeth A F (__), 2) (Petersburg Co, 2 Nov 1865) Jane A E Rolfe Ellyson (1832-1894), she pen $12 mo Feb 1894 **VI:** Son of Mary Cocke Wilcox (1762-21 Aug 1846). Recd BLW **P:** Widow **BLW:** Yes **PH:** No **SS:** B pg 169; M pg 127; O- pen files Fold3 **BS:** 245

Key: *Additional veteran entry   **Corrected veteran entry   ***Deleted veteran entry
RU=Rank/Unit   CEM=Cemetery   GS=Gravestone   SP=Spousal Information   VI=Other Veteran Info   P=Pension
BLW=Bounty/Land Warrant   PH=Photo   SS=Service Source   BS=Burial Source   VMR= VA Military Regiment
LNR= Last Known Residence

35

*DULIN, John; b 23 May 1776, Upperville, Fauquier Co, d 22 Oct 1828 **RU**: Private, 57th VMR, Captain Martin Kitzmiller's Co, Loudoun County, **CEM**: Dulin-Kenne; Loudoun; Sterling **GS**: Yes **SP**: Rebecca Elgin (1775-1862) **VI**: Son of Thaddeus Dulin and Elizabeth Powell; was injured in the war and confined to a wheelchair **P**: No **BLW** Yes **PH**: No **SS**: B pg 120; BD pg 626; M pg 129 **BS**: 245

*DUNCAN, Greenberry (Greenbury); b1791 Montgomery Co, d 12 May 1860 **RU**: Private 78th VMR, Capt Timothy Dalton's Co, Grayson County, attached to 4th VMR (McDowell-Koontz-Chilton) **CEM**: Greenberry Duncan Family; Carroll; W of Rt 613 Greasy Creek Rd; Dudspur **GS**: Fieldstone **SP**: Mar (Grayson Co, 1809) Nancy Phillips (1793-19 Feb 1882), daug of Tobias Phillips. Recd two widow pen **VI**: Son of John Duncan and Elizabeth (Betty) Holtzclaw. Recd BLW 160 acres, 1855 **P**: Widow **BLW**: Yes **PH**: No **SS**: B pg 86; M pg 129; O- pen files Fold3 **BS**: 245

*DUNFORD, Philip Tennyson: b 19 Aug 1794 Cumberland Co, d 1 Feb 1872; **RU**: Private 17th VMR, Capt Benjamin Allen's Co, Cumberland County, attached to 1st VMR (Truehart) **CEM**: Hollywood; Richmond City; 412 So Cherry St **GS**: Govt **SP**: No spousal info **VI**: Recd pen $8.00 mo and two BLW 40 acres 1850, 160 acres 1855, LNR 511 West Clay St, Richmond City **P**: Yes **BLW**: Yes **PH**: No **SS**: B pg 64; M pg 130; O- pen files Fold3 **BS**: 245

*DUNN, John; b unk, d Aug 1859 **RU**: Private, 33rd VMR, Captain William Allen's Co, Henrico County **CEM**: Hollywood; Richmond; 412 So Cherry St **GS**: No **SP**: No spouse info **VI**: No further data **P**: No **BLW**: No **PH**: No **SS**: K pg 90 **BS**: 245

*EAKLE, John: bef 1800, d 23 Mar 1883, Wilsonville **RU**: Private, 32d VMR Capt Alexander Given's Co, Augusta Co, attached to 5th VMR (McDowell) Served as a substitute **CEM**: Eakle Family; Bath; Little Valley Rd; Bolar **GS**: Yes Govt **SP**: Mar (Bath Co, 25 Sep 1825) Sarah Carpenter **VI**: Recd two pen and two BLW, 1850 -1855 -120 acres **P**: Yes **BLW**: Yes **PH**: No **SS**: B pg 39; M pg 132; O- pen files Fold3 **BS**: 245

*EAST, David C; b 12 Feb 1788, d 23 Mar 1881 **RU**: Private, 8th VMR, in company, Rockbridge County attached to 5th VMR (McDowell) **CEM**: New Providence; Rockbridge; Raphine **GS**: Yes **SP**: Elizabeth (__) **VI**: No further data **P**: No **BLW**: No **PH**: No **SS**: A rec 111; B pg 179 **BS**: 245

*EASTWOOD, William; b 1727, d 20 May1868 **RU**: Private 7th VMR, Capt John Thompson's Co, Norfolk Co, attached to 9th VMR (Sharp) **CEM**: Bellamy United Methodist Church; Gloucester; Rt 615 Chestnut Fork Rd; Bellamy **GS**: Yes **SP**: Mar (Gloucester Co, 16 Jan 1827) Susan Jarvis (1797-bef 28 Jul 1885). LNR P.O. Gloucester CH 1879. She recd pen Act of 1878 having applied (26 May 1879, age 82) **VI**: Was a minister. Recd BLW Act of 1855. War of 1812 gr marker **P**: Widow **BLW**: Yes **PH**: No **SS**: B pg 149; M pg 133; O- pen files Fold3 **BS**: 245

*EBERLY/EBERLEY, George; b 1793, d1874 **RU**: Private, 13th or 97th VMR in a company Shenandoah Co. attached to 6th VMR (Coleman) **CEM**: Saint Pauls Lutheran Church; Shenandoah; Strasburg **GS**: Yes **SP**: Mary A (__) (1796-1828) **VI**: No further data **P**: Yes **BLW**: Yes **PH**: No **SS**: A rec 964; B pgs 184-185; M pg 133 **BS**: 245

*EBERLY, Jacob; b 16 Jan 1795- d 23 Feb 1854 **RU**: Private, 13th or 97th VMR, Capt Samuel Colville's Co, Shenandoah Co, attached to 6th VMR (Coleman) **CEM**: St Pauls Lutheran Church; Shenandoah; 12180 Back Rd; Toms Brook; Strasburg **GS**: Yes **SP**: Mar (20 Dec 1839) Elizabeth Hockman (14 Dec 1808-20 Jun 1890), daug of Henry Hockman (1774-1828) and Rebecca Dillinger (1778-1828). Recd widows pen Act of 1878, LNR Mt Olive, Shenandoah Co, 1878 **VI**: Recd BLW 1850 of 40 acres, 1855 120 acres **P**: Widow **BLW**: Yes **PH**: No **SS**: B pg 184; M pg 133; O- pen files Fold3 **BS**: 245

EDMONDS, Elias; b 1778, d 1871 **RU: Lieutenant, Maj Kemper's Command **CEM**: Ivy Hill; Fauquier; Warrenton **GS**: Yes **SP**: Mar 1) Adeline (__) (1802-1837), 2) Emma James (1828-1913) **VI**: Father of Elias Edmonds (1832-1900) **P**: No **BLW**: No **PH**: No **SS**: A rec 1974 **BS**: 3 pg 17

*EDMONDS, James; b 16 Mar 1779; d aft 1812 **RU**: Private,85th VMR, Capt Seth Comb's Co, Fauquier Co, attached to 41st VMR (Branham) **CEM**: Oak Springs; Fauquier; 770 Fletcher Drive **GS**: Y **SP**: No spouse info **VI**: Son of Col William Edmonds (1725-19 Feb 1816) and Elizabeth Blackwell (1742-28 Feb 1817) **P**: No **BLW**: No **PH**: No **SS**: A rec 1987; B pg 73 **BS**: 30

Key: *Additional veteran entry   **Corrected veteran entry   ***Deleted veteran entry
RU=Rank/Unit   CEM=Cemetery   GS=Gravestone   SP=Spousal Information   VI=Other Veteran Info   P=Pension
BLW=Bounty/Land Warrant   PH=Photo   SS=Service Source   BS=Burial Source   VMR= VA Military Regiment
LNR= Last Known Residence

EDMONDSON, Andrew; b 25 Jan 1794; d 18 Jul 1852 **RU: Sergeant, Bradley's Regiment **CEM**: Rock Spring; Washington; vic jct Rts 803 & 91 **GS**: Y **SP**: Elizabeth Steele (1803-1827) **VI**: No further data **P**: No **BLW**: No **PH**: No **SS**: A rec 2045 **BS**: 116 pg 206.

*EDMONDSON, Richard H; b 1787, Mecklenburg Co, d 6 Nov 1859 **RU**: Private 69th VMR Capt Joseph Sanford's Trp of Calvary, Halifax Co, assigned to Detachment of Calvary **CEM**: St Johns Episcopal Church; Halifax; 13953 Halifax Rd, Java **GS**: Unk **SP**: Mar (Halifax Co, 15 May 1823) Susan Howell Chastain (23 Dec 1803-14 Nov 1891), her LNR Halifax Co 1878, Recd pen $12.00 mo **VI**: Recd pen and BLW 80 acres 1852 **P**: Both **BLW**: Yes **PH**: No **SS**: B pg 90; M pg 134; O- pen files Fold3 **BS**: 245

*EDWARDS, James; b unk, d 31 Dec 1857 **RU**: Fifer, 19th VMR, Capt Andrew Stevenson's Artillery Co, Richmond City, attached to 2d VMR (Ballowe) **CEM**: Hollywood; Richmond City; 412 So Cherry St **GS**: Unk, Sec B, lot 125 **SP**: Mar (24 May 1826) Sophia Thomas (1803-28 Dec 1892) LNR 1410 Clay St, Richmond. Recd widows pen $12 mo 4 Apr 1887 **VI**: Recd BLW, was Fifer also Public Guard **P**: Both **BLW**: Yes **PH**: No **SS**: B pg 175; M pg 134; O- pen files Fold3 **BS**: 245

*EDWARDS, Joseph; b Aug 1784, d 4 Oct 1859 **RU**: Private, 56th VMR, Loudoun Co. **CEM**: Saint Pauls Lutheran Church; Loudoun; Neersville **GS**: Yes **SP**: Elizabeth Conrad (6 Jun 1795-22 Dec 1871) **VI**: No further data **P**: No **BLW**: No **PH**: No **SS**: A rec 452; B pgs 119-121 **BS**: 245

*EDWARDS, William; b 1780, Surry Co, d 22 Jan 1827 **RU**: Private. 71st VMR, Capt Sampson White or Sampson Wilson's Co, Surry Co **CEM**: Cedar Grove; Portsmouth City; Effington St and Fort Lane **GS**: Unk **SP**: Frances (Fanny) Green Seawall (__-Jan 1857) **VI**: Son of William Edwards and Susannah Edmunds **P**: No **BLW**: No **PH**: No **SS**: B pg 193; L pg 841 **BS**: 245

*EFFINGER, John Frederick; b 5 Jun 1786, Shenandoah Co, d 6 Feb 1840 **RU**: Private, 116th VMR, Capt Daniel Matthews's Co, Rockingham County, attached to Flying Camp (McDowell) **CEM**: Woodbine; Harrisonburg City; at jct with E Market St and Reservoir St **GS**: Yes **SP**: Mar (Harrisonburg, 18 Apr 1811) Mary Hite (1793-18 Dec 1872). Recd BLW 80 acres 1855 and pen 4 Oct 1872 **VI**: Son of Ignatius Effinger **P**: Widow **BLW**: Widow **PH**: No **SS**: B pg 185; M pg 133; O- pen & BLW files Fold3 **BS**: 245

*ELGIN, Charles West; b 11 Jul 1788, d 9 Mar 1824 **RU**: Captain, 57th VMR, Company commander, Loudoun Co **CEM**: Leesburg Presbyterian Church, Loudoun; 207 W Market St **GS**: No **SP**: Rowena Drish Gore (8 Oct 1795-25 Dec 1855), daug of John Drish and Eleanor (__) **VI**: Son of Gustavus Elgin (1754-1834) and Rebecca Thrift (1767-1822) **P**: No **BLW**: No **PH**: No **SS**: B pg 120 **BS**: 245

*ELGIN, Robert; b 28 Jun 1783, d 13 Mar 1844 **RU**: Sergeant 57th VMR Capt Charles Elgin's Co, Loudoun Co, attached to (Mason-Minor) **CEM**: Elgin Family on Kingdom Farm; Loudoun; Evergreen Mill Rd, Sycolin **GS**: No **SP**: Elizabeth Elgin (17 Apr 1796-24 Mar 1841), daug of Gustavus Elgin (1754-1834) and Rebecca Thrift (1767-1822) **VI**: Son of Francis Elgin (1758, MD-6 Dec 1813) and Jane Adams (1759-aft 1820), The family gr st from cem moved to Union Cem in Leesburg. Recd BLW 160 acres 1855 **P**: No **BLW**: Yes **PH**: No **SS**: B pg 120; M pg 136; O- BLW files Fold3 **BS**: 245

** ELLIOTT, William Aquila, Sr, b 1785, MD, d 1875 **RU**: Captain, Naces Regt, and 41st Regt MD Militia **CEM**: Elliott Family; Frederick; off Rt 50, short distance W of Hayfield on private property **GS**: Unk **SP**: Nancy Ann Wright (1795-1875), daug of George Wright **VI**: Recd BLW 160 acres Act of 1855 **P**: No **BLW**: Yes **PH**: No **SS**: O- Serv Index and BLW files Fold3 **BS**: 79 pg 103; 245

*ELLIS, William; b 1784, d 1829 **RU**: Private 71st VMR One of four companies of Lunenburg Co, under command of Lt Col Grief Green and at Norfolk attached to 6th VMR (Sharp) **CEM**: Ellis Family; Lunenburg; Rehoboth **GS**: Yes **SP**: Elizabeth (__) (1786-1848) **VI**: No further data **P**: No **BLW**: No **PH**: No **SS**: A rec 1288; B pg 125; **BS**: 245

*ELLMORE, John; b 20 Apr 1781, d 02 Apr 1871, Loudoun Co **RU**: Private, 57th VMR Capt Charles Veale's Co, Loudoun Co **CEM**: Chestnut Grove; Fairfax; 831 Dranesville Rd, Herndon **GS**: Yes **SP** Mar (14 Apr 1833) Elizabeth Ann Rose (__-14 Dec 1896). Recd pen $12.00 mo. LNR Farmwell Station, Loudoun Co, 1878 **VI**: Recd BLW **P**: Widow **BLW**: Yes **PH**: No **SS**: B pg 123; M pg 137; O- pen files Fold3 **BS**: 245

Key: *Additional veteran entry   **Corrected veteran entry   ***Deleted veteran entry
RU=Rank/Unit   CEM=Cemetery   GS=Gravestone   SP=Spousal Information   VI=Other Veteran Info   P=Pension
BLW=Bounty/Land Warrant   PH=Photo   SS=Service Source   BS=Burial Source   VMR= VA Military Regiment
LNR= Last Known Residence

37

*EMBREY, Daniel; b 1792 Fauquier Co, d 25 Dec1878 **RU**: Private, 85th VMR Capt Seth Combs Co, and Capt Thomas Brooks Co, Fauquier Co, attached to 2d VMR (Ballowe) **CEM**: Daniel Embrey Family (AKA Embrey Cem #3); Stafford; S side of Ramouth Church Rd; nr jct I-95 **GS**: No **SP**: Winnifred Brown (Fauquier Co, 1801-15 Jan 1858) **VI**: Recd two pen and two BLW of 80 acres each 1855 **P**: Yes **BLW**: Yes **PH**: No **SS**: B pg 73; M pg 137; O- pen files Fold3 **BS**: 79 pg 103; 245

EMBREY, Robert; b 1768, d May 1857RU**: Sergeant, 44th VMR, Capt William O'Bannon's Company, Fauquier Co, attached to 36th VMR (Reno) **CEM**: Embrey Family; Fauquier; 1 mi NE of Summerduck **GS**: Unk **SP**: Sarah D Glass (Morrisville Co, 4 Oct 1778-aft May 1857) **VI**: Son of Thomas Embrey (1731-1838), Pineview & Mary (__) (1736-__) **P**: No **BLW**: No **PH**: No **SS**: A rec 2144 **BS**: 105

*ENGLAND, John; b 1795, d bef 12 Feb 1879 **RU**: Private, 45th & 16th VMRs Capt Lewis Alexander's Co Stafford County and Capt John Quarles Co, Spotsylvania County **CEM**: Hollywood; Richmond City; 412 So Cherry St **GS**: Yes Section K, lot 33 **SP**: Not married **VI**: Recd two pen and two BLW 40 acres 1850 and 120 acres 1855. LNR Atlees Station, Hanover Co, 1878 **P**: Yes **BLW**: Yes **PH**: No **SS**: B pgs 189; 190; M pg 138; O- pen files Fold3 **BS**: 245

EPPES, Francis Alexander; b 8 Jun 1773; d 13 May 1844 **RU: Private, 62nd VMR, Capt William Harrison, Prince George Co, attached to William Allen's 1st VMR **CEM**: Lewis / Bland; Prince George; Rt 625, 3.2 mi S of Disputanta **GS**: Unk **SP**: 1) Mildred Warmark (__-1823), 2) Ann (__), 3) unk **VI**: No further data **P**: No **BLW**: No **PH**: No **SS**: A rec 783; B pgs 170; 223 **BS**: 148

*EVANS, George; b unk, d 10 Jan 1816 **RU**: 1/LT, 20th US Army Inf **CEM**: Masonic; Fredericksburg; 900 block Charles St **GS**: Yes **SP**: No spouse info **VI**: Commissioned 2/LT (14 May 1812), 1/LT (2 Mar 1814), discharged (15 Jun 1815); son of Dr George Evans of Chesterfield Co **P**: No **BLW**: No **PH**: No **SS**: AF **BS**: 245

EVANS, John; b 1776; d 28 Mar 1850 **RU: Sergeant, 4th VMR (Beatty) **CEM**: Bethel Church; Frederick; Bethel Church Rd (Rt 610), Gore **GS**: No **SP**: Frances Hardesty (1787-12 Sep 1864, IL) **VI**: No further data **P**: No **BLW**: No **PH**: No **SS**: A rec 384 **BS**: 79 pg 104

*EVANS, Meredith Nathaniel; b 30 Nov 1789, Pittsylvania Co, d 26 Apr 1880, Dwina **CEM**: Hamm; Wise; jct Rts 657 & 658, Dry Fork **GS**: Yes Govt **SP**: Mar (Russell Co, 2 Sep 1815) Sarah Nancy Skeens (1796-26 Apr 1878) **VI**: Recd two pen and two BLW, one 1850-40 acres, 1855-120 acres. Was a Baptist minister. LNR P.O. Bickley's Millis, Russell Co, 1871 **P**: Yes **BLW**: Yes **PH**: No **SS**: B pgs 167; M pg 141; O- pen files Fold3 **BS**: 245

*EVANS, William; b, 1773, d 1835 **RU**: Lieutenant, 98th VMR, Capt Richard Daley's Troop of Cav or Capt John K Moore's Co, Mecklenburg County, attached to 1st VMR (Byrne) **CEM**: Oakwood; Mecklenburg; South Hill **GS**: Yes **SP**: Mary Walker (1785-1820) **VI**: Son of Anthony Evans & Mary Davis **P**: No **BLW**: No **PH**: No **SS**: A rec 737; B pgs 130-131 **BS**: 245

*EWING, Alexander; b 1773, d 1869 **RU**: Corporal, 1st Regt (Wears) East TN Volunteers **CEM**: Ewing-McClure; Lee; Jonesville **GS**: Yes **SP**: No spouse info **VI**: No further data

P: No **BLW**: No **PH**: No **SS**: A rec 1767 **BS**: 245

*EWING, Mitchell; b 1775, d 1 Aug 1825 **RU**: Ensign, 10th & 91st VMR, Bedford Co, attached to 5th VMR (Mason-Preston) **CEM**: Ewing-Patterson; Bedford; Penicks Mill **GS**: Yes **SP**: No spouse info **VI**: No further data **P**: No **BLW**: No **PH**: No **SS**: A rec 1849; B pgs 42-43 **BS**: 245

*FADELY, Henry; 14 Feb 1790, d 22 Mar 1833, MT Clifton **RU**: Private 13th or 97th VMR, Capt Jacob Fry's Co, Shenandoah Co, attached to 4th VMR (Wooding) **CEM**: Hudsons Crossroads Community (AKA St Johns United Church of Christ); Shenandoah; vic jct Rts 720 & 42 **GS**: Unk **SP**: Mar (1 Jan 1816) Sarah Heaton (1799-1861) **VI**: Rec pen & BLW 1871 **P**: Yes **BLW**: Yes **PH**: No **SS**: B pgs 185; M pg 142; O- pen files Fold3 **BS**: 245

Key: *Additional veteran entry   **Corrected veteran entry   ***Deleted veteran entry
RU=Rank/Unit   CEM=Cemetery   GS=Gravestone   SP=Spousal Information   VI=Other Veteran Info   P=Pension
BLW=Bounty/Land Warrant   PH=Photo   SS=Service Source   BS=Burial Source   VMR= VA Military Regiment
LNR= Last Known Residence

FINNEL (FINNALL), Jonathan; b 1782; d 18 Nov 1852 **RU: Private, 45th VMR, Capt John C Edrington's Company, Stafford Co **CEM:** Finnall Family; Stafford; Rt 600 abt 0.6 mi from jct with Rt 218, Stafford C. H. **GS:** Unk **SP:** No spouse info **VI:** Son of Jonathan Finnell and Magdalen (__), assistant inspector of tobacco 1778 **P:** No **BLW:** No **PH:** No **SS:** L pg 301 **BS:** 26 pg 199

FISHBACK, Martin; b 12 Oct 1763; d 24 Jan 1842 **RU: 2nd Lieutenant, US Army, 5th Regt serving 2 Jul 1813 to 15 Jun 1815 **CEM:** Fleetwood; Culpeper; Jeffersonton **GS:** Yes **SP:** Mar (23 Mar 1783) Lucy Amiss (8 Jul 1763-12 Sep 1843), daug of William Amiss and Annie (__) **VI:** Son of Johann Freidrich Fishbach and Eve Martin. Furnished land and buildings for Jeffersonton Academy. Also served as private in the Rev War (stone marked by SAR). *Records of Men Enlisted in the U.S. Army Prior to the Peace Establishment, May 17, 1815* states he died in service on 15 Mar 1815. This death record is for his son, Martin Fishback, Jr. who was born in 1791. It is not clear why this was entered with the service record of Martin Fishback, Sr. It is possible that the son was also in the U.S. Army, but if so no record of his service has been found **P:** No **BLW:** No **PH:** Yes **SS:** AF; D pg 421 **BS:** 31

*FISHER; William: b 5 Aug 1772, Wythe Co, d 30 Jan 1840 **RU:** Private, 35th USA Regt, Capt Isaac Preston's Co, at Fort Norfolk **CEM:** Northern Methodist; Rockingham; Mount Crawford **GS:** Yes Row 8, Plot 6 **SP:** Mary A (__) (1791-1874) **VI:** Blacksmith **P:** Yes **BLW:** No **PH:** No **SS:** C pg 62; O- pen files Fold3 **BS:** 245

*FITTS, Cornelius: b 1798, d 7 Jan 1865 **RU:** Fifer, 29th VMR Capt Hamilton Shields Co, Isle of Wight County, attached to 7th VMR (Saunders) **CEM:** Fitts Family; Lee; Rt 758 to Fitts Gap, at bottom of hill behind barn on right, 300 yds up hill, Jonesville **GS:** Yes **SP:** Mar (1 Feb 1814) Susannah Coleman (1823-1848) Recd pen **VI:** Recd BLW **P:** Widow **BLW:** Yes **PH:** No **SS:** B pg 104; M pg 149; O- Serv Index Card & pen files Fold3 **BS:** 245

FITZGERALD, James Henderson; b Dec 1793, Nottoway Co; d 6 May 1852 **RU: 49th VMR, Capt William Fitzgerald's Co, Nottoway County, attached to 4th VMR (Huston / Wooding) **CEM:** Thornton / Forbes / Washington; Fredericksburg; off Hunter St behind Mary Washington Hospital Home Health Agency **GS:** Yes **SP:** Mar (1810) Elizabeth A (__) (1793-18 Feb __) **VI:** Probably a memorial gravestone **P:** No **BLW:** No **PH:** Yes **SS:** B pg 154 **BS:** 26 pg 371

FITZHUGH, William Henry; b 1788, d 15 Apr 1854 **RU: Captain, 45th VMR, Company Commander, Stafford Co **CEM:** Fitzhugh Family; Stafford; 37 King George Grant Rd (Rt 608); Falmouth **GS:** Yes **SP:** Eliza Churchill Darby (1795-31 Jan 1850); "Consort of William H Fitzhugh" **VI:** Son of Thomas Fitzhugh (1760-1820) of Boscobel, Justice of Peace, Stafford Co, 1824; 1825; 1834; Sheriff 1833 **P:** No **BLW:** No **PH:** No **SS:** A rec 442 **BS:** 80; 26 pg 200

*FLANAGAN, William; b 25 Dec 1785, d 18 Aug 1878 **RU:** Private, 32d VMR, Capt John C Sower's Co, "Staunton's Artillery" Augusta Co, attached to a Battalion of Artillery. Also served in Maj Moses McCue's Detachment **CEM:** Sunset (AKA Christiansburg Municipal); Montgomery; jct So Franklin St & Elliott Dr, SE, Christiansburg **GS:** Yes **SP:** Mar (20 Jul 1823) Peggy Wall (1815-bef 29 Aug 1871) **VI:** Recd two pensions and two BLWs one 40 and one 120 acres **P:** Yes **BLW:** Yes **PH:** No **SS:** B pgs 39- 40; M pg 149; O- pen files Fold3 **BS:** 245

*FLEENOR, Isaac Blackhawk; b 19 Jan 1799, d 26 Feb 1879 **RU:** Private, 70th or 105th VMR, Capt Abram Fulkerson's Co, Washington County, attached to 5th VMR (Mason-Preston) **CEM:** Fleenor Memorial Baptist; Washington; nr 11547 Rich Valley Rd **GS:** Yes **SP:** Mar (18 Dec 1817) Susannah Andis, daug of Matthias Andis and Susannah Thomas **VI:** Recd two pensions and two BLW, both 80 acres each, LNR Craig Mills, Washington Co **P:** Yes **BLW:** Yes **PH:** No **SS:** B pg 198; M pg 149; O- pen files Fold3 **BS:** 245

*FLEMING, William; b 1 Aug 1797, d 15 Jan 1863 **RU:** Private, 56th VMR, Captain Thomas Gregg's Company of Loudoun Co, attached to 4th VMR (Beatty) **CEM:** South Fork Meeting House; Loudoun; Unison **GS:** Yes **SP:** No spouse info **VI:** No further data **P:** No **BLW:** No **PH:** No **SS:** A rec 1351; B pg 120 **BS:** 245

Key: *Additional veteran entry   **Corrected veteran entry   ***Deleted veteran entry
RU=Rank/Unit   CEM=Cemetery   GS=Gravestone   SP=Spousal Information   VI=Other Veteran Info   P=Pension
BLW=Bounty/Land Warrant   PH=Photo   SS=Service Source   BS=Burial Source   VMR= VA Military Regiment
LNR= Last Known Residence

*FLETCHER, Andrew; b 1787, d 18 Feb 1850 **RU:** Private, 13th VMR, Captain, Samuel Hawkins's Company, Shenandoah Co, attached to 4th VMR (Boyd) **CEM:** Rileyville; Page; end of cem Rd, Luray **GS:** Yes **SP:** No spouse info **VI:** No further data **P:** No **BLW:** No **PH:** No **SS:** A rec 1184; B pg 184 **BS:** 245

*FLETCHER, John Goldsmith; b 17 Apr 1792, d 4 Mar 1879 **RU:** Private, 94th VMR, Capt Jeremiah Neill's Co, Lee County, attached to 7th VMR (Saunders) **CEM:** Fletcher Family; Lee; Jonesboro **GS:** Yes **SP:** Mary Rebecca Randolph (1799-1893) **VI:** No further data **P:** No **BLW:** No **PH:** No **SS:** A rec 1676; B pg 118 **BS:** 245

*FLETCHER, Thomas; b 1 Feb 1776, d 16 Jan 1844 **RU:** Sergeant, 99th VMR; Capt John Staton's Co, Accomack Co **CEM:** Fletcher Family; Accomack; nr Jenkin's Bridge, Grotons **GS:** Unk **SP:** No spouse info **VI:** No further data **P:** No **BLW:** No **PH:** No **SS:** B pg 34; L pg 738 **BS:** 245

*FLOOD, William; b 1776, Appomattox Co, d 1818 **RU:** Captain, 100th VMR, Company commander, Buckingham Co, attached to 5th VMR (Mason & Preston) **CEM:** Red Oak Hill, Buckingham, David's Creek **GS:** Yes **SP:** Martha Guerrant, daug of Peter Guerrant, Jr and Mary Perreau **VI:** Son of John Flood and Mary Fuqua **P:** No **BLW:** No **PH:** No **SS:** A rec 2189; B pg 50 **BS:** 245

*FORD, Tipton; b bef 1800, d 27 Mar 1869 **RU:** Private, 12th VMR, Capt Horace Timberlake's Company, Fluvanna Co, attached to 7th VMR (Grays) **CEM:** Reedy; Rockingham; Rt 878, Wenger Mill **GS:** Unk **SP:** Mar (Dec 1825), Mary Smith (__-Jun 1887) who recd pen $12 mo, LNR Coots Store, Runion Creek, Rockingham Co **VI:** Died in Poorhouse. Recd BLW 1855 **P:** Widow **BLW:** Yes **PH:** No **SS:** B pg 104; M pg 149; O- Serv Index Card & pen files Fold3 **BS:** 245

*FOLTZ, George; b 4 Mar 1796, d 5 May 1868 **RU:** Private, 13th VMR, in Shenandoah Co, company attached to 5th VMR (Mason & Preston) **CEM:** Zion Lutheran Church: Shenandoah; Hamburg **GS:** Yes **SP:** Mary Wetzel (1795-1876) **VI:** No further data **P:** No **BLW:** No **PH:** No **SS:** A rec 609; B pgs 185-6 **BS:** 245

*FORTUNE, Thomas Eubank; b 1782, d 14 Mar 1857 **RU:** Captain 28th VMR, company commander of Artillery, Nelson Co. **CEM:** Fortune; Nelson; Lovingston **GS:** No **SP:** Mar 1) Cynthia Roseanne Loving (1785-1807), 2) (16 Sep 1810) Jane McAlexander (1787-1874) **VI:** Son of Thomas Fortune (1740-1804) and Elizabeth Eubank (1745-1821) **P:** Both He applied and 2nd wife recd **BLW:** Yes **PH:** No **SS:** B pg 142; BD pg 744; K pg 192 **BS:** 245

FOSTER, Isaac; b 1786; d 26 Sep 1854 **RU: Sergeant, 61st VMR, Lt Thomas Tabb's Co, Mathews County **CEM:** Pear Tree; Mathews; Rt 609; Onemo Post Office **GS:** Yes **SP:** No spouse info **VI:** Sea Captain and later keeper of New Point Lighthouse **P:** No **BLW:** No **PH:** No **SS:** K pg 304 **BS:** 54 pg 4

*FOSTER, James Coulter; b 13 Feb 1775, d 6 Apr 1838 **RU:** Corporal, 8th VMR, Capt Robert Davidson's or Capt John Elliott's Co Rockbridge County, attached to 6th VMR (Dickinson, Scott, Coleman) **CEM:** Walkerland; Rockbridge; Rt 602 Walker Creek Rd, top of hill behind Maxwelton's cabins **GS:** Yes **SP:** Sarah Sally McCray (10 May 1777-17 Aug 1858) **VI;** No further data **P:** Unk **BLW:** Unk **PH:** No **SS:** A rec 513; B pg 179; O-Serv Rec Fold3 **BS:** 245 cites Rockbridge Area Gen Soc; *Rockbridge Co, VA Cem;* pg 229

*FOUSHEE, Griffin Henry; b c1793, d 3 May 1844 **RU:** Ensign, 37th VMR, Captain William Jett's Co, Northumberland County **CEM:** Cralle's Banks (AKA Barnes-Nokomis); Northumberland; nr head of Coan River Barnes-Nokomis **GS:** Yes **SP:** Not married **VI:** Son of John H Foushee and Nancy Cralle; Owned Hughlett's tavern in Heathsville **P:** No **BLW:** No **PH:** No **SS:** K pg 396 **BS:** 269 pg 135

*FOWLKES, John Hall; b 1782, d 1838 **RU:** Private, 49th VMR, Capt Samuel Thomas's Co, Nottoway County, attached to 6th VMR (Sharps) **CEM:** Fowlkes Family; Nottoway; 6808 W. Courthouse Rd **GS:** Unk **SP:** Mar 1797, Elizabeth Dickenson Jennings (1779-1821), daug of James Jennings (24 Dec 1756-__) **VI:** Son of John Fields Fowlkes (1745-1824) and Dicey Hall (1744-__) His grandfather built Hyde Park, where he lived which has a VA historical Road sign **P:** No **BLW:** No **PH:** No **SS:** A rec 1858; B pg 154 **BS:** 245

Key: *Additional veteran entry · **Corrected veteran entry · ***Deleted veteran entry

RU=Rank/Unit	CEM=Cemetery	GS=Gravestone	SP=Spousal Information	VI=Other Veteran Info · P=Pension
BLW=Bounty/Land Warrant	PH=Photo	SS=Service Source	BS=Burial Source	VMR= VA Military Regiment
LNR= Last Known Residence				

40

*FRANCIS, Joseph E; b unk, d Jul 1868 **RU:** Private 33rd VMR, Captain Francis Wicker's Co, Henrico County **CEM:** Hollywood; Richmond City; 412 S Cherry St **GS:** No **SP:** No spouse info **VI:** No further data **P:** No **BLW:** No **PH:** No **SS:** K pg 829 **BS:** 245

*FRANCIS, Miles; b 20 Sep 1795, Buckingham Co, d 21 Oct 1875 **RU:** Private, 75th VMR, Capt William Currin's Co, Montgomery County **CEM:** Francis Family; Montgomery; jct Rts 666 & 1245, Christiansburg **GS:** Yes **SP:** Mar 1) Jane R Hall (27 Dec 1807-30 Jun 1846), daug of David Hall (1780-1849) and Elizabeth Pate (1783-1858), 2) Malvina L Simpkins (6 Mar 1816-9 Mar 1880), daug of John Thomas Simpkins and Delilah Akers, applied for pen **VI:** Pen rejected but recd BLW **P:** Widow **BLW:** Yes **PH:** No **SS:** B pg 138; O- pen files, Fold3 **BS:** 245

*FRANCISCO, Peter; b 9 Jul 1760, d 16 Jan 1831 **RU:** Private, 5th Regt U.S. Infantry **CEM:** Shockoe Hill; Richmond; 100 Hospital St **GS:** Yes **SP:** Mar as 3rd wife, Buckingham Co, 3 Jun 1823, Mary P West **VI:** A Rev War veteran whose widow was awarded Rev War pension and War of 1812 pension, Served as a Sergeant-of-Arms VA House of Delegates **P:** Both **BLW:** No **PH:** Yes **SS:** O- Commissioner's Ltr Rev War files **BS:** 31

*FRAZIER, James A; b 10 Aug 1780, Co Armagh, Ireland, d 3 Jan 1863 **RU:** Private, 81st VMR in a company of Bath Co **CEM:** Frazier Family; Augusta; W of Lone Fountain; Rt 250 at Jennings Gap **GS:** Yes **SP:** Mar (Augusta Co, 12 Feb 1809) Patsy Rankin **VI:** Operated a General Store **P:** No **BLW:** No **PH:** No **SS:** A rec 1315; B pg 41 **BS:** 245

*FREEMAN, David; b 1789, d bef 18 Aug 1886 **RU:** Private, 10th or 91st VMR, Capt Willie Jones Co, Bedford County, attached to 5th VMR, (Francis Preston) **CEM:** Freeman Family; Scott; N side Rt 614, 1 mi E of jct with Rt 636 **GS:** Yes Govt **SP:** Mar (Scott Co, Nov 1825) Salitha Rhatuss (__-bef 1871) **VI:** Recd pen 1871 age 81 as widower, also BLW. LNR P.O. Estillville, Scott Co **P:** Yes **BLW:** Yes **PH:** No **SS:** B pgs 43,229; M pg 157; O- pen files Fold3 **BS:** 245

*FREEMAN, William Nance; b 1786, d 5 Oct 1879 **RU:** Private 52d VMR, Capt Robert Perkin's Co, New Kent County **CEM:** St Clair Bottom Primitive Baptist Church; Smyth; jct Rts 600 and Riverside Rd, Chilhowie **GS:** No **SP:** Sarah Belsher **VI:** Recd BLW and pen $8.00 mo but pen rejected 30 Jan 1782 for insufficient serv of only 15 days. LNR Seven Mile Ford, Smyth Co, 1878 **P:** Yes **BLW:** Yes **PH:** No **SS:** B pg 144; M pg 157; O- pen files Fold3 **BS:** 245

FRITTER, Barnett; b 27 May 1792; d 23 Apr 1872 **RU: Private, 45th VMR, Capt Daniel Mason & Capt Elijah Hardin's companies Stafford Co **CEM:** Fritter Family; Stafford; Jct Rts 627 & 648, outside wall of Master's Cem **GS:** Unk **SP:** Mar 1) (12 Aug 1820) Betsy Faut, (1774-abt1827, 2) Mary L Fant, (19 Aug 1792-1861) **VI:** Likely son of John Fritter (1792-__). LNR PO Stafford Courthouse, 1871 **P:** Yes **BLW:** Yes **PH:** N **SS:** A rec 386; BD pg 768 **BS:** 26 pg 212; 49

FRY, Jacob; b 10 Mar 1769; d 23 Dec 1838 **RU: Captain, commanded a company, Frederick County, attached to 4th VMR **CEM:** St John's Lutheran; Frederick; 3623 Back Mountain Rd, Hayfield **GS:** Yes **SP:** Mar (Frederick Co, 6 Jan 1799) Elizabeth Linn, (25 Apr 1776-16 Mar 1838). Stone styles her "consort of Jacob Fry" **VI:** No further data **P:** No **BLW:** No **PH:** N **SS:** A rec 878 **BS:** 79 pg 117; 151

*FRIES/FRY, Michael; b 1777, d 2 May 1828 **RU:** Private, 31st or 51st VMR, in a company of Frederick Co attached to 4th VMR **CEM:** Old Stone Church: Frederick; Green Spring **GS:** Yes **SP:** No spouse info **VI:** No further data **P:** No **BLW:** No **PH:** No **SS:** A rec 937; B pgs 78-80 **BS:** 245

*FRYE, Jacob, b unk, d 16 Aug 1814 **RU:** Captain, 13th or 97th VMR, commanded company, Shenandoah Co. attached to 4th VMR (Huston and Wooding) **CEM:** Frye Family; Shenandoah; Wheatfield **GS:** No **SP:** Elizabeth (__) **VI:** After his death, command of company taken over by Captain John Sloan **P:** No **BLW:** No **PH:** No **SS:** B pg 185 **BS:** 245

Key: *Additional veteran entry   **Corrected veteran entry   ***Deleted veteran entry
RU=Rank/Unit   CEM=Cemetery   GS=Gravestone   SP=Spousal Information   VI=Other Veteran Info   P=Pension
BLW=Bounty/Land Warrant   PH=Photo   SS=Service Source   BS=Burial Source   VMR= VA Military Regiment
LNR= Last Known Residence

41

FULKERSON, Adam, b 3 Apr 1789, Lee Co, d 2 Oct 1859 **RU:** Captain, 70th VMR, commanded a company Washington County, attached to 5th VMR **CEM:** Sinking Spring Presbyterian; Washington; Blackfield Rd, one block fr Main St, Abington **GS:** Yes **SP:** Margaret Laughlin Vance (12 Sep 1794-22 Jun 1864), daug of Samuel Vance (1749-1838) and Margaret Laughlin (1775-1814) **VI:** Son of James Fulkerson (22 Jun 1737, PA-22 Sep 1799) and Mary Van Hook (19 Sep 1747-12 Jul 1830). Later became Colonel, VA Militia **P:** No **BLW:** No **PH:** No **SS:** B pg 198 **BS:** 116 pg 82; 245

FULKERSON, Peter; b 26 Sep 1764, Lancaster Co, PA, d 23 Jun 1847, Loudoun Co **RU:** Colonel, 94th VMR, Commander of regiment, Lee Co **CEM:** Powell Valley; Lee; Dryden **GS:** Unk **SP:** Mar (Washington Co) Margaret Craig (22 Jan 1773-28 Oct 1838) daug of Robert Craig, (Lancaster Co. PA, 28 Dec 1744-Abingdon, 4 Feb 1834) **VI:** Son of James Fulkerson (1737-1799) and Mary Van Hook (1747-1830). Recommended Captain 94th VMR 8 Nov 1796, appointed Colonel (25 Dec 1799). Served in VA House of Delegates, (1800-1802) **P:** No **BLW:** No **PH:** No **SS:** B pg 118 **BS:** 245

FULLER, John W; b 3 Mar 1797, d 7 Nov 1876 **RU:** Private, Capt John Lyle's Co, Rockbridge County, attached LTC Griffin Taylor's 1st VMR **CEM:** Stonewall Jackson Memorial; Lexington City; jct Rts 11 (Bus) & White St **GS:** Yes, Plot White 24, 1 **SP:** Ann F (__) (Alexandria, 1795-aft 1880) **VI:** Sadler, Librarian Franklin Hall **P:** No **BLW:** No **PH:** No **SS:** A rec 1693; B pg 247 **BS:** 245

FULTON, James; c1790, d 4 Apr 1855 **RU:** Private, 58th or 116th VMR, Capt John Snapp's and Capt Ralph Loftus's Trp of Cav, Rockingham Co **CEM:** Greenwood (AKA Bridgewater or Greenwood Ames); Rockingham; jct Green and N Grove Sts, Bridgewater **GS:** Yes **SP:** No spouse info **VI:** No further data **P:** No **BLW:** No **PH:** No **SS:** B pg 182; O- Serv Index card **BS:** 245

FUNK, Michael; b bef 1800, d 12 Nov 1875 **RU:** Private 51st VMR, Capt James H Sowers Co, Frederick Co, attached to McDowell's Flying Camp and 4th VMR **CEM:** Reliance; Warren: Rt 627 N of Reliance Ln **GS:** Yes **SP:** Mar (Fort Shenandoah, 23 Sep 1807) Catherine Ritenour **VI:** Recd two pen and two BLW of 80 acres each **P:** Yes **BLW:** Yes **PH:** No **SS:** B pg 80; M pg 160; O- pen files Fold3 **BS:** 245

FURR (FUR), Moses; b 5 Jan 1796, d 5 Jan 1867 **RU:** Private 44th or 85th VMR Capt George Love's Co, Fauquier County, attached to 2d VMR (Ballowe) **CEM:** Franks; Clarke; Rt 607 Ebenezer Rd **GS:** Yes **SP:** Mar (Warren Co, Nov 1818) Margaret Tracy (1 Jun 1798-18 Dec 1881) Recd two pen; LNR Paris, Fauquier Co, 1878 **VI:** Recd two BLW both 80 acres **P:** Widow **BLW:** Yes **PH:** No **SS:** B pg 74; M pg 161; O- pen files Fold3 **BS:** 245

GARDNER, James; b 1784, d 12 May 1858 **RU:** Private 16th VMR Capt John Quarles's and Capt Lewis Halladay's(Holliday)(Holladay) Co Spotsylvania County, attached to 1st VMR (Crutchfield) **CEM:** Rose Valley Farm; Spotsylvania; Rose Hill Subdivision **GS:** No **SP:** Mar 1) Mary Young (Louisa Co__-16 Sep 1847), 2) Lucy Tate Quisenberry (27 Aug 1810-25 Jun 1897), daug of Elijah Quisenberry (1781-1845) and Lucy Nelson (1783-1848). LNR Twyman's Store, 1878. Recd pen $12.00 mo **VI:** Burials are re-internments fr orig farm loc. Recd BLW of 40 and 120 acres **P:** Yes **BLW:** Yes **PH:** No **SS:** B pg 189; M pg 163; O- pen files Fold3 **BS:** 245

GARDNER, John; b 3 Jun 1780, PA, d 15 Nov 1853 **RU:** Private 75th VMR, Capt William Pepper's Co, Montgomery County, attached to 4th VMR (McDowell, Koontz, Chilton) **CEM:** Sunset (AKA Christiansburg Municipal); Montgomery; jct S Franklin St & Ellett Dr, SE **GS:** Yes **SP:** Mar (21 Jan 1813) Elizabeth (Betsey) Page (NJ, 12 May 1795-10 Jul 1875). Recd two widow's pen **VI:** Recd two BLW of 80 acres ea. Photo of vet-FindaGrave **P:** Widow **BLW:** Yes **PH:** No **SS:** B pg 138; M pg 163; O- pen files Fold3 **BS:** 245

GARLAND, James; b 6 Jun 1791, Ivy Depot, Albemarle Co; d 8 Aug 1885 **RU:** 1st Lieutenant, 53rd or 117th VMR, Capt William Cocke's Trp of Calvary, Campbell Co, attached to1st VMR (Holcombe) **CEM:** Spring Hill; Lynchburg; 3000 Fort Ave **GS:** Unk **SP:** Mar Mary (__) (21 Jul 1786-5 Jul 1854) **VI:** Lawyer, member VA House of Delegates 1829-31; U.S. Congress 1835-1841 **P:** No **BLW:** No **PH:** No **SS:** A rec 958 **BS:** 168

Key: *Additional veteran entry   **Corrected veteran entry   ***Deleted veteran entry
RU=Rank/Unit   CEM=Cemetery   GS=Gravestone   SP=Spousal Information   VI=Other Veteran Info   P=Pension
BLW=Bounty/Land Warrant   PH=Photo   SS=Service Source BS=Burial Source   VMR= VA Military Regiment
LNR= Last Known Residence

*GARNER, William; b c1781, England, d 15 Nov 1819 **RU**: Private, 89th VMR, Capt Benjamin Tyler's Co, Prince William County **CEM**: Presbyterian; Alexandria City; Wilkes St **GS**: Yes **SP**: No spouse info **VI**: No further data **P**: No **BLW**: No **PH**: No **SS**: B pg 172; L pg 791 **BS**: 245

*GARNETT, James Mercer; b 30 Oct 1794, d 14 Jul 1824 **RU**: Private 6th VMR, Captain William Garnett's Company, Essex Co. **CEM**: Garnett Family; Essex; Elmwood; Loretto **GS**: Unk **SP**: No spouse info **VI**: Son of James M Garnett (1770-1843) and Mary Eleanor Mercer (1774-1837) **P**: No **BLW**: No **PH**: No **SS**: A rec 1332; B pgs 69; 70 **BS**: 245

GARNETT, Robert; b 17 Jun 1770, Culpeper Co, d 11 Sep 1854, Syria, Madison Co **RU: Private, Major Thomas Hill's Detachment of Inf, King William Co **CEM**: Old Garnett; Madison; Rt 600; Nr Syria P.O. **GS**: Yes **SP**: Rhoda Sampson (24 Apr (1773-28 Apr 1855) **VI**: An Elder in church. Son of James Garnett, (1743-1830) **P**: No **BLW**: No **PH**: No **SS**: A rec 1356; B pg 115; K pg 49 **BS**: 191; 245

*GARNETT, Robert Selden, Esq; b 26 Apr 1789, d unk **RU**: Private, 6th VMR, Capt William Garnett's Company, Essex Co **CEM**: Champlain Estate; Essex; Lloyds **GS**: Yes **SP**: Charlotte Olympia De Gougea, (__-1856) **VI**: Attended Princeton University, recd degree as lawyer; served VA House of delegates, then US Congress 1817 to 1827 **P**: No **BLW**: No **PH**: No **SS**: A rec 1358; B pg 70; **BS**: 245

*GARNETT, Thomas; b unk, d 5 Apr 1826 **RU**: Private, 24th VMR Captain John Gannaway's Co, Buckingham County, attached to 8th VMR (Walls) **CEM**: Garnett Family; Buckingham; loc on his property **GS**: Yes **SP**: No spouse info **VI**: No further data **P**: No **BLW**: No **PH**: No **SS**: B pg 50; K pg 96 **BS**: 245

*GARRETT, Ira; b 3 Jul 1791, Louisa Co, d 26 Jul 1870 **RU**: Private, 28th VMR Capt Thomas Fortune's Artillery Co, Nelson County **CEM**: Maplewood; Charlottesville City; jct Lexington Ave & Maple St **GS**: Yes **SP**: Mar (Milton, Albemarle Co, 14 Mar 1817) Elizabeth J Watson (14 Oct 1796-22 Apr 1880). LNR Alexandria at age 81. Recd pen **VI**: Deputy Sheriff, County & Circuit Clerk, Albemarle Co. Recd BLW 1855 of 160 acres **P**: Widow **BLW**: Yes **PH**: No **SS**: B pg 142; M pg 164; O- pen files Fold3 **BS**: 245

*GARY, James; b 1764, d Apr 1831, Centerville **RU**: Private, 39th VMR, Captain Richard McRae's Co, Petersburg Volunteers **CEM**: Gary Family; Fairfax; Centerville **GS**: Unk **SP**: No spouse info **VI**: Enlisted at age 16, 1st VA Regt during the Rev War **P**: No **BLW**: No **PH**: No **SS**: B pg 30 **BS**: 30

*GATHRIGHT (GARTHWRIGHT) , William Jr; b 1 Mar 1778, d 1850 **RU**: Private, 33rd VMR, Captain William Allen's Company, Henrico Co **CEM**: Woodson Family #2; Goochland; Rt 609, 1.5 mi fr East Lake **GS**: N **SP**: Jane (__) (4 Mar 1776-1825) **VI**: No further data **P**: No **BLW**: No **PH**: No **SS**: K pg 10 **BS**: 78; 245

*GIANNINY, Nicholas; b 1779, d 19 Jan 1854 **RU**: Private, 47th VMR, Capt John Cole's Company, Albemarle Co, attached to 6th VMR (Coleman) **CEM**: Gianniny; Albemarle; Buck island creek area, betw Monticello and Carter's bridge **GS**: Yes **SP**: Amanda Polly Pace **VI**: Neighbor to Thomas Jefferson, nephew of President Thomas Jefferson **P**: No **BLW**: No **PH**: No **SS**: B pg 25 **BS**: 31

*GIBSON, William; b unk, d 1851 **RU**: Private, 56th VMR, Capt Thomas Gregg's Company Loudoun Co, attached to 4th VMR (Beatty) **CEM**: Goose Creek Burying Ground; Loudoun; Rt 722 Lincoln **GS**: Yes **SP**: No spouse Info **VI**: No further data **P**: No **BLW**: No **PH**: No **SS**: A rec 1242- 4; B pg 119-121 **BS**: 245

***GILL, Richard H: Deleted as service identified pertains to another individual

*GILMER, George; b 11 Mar 1778, d 6 Oct 1836 **RU**: Private, 88th VMR, Captain Samuel Carr's Troop of Calvary, Albemarle Co, attached to Randolph's 1st Corps DeElite **CEM**: Mount Air; Albemarle; Keene **GS**: No **SP**: Elizabeth Anderson Hudson (30 Apr 1784-8 Aug 1820), daug of Christopher Hudson (1758-1825) and Sarah Anderson (1758-1807) **VI**: Son of Dr George Gilmer (1742-1793) and Lucy Walker (1751-1825); Doctor and friend of Thomas Jefferson **P**: No **BLW**: No **PH**: No **SS**: B pg 85; L pg 196 **BS**: 245

Key: *Additional veteran entry **Corrected veteran entry ***Deleted veteran entry
RU=Rank/Unit   CEM=Cemetery   GS=Gravestone   SP=Spousal Information   VI=Other Veteran Info   P=Pension
BLW=Bounty/Land Warrant   PH=Photo   SS=Service Source   BS=Burial Source   VMR= VA Military Regiment
LNR= Last Known Residence

43

*GLAIZE (GLAZE), Sampson; b 3 Oct 1791, d 6 Feb 1850 **RU:** Private, Frederick Co in company attached to 1st VMR (Taylor) **CEM:** Hebron; Winchester; GPS 39.18170, -78.1572, 305 E Boscawen St **GS:** Yes **SP:** Elizabeth Rachel Renner (13 May 1798-1 Oct 1856) **VI:** Son of George and Catherine (Hetzal) Glaize **P:** No **BLW:** No **PH:** No **SS:** A rec 2721 **BS:** 56 pg 180

GLASCOCK, John; b unk, d 29 Jan 1871 **RU: Ensign, 44th VMR, Capt Enoch Jeffrie's Co, Fauquier County, attached to 4th VMR (Beatty) **CEM:** Rector town; Fauquier; Rt 624 NE Rector town **GS:** Yes **SP:** No spouse info **VI:** No further data **P:** No **BLW:** No **PH:** No **SS:** A rec 2389; B pg 74 **BS:** 4 pg 67

GLASCOCK, William; b 20 May 1785, d 17 Feb 1857 **RU: Fifer, 44th VMR, Capt Nathaniel Grigsby's, Troop of Cavalry, Fauquier Co **CEM:** Glascock Family; Fauquier; "Glenmore," Rt 624 NE of Rector town **GS:** Yes **SP:** Mar (Fauquier Co, 14 Mar 1798) Mahalia Alice Cole **VI:** Son of Lt Thomas Glascock and Agnes / Agatha Rector **P:** No **BLW:** No **PH:** No **SS:** A rec 2654 **BS:** 4 pg 67

*GOGGIN, Pleasant Moorman; b 11 Feb 1777, d 3 Feb 1831 **RU:** Captain, Company commander, 10th or 91st VMR, Bedford Co, attached to 7th VMR (Saunders) **CEM:** Goggin Family; Bedford; Bunker Hill **GS:** Yes **SP:** Mary Otey Leftwich (1789-1854) **VI:** No further data **P:** No **BLW:** No **PH:** No **SS:** A rec 1298; B pg 42 **BS:** 245

GOOD, Felix; b 30 Aug 1794; d 26 Sep 1875 **RU: Captain, 31st VMR, Capt Isaac Van Horn's Company, or his own company, Frederick Co, attached to 4th VMR (Boyd) **CEM:** Good Family; Frederick; Pinetop Rd **GS:** Unk **SP:** Mar (Frederick Co, 14 Dec 1820) Rachel Orndorff (22 Oct 1798-3 Jun 1857), daug of John Orndorff and Margaret Renner **VI:** Son of Felix Good, Sr. and Margaret Delong, LNR Back Creek, Frederick Co **P:** Yes **BLW:** Yes **PH:** No **SS:** BD pg 833; B pg 80 **BS:** 79 pg 126

GOOD, Jacob; b 6 Apr 1799; d 12 Mar 1881 **RU: Fifer, 31st, 51st or 122nd VMR, in a company Frederick Co, probably his fathers, attached to 4th VMR **CEM:** Gainesboro; Frederick; 166 Siler Ln, Gainesboro **GS:** Yes **SP:** Mar (Frederick Co, 5 Jan 1829) (bond), Lucy Wigginton, James Walls surety **VI:** Son of Felix Good, Sr and Margaret Delong **P:** No **BLW:** No **PH:** No **SS:** A rec 37 **BS:** 79 pg 126

*GOODALL, Philander; b Jan 1794, d 27 May 1890 **RU:** Private, 51st VMR, Capt John Pitman's Co, Frederick County, attached to 1st VMR (Taylor) **CEM:** Goodall; Madison; Philander, 1.1 miles upriver from Graves Mill on Rt 662, Locust Dale **GS:** Yes **SP:** Mar 1) (Madison Co, 3 Aug 1818) Mourning Marshall (1802-bef 1870), 2) (27 Nov 1872) Jemima Catherine Gallehugh (1840-bef 1890), daug of Elijah Gallehugh **VI:** Son of Joseph Goodall & Edney Grimsley. Recd pen and BLW of 160 acres **P:** Yes **BLW:** Yes **PH:** No **SS:** A rec 163; B pg 79; M pg 173; O- Pen files, Fold3 **BS:** 30, Family Cem; 245

*GOODE, Jacob; b 1793, d 1887 **RU:** Private, 110th VMR, Lt Smith Webb's Co, Franklin County **CEM:** Goode Home Place; Henry; Oak Level **GS:** Unk **SP:** Alice Mullins (1796-__), daug of Booker Mullins **VI:** Son of David Goode (1762-1851) & Elizabeth Waltz (1768-1848), Recd pen and BLW of 160 acres **P:** Yes **BLW:** Yes **PH:** No **SS:** B pg 77; M pg 173; O- Serv index card and pen files Fold3 **BS:** 245

*GOODWIN, Thomas Cary; b 1 Sep 1794, d 6 Jan 1864 **RU:** Private, 47th VMR, Capt John Field's Co, Albemarle Co, attached to 8th VMR (Walls) **CEM:** Goodwin Family; Amherst; off Rt 60, West of Amherst at forks of Buffalo River **GS:** Yes Govt **SP:** Lucinda Montgomery **VI:** Son of Thomas Goodwin & Temperance Harris **P:** No **BLW:** No **PH:** No **SS:** B pg 35, K pg 91 **BS:** 245

*GOODWIN, William; b 21 Jan 1791, d 5 Apr 1863 **RU:** Private' 40th VMR. Capt David Watson's Troop of Cavalry, Louisa Co, attached to Cocke's Detachment **CEM:** Goodwin Family; Louisa; Ellisville Dr, Rt 669 **GS:** Yes **SP:** Mar (8 May 1823) Frances Jane Goodwin (1805-1891). Recd pen **VI:** Recd two BLW of 80 acres ea **P:** Widow **BLW:** Yes **PH:** No **SS:** B pg 124; M pg 175; O- pen files Fold3 **BS:** 245

*GORDON, Alexander B; b 1797, d 19 Mar 1832 **RU:** Private 16th VMR serving in one of several companies of Spotsylvania County, attached to the 1st VMR commanded by Lt Col Robert Crutchfield **CEM:** Gordons of Kenmore; Fredericksburg City; GPS 38.30560,-77.46843; on Washington Ave nr Hitchcock St **GS:** Yes **SP:** No spousal info **VI:** Son of John Gordon of Lochdougan, Scotland **P:** No **BLW:** No **PH:** No **SS:** A rec 1721; B pgs 188; 189; 247 **BS:** 245

Key: *Additional veteran entry   **Corrected veteran entry   ***Deleted veteran entry
RU= Rank/Unit   CEM=Cemetery   GS=Gravestone   SP=Spousal Information   VI=Other Veteran Info   P=Pension
BLW=Bounty/Land Warrant   PH=Photo   SS=Service Source   BS=Burial Source   VMR= VA Military Regiment
LNR= Last Known Residence

*GORDON, James; b 22 Jul 1787, d 15 Jun 1824 **RU**: Prob Sergeant 60th VMR in a company of Fairfax County or as a Private in a company of unk county, attached to 4th VMR (Boyd) **CEM**: Maplewood; Orange; Spotswood Trail, Gordonsville **GS**: Yes **SP**: No spousal info **VI**: No further data **P**: No **BLW**: No **PH**: No **SS**: A rec 1845 **BS**: 245

*GORDON, Samuel; b 1779, d 1845 **RU**: Sergeant, 100th VMR, Capt William Moseley's Troop of Cavalry, Buckingham Co, attached to 1st VMR of Calvary (Holcombe) **CEM**: Gordons of Kenmore; Fredericksburg City; GPS 38.30560,-77.46843; on Washington Ave nr Hitchcock St **GS**: Yes **SP**: Elizabeth Cole Fitzhugh (1775-Jul 1869), daug of William Fitzhugh (1741-1809) and Sarah Diggfes (1757-1804) **VI**: No further data **P**: No **BLW**: No **PH**: No **SS**: A rec 2003; B pgs 51; 238; L pg 602 **BS**: 245

GORDON, William Richards; b 1780; d 1855 **RU: Corporal, 36th VMR, Capt James Fox's Co, Prince William County **CEM**: Gordon Family; Stafford; Rosepetal St (Rt 1266) **GS**: Yes, lot 49 **SP**: Mary A M (__), (1808-1874) **VI**: Justice of Peace, 1825, Stafford Co **P**: No **BLW**: No **PH**: No **SS**: A rec 2077 **BS**: 26 pg 221

*GRABILL (GRABEEL), Daniel; b 11 Jun 1787, d 16 Aug 1854 **RU**: Ensign, 13th VMR, Capt George Shrum's Co, Shenandoah County attached to 1st VMR (Truehart) and to Maj Perkin's Garrison **CEM**: Neff-Kagey; Shenandoah; GPS: 38.69443, -78.65895; Old Bridge Rd, New Market **GS**: Yes **SP**: Mar (14 Feb 1814) Hannah Swartz **VI**: Son of George Grabill (1750-1825) and Anna Neff (1765-1848) **P**: No **BLW**: No **PH**: No **SS**: B pg 185; K pg 432 **BS: 245**

*GRAHAM, George; b˜16 May 1770, Dumfries, Prince William Co, d 9 Aug 1830, Washington, DC **RU**: Captain; 60th VMR, Commander, Troop of Calvary "Fairfax Light Dragoons", 1st Cav, Fairfax Co, attached to 1st Corps D'Elite (Randolph) **CEM**: Arlington National; Arlington; off George Washington Parkway **GS**: Yes **SP**: 1) Elizabeth Mary Anne Hooe, 2) (__) Watson **VI**: Originally buried Congressional Cem, DC, reinterred Arlington National Cemetery section 3, lot 1989. His military unit was on duty (23 Aug-4 Oct 1814), thus believed to have been in the battle in Bladensburg, MD. Served as Acting Secretary of War (1815-1817) and was sent on a special mission to Galveston Island, TX. He was President of the Washington Branch of the Bank of US from 1817 to 1823. Served as Commissioner of the General Land Office of the US. Son of Richard Graham and (__) Brent, daug of George Brent, Esq. Graduated from Columbia College 1790 and practiced law first in Dumfries, then in Fairfax Co. **P**: No **BLW**: No **PH**: No **SS**: A rec 1046; B pg 71 **BS**: 53 pg 19; 55

*GRANT, Gardner; b 29 Apr 1791, d 10 Dec 1856 **RU**: Private, 105th VMR, Capt Henry St John Dixon's Co, Washington County, attached to 5th VMR (Mason, Preston) **CEM**: Zion Methodist Church; Washington; Rt 712 nr jct with Rt 58, Abingdon **GS**: Yes **SP**: Mar (23 Jan 1815), Hannah Gilliam (__-3 Jan 1877, Verona, MO) She recd pen,1871 **VI**: Recd BLW of 80 acres **P**: Widow **BLW**: Yes **PH**: No **SS**: B pg 198; M pg 177; O- pen files Fold3 **BS**: 245

*GRAVELLY (GRAVELY), Joseph Jefferson; b 1778, Pittsylvania Co, d 2 Feb 1871 **RU**: Private, George Vashon's Co, 10th Inf USA Enlisted 26 Jan 1813, discharged 27 Jan 1814 **CEM**: William Marshall Gravely; Henry; one tenth mi East of jct Rt 620 & Pebble Brook Rd, Axton **GS**: Unk **SP**: Mar (Henry Co, 30 Dec 1799) Mary Helen King (10 Jan 1780-1 Apr 1860), daug of John King (1758-1821) & Mary Elizabeth Seward **VI**: No further data **P**: No **BLW**: No **PH**: No **SS**: C pg 75; O-Army Reg; pg 422 **BS**: 245

*GRAVELY, Lewis; 1796, d 11 May 1884 **RU**: Private, 64th VMR Capt Thomas Graves Co, Henry County **CEM**: Dyer Family; Henry; nr jct Foxfire & Dyers Store Rd, Leatherwood; **GS**: Unk **SP**: Rachel Martha Dyer (1800-24 Oct 1876), daug of George Dyer and Rachel Dalton **VI**: Recd two pen and two BLW, one 40 acres in 1850 and 160 acres 1855 **P**: Yes **BLW**: Yes **PH**: No **SS**: B pg 101; M pg 178; O-pen files Fold3 **BS**: 245

GRAVES, George; b 1783, d 1823 **RU: Private, 15th VMR, Capt Isaac Mitchell's Company, Sussex Co **CEM**: Shockoe Hill; Richmond City; 100 Hospital St **GS**: Unk **SP**: No spouse info **VI**: No further data **P**: No **BLW**: No **PH**: No **SS**: L pg 594 **BS**: 38 pg 1.

GRAY, James; b 15 Apr 1789, Amelia Co, d 3 Apr 1655 **RU: Private, 19th VMR, Capt George Booker's or Capt Samuel Jones's Co, Richmond City **CEM**: Hollywood; Richmond City; 412 S Cherry St **GS**: Yes **SP**: Harriet Ann Wherry (1796-1863) **VI**: No further data **P**: No **BLW**: No **PH**: No **SS**: A rec 314; L pgs 155, 498 **BS**: 260; 245

*GRAY, James; b 1783, d 21 Apr 1844 **RU**: Private, enlisted Alexandria 15 Jul 1814 in the 14th Inf U.S. Army **CEM**: St James Episcopal Church; Loudoun; Church St NE, across st fr Loudoun Co Circuit Ct bldg, Leesburg **GS**: Yes **SP**: No spousal info **VI**: No further data **P**: No **BLW**: No **PH**: No **SS**: C pg 75 **BS**: 245

***GRAY, John: Deleted as service or burial not adequate

*GRAY, Zebedee; b 1 Nov 1795, d 15 Jul 1881 **RU**: Private, 55th VMR, Capt Thomas Cockrell's Co, Jefferson County (now WVA), attached to 1st VMR (Lt Col Griffin Taylor) **CEM**: Berryville Baptist Church; Clarke; jct Academy St & Rice St, Berryville **GS**: Yes **SP**: Mar (18 Mar 1823) Nancy M (or A) Dowell (27 Sep 1795-13 May 1866) **VI**: Recd pen and BLW of 160 acres **P**: Yes **BLW**: Yes **PH**: No **SS**: B pg 107; M pg 179; O- pen files Fold3 **BS**: 245

*GREANER, William; b 18 Aug 1793, MD, d 31 Dec b1868 **RU**: Private Capt Joseph K Stapleton's and Capt William Meyer's Cos, MD Militia at battle of North Point, MD **CEM**: Hollwood; Richmond City; 412 S Cherry St **GS**: Yes **SP**: 1) (Baltimore) Temperance Temple (__-c1840), 2) (9 Mar 1845) Sarah (__) widow Talbert, pen Richmond 1878 **VI**: Owned tobacco factory used as Castle Thunder Prison in War. Recd two BLW, 40 acres in 1850; 120 acres 1855 acres **P**: Widow **BLW**: Yes **PH**: No **SS**: M pg 179; O- pen files Fold3 **BS**: 245

***GREEN, Duff: Deleted as service or burial not adequate

*GREEN, Theophilus F; b 2 Oct 1789, d 14 Apr 1862 **RU**: Corporal 30th VMR, Capt William F Gray's Co, Caroline County attached to 9th VMR (Boyd) **CEM**: Old Salem Baptist Churchyard; Caroline; Sparta **GS**: No **SP**: Mar (24 Dec 1816) Patsy E. Walden (1796-aft 1878), LNR Sparta, Caroline Co, recd pen 1878 **VI**: Son of George Green and Eliza Walden. Recd BLW 40 acres 1850 **P**: Widow **BLW**: Yes **PH**: No **SS**: B pg 55; M pg 179; O- pen files Fold3 **BS**: 245

*GREEN, William; b 1774, England, d 01 Jun 1824 **RU**: Private, Major King's Detachment, DC Militia **CEM**: Saint Paul's Episcopal Church; Alexandria; 228 S Pitts St **GS**: Yes **SP**: Mary (__), (1782-1818) **VI**: No further data **P**: No **BLW**: No **PH**: No **SS**: A rec 2107; **BS**: 245

*GREEN, William; b 12 Jan 1793, d 10 Jun 1867 **RU**: Private, 52nd VMR, Captain Robert Perkins's Co, New Kent County **CEM**: Old Graveyard: Henrico; 5.6 mi SE of Richmond on Darbytown Rd, then N dirt Rd 4 mi **GS**: Unk **SP**: No spouse info **VI**: No further data **P**: No **BLW**: No **PH**: No **SS**: A rec 2183; B pg 144; L pg 632 **BS**: 245

GREENLAW, William P; b by 1799, d aft July 1813 (War of 1812 service) **RU: Private, 25th VMR, Capt John Ashton's Co, King George County **CEM**: Hollywood; Stafford; Hollywood Farm, west of Caisson Road; Rt 601; abt 600 feet fr Hollywood farmhouse **GS**: Yes **SP**: No spouse info **VI**: No dates on stone **P**: No **BLW**: No **PH**: No **SS**: L pg 111 **BS**: 26 pg 248

*GREGG, James B; b Unk, d 1847 **RU**: Private, 32nd PA Militia, Capt Squire's Co. & 5th PA Militia (Fenton's) **CEM**: Goose Creek Burying Ground; Loudoun; Rt 722 Lincoln **GS**: Yes **SP**: No spouse info **VI**: No further data **P**: No **BLW**: No **PH**: No **SS**: A rec 2751 & 2753 **BS**: 245

*GREGORY, William; b 3 Mar 1789, Kilmarnock, East Ayrshire, Scotland, d 13 Jun 1875 **RU**: Private, Captain Charles Knight and Captain Lewis Hopkin's Co, DC Militia **CEM**: Presbyterian Cem; GPS: 38.80015,-77.05791; Alexandria **GS**: Yes **SP**: Mar (6 Mar 1838) Mary Donaldson (1809-1896) **VI**: No further data **P**: Spouse (WC-8556) **BLW**: Yes **PH**: No **SS**: A rec 2445; BD pg 873 **BS**: 245

Key: *Additional veteran entry   **Corrected veteran entry   ***Deleted veteran entry
RU=Rank/Unit   CEM=Cemetery   GS=Gravestone   SP=Spousal Information   VI=Other Veteran Info   P=Pension
BLW=Bounty/Land Warrant   PH=Photo   SS=Service Source   BS=Burial Source   VMR= VA Military Regiment
LNR= Last Known Residence

46

*GRESHAM, Thomas; b 5 Apr 1774, d 2 Dec 1837 **RU:** Ensign, 9th VMR, commander of unit of King and Queen Co, attached to Major Campbell of Caroline Co. **CEM:** Dew Family; King & Queen; Owenton **GS:** Yes **SP:** Mary Ellen Dew (1786-Feb 1836), daug Thomas Roderick Dew and Lucy Ellen **VI:** Son of Samuel Gresham and Hannah Farmer. Was a lawyer, Essex Co. Owned Woodlawn, King and Queen Co **P:** No **BLW:** No **PH:** No **SS:** B pg 113 **BS:** 245

GRIFFIN, James; b 1797, d 2 Jul 1852 **RU: Private, 65th VMR Capt James Clayton's Co, Southampton County **CEM:** Hollywood; Richmond City; 412 S Cherry St **GS:** Yes, Sec D, lot 144 **SP:** 1) (_) Bowles, 2) (16 Oct 1844) Jane A Barrett (_-10 Aug 1888) LNR Henrico Co, 1888. Recd two pen **VI:** Resided Southampton Co 1850, moved to Henrico Co by 1855. Recd BLW 40 acres (1850); 120 acres (1855) **P:** Widow **BLW:** Yes **PH:** No **SS:** B pg 186; O- pen files **BS:** 237; 245

*GRIFFITH, John M; b unk, d 10 Nov 1816 **RU:** Private, 31st or 58th VMR, a company of Frederick Co, attached to 4th VMR (Boyd) **CEM:** Gainesboro Methodist; Frederick; Gainesboro **GS:** Yes **SP:** No spouse info **VI:** No further data **P:** No **BLW:** No **PH:** No **SS:** A rec 116; B pgs 78-80 **BS:** 245

*GRIFFITH, Thomas; b 1797, d 22 Dec 1863 **RU:** Private, 56th VMR, Capt Jacob Price's Co, Loudoun County **CEM:** North Fork Baptist Church; Loudoun; 38130 North Fork Rd, North Fork **GS:** Yes **SP:** Mar (23 May 1822) Sarah Elizabeth Van Horn (1805-18 Feb 1896). Recd pen 1879, LNR Charlestown, WVA **VI:** Son of Thomas Griffith (1757-1835) and Mary March (1758-1842). Recd BLW 1850 **P:** Widow **BLW:** Yes **PH:** No **SS:** B pg 120; O- pen files **BS:** 245

GROVE, John W; b 23 May 1787; d 8 Mar 1863 **RU: 1st Sergeant, 51st VMR, Capt John Gilkerson's Co, Frederick County, attached to McDowell's Flying Camp **CEM:** Hironimus; Frederick; Old Mill Ln, Whitacre **GS:** Unk **SP:** Mar (Frederick Co, 31 Jan 1813) (return by John B Tilden) Jane Young, (_-bef 9 Sep 1882), LNR PO New Town Stevensburg, (Frederick Co, 1879) **VI:** Methodist preacher **P:** Widow **BLW:** No **PH:** No **SS:** BD pg 884; B pg 79 **BS:** 79 pg 132

*GUNN, Burwell; b 1795 d aft 1850 Census, Nottoway Co **RU:** Corporal, 1st VMR, Capt Moses Overton's Co, Amelia County, attached to 5th VMR (Mason & Preston) **CEM:** Butterwood Church; Dinwiddie; Darvills **GS:** Yes **SP:** Mary S (_) (1796-aft 1870 Census, Dinwiddie Co) **VI:** Dis-internment fr Roger's Fam Cem on Fort Pickett, 1941-2 period **P:** Widow applied **BLW:** No **PH:** No **SS:** A rec 1797; B pg 37; BD pg 889; M pg 184 **BS:** 245

*GUNNELL, George West; b 1789, d 8 Jan 1878 **RU:** Captain, 60th VMR, Company commander, Fairfax Co **CEM:** Andrew Chapel Methodist Church; Fairfax; 9201 Leesburg Pike, Vienna **GS:** Yes **SP:** 1) Lucy (Louisiana) Ratcliffe, 2) (Fairfax Co, 3 Oct 1839) Emmaline Young Adams,(_-19 Nov 1887), LNR Vienna, 1878 **VI:** Attained rank of Colonel after war **P:** 2d Spouse **BLW:** Yes **PH:** No **SS:** A rec 1870; B pg 71; BD pg 890; M pg 184 **BS:** 245

*GUY, Robert; b 23 Mar 1792, Ireland, d 15 Jan 1863 **RU:** Sergeant, Capt Jesse Dold's Troop of Calvary, Augusta Co attached to Woolford's Squadron **CEM:** Augusta Stone Presbyterian Church; Augusta; Fort Defiance **GS:** Yes **SP:** Mar Gilly S (_) (1804-1841) **VI:** No further data **P:** No **BLW:** No **PH:** No **SS:** B pg 39 **BS:** 245

*HAIRSTON, George; b 20 Sep 1750, d 5 Mar 1825 **RU:** Lt Col, Hairston's Brigade, VA Militia **CEM:** Hairston Family; Henry; Beaver Creek Plantation **GS:** Yes **SP:** Mar (Henry Co, 1 Jan 1781) Elizabeth Perkins Letcher (1759-1818) **VI:** Son of Robert and Ruth (Stovall) Hairston; was Revolutionary War Captain company commander in Henry Co **P:** No **BLW:** No **PH:** No **SS:** A rec 2692 **BS:** 245

*HALE, Lewis II; born 14 May 1781 d 24 May 1842 **RU:** Captain, 78th VMR, Co Commander, Grayson Co, attached to 4th VMR (McDowell's, Koontz, Chilton) **CEM:** Hale Family; Grayson; on W side Elk Creek; N of Window Shade Ln **GS:** Yes **SP:** Elizabeth Bourne (20 Mar 1785-22 Aug 1866), daug of William Bourne (1743-1836) and Rosemond Jones (1750-1821) **VI:** Son of Lewis Hale (1746-1802) and Mary Burwell (1744-1809) **P:**No **BLW:** No **PH:** No **SS:** B pg 86 **BS:** 245

HALL, David; b 1777; d 1849 **RU: Corporal, 12th VMR of Cav Boughton, NY Militia **CEM:** Hall Family; Pulaski; Mack Creek Village; Little Dam Rd; Elliston **GS:** Yes **SP:** No spouse info **VI:** Son of Asa Hall of Dutchess Co, NY, a Rev War soldier **P:** No **BLW:** No **PH:** No **SS:** A rec 243 **BS:** 245

Key: *Additional veteran entry   **Corrected veteran entry   ***Deleted veteran entry
RU=Rank/Unit   CEM=Cemetery   GS=Gravestone   SP=Spousal Information   VI=Other Veteran Info   P=Pension
BLW=Bounty/Land Warrant   PH=Photo   SS=Service Source   BS=Burial Source   VMR= VA Military Regiment
LNR= Last Known Residence

*HALL, John Byrd; b 1787, d 1 Sep 1862 **RU:** Private, 109th VMR, Capt Carter Berkley's Co, Middlesex County **CEM:** Fredericksburg City; Fredericksburg; jct William St and Washington Ave **GS:** Yes **SP:** Mar (23 Oct 1817) Harriet Stringfellow (1801-1888), recd pen1878 **VI:** A doctor, recd BLW **P:** Widow **BLW:** Yes **PH:** No **SS:** B pg 132; 171 pg 133; 596; M pg 186 **BS:** 245

HALL, Robert; b 1784; d 1824 **RU: Private, 93rd VMR, Capt Samuel Steele's Company, Augusta Co, attached to Cocke's Detachment **CEM:** Old Providence Church; Augusta; 1005 Spotswood Rd; Spotswood **GS:** Yes **SP:** No spouse info **VI:** Son of Patrick Hall (1751-1814) & Susanna (McChesney) Hall **P:** No **BLW:** No **PH:** Yes **SS:** A rec 1174 **BS:** 2 pg 57; 31; 245

*HAMILTON, James; b 1769, d 14 Jul 1837 **RU:** Private, Capt John McCullough's Co, PA Militia **CEM:** Tinkling Spring Presbyterian Church; Augusta; Fisherville **GS:** Yes **SP:** Isabella Helena Hamilton (22 Feb 1772-8 Nov 1855), daug of William Hamilton (1748-1795) and Patience Craig (1752-1822) **VI:** No further data **P:** Spouse (#WC-21671) **BLW:** No **PH:** No **SS:** B pg 40; BD pg 910 **BS:** 245

*HANCOCK, Benjamin Peter; b 16 Jun 1782, Fluvanna Co, d 25 Mar 1860 **RU:** Private, 12th VMR, Capt Horace Timberlakes Co, Fluvanna County, attached to 7th VMR (Gray) **CEM:** Hancock Family; Franklin; GPS 36.97299, -79.68562; East of Rt 946 on Noveltry Rd, Union Hall **GS:** No **SP:** Elizabeth Booth (29 Jul 1801-31 Mar 1860) **VI:** Findagrave shows home and portrait **P:** No **BLW:** No **PH:** No **SS:** A pg 1887; B pg 75 **BS:** 245

*HANCOCK, John; b 6 May 1793, d 18 Jun 1831 **RU:** Private, Capt John Pollock's Company, Charlotte Co, attached to 7th VMR **CEM:** Martin Hancock; Charlotte; Red House **GS:** No **SP:** No spouse info **VI:** Son of Martin Hancock (1767-1838) and Sarah Harvey (1774-1845) **P:** No **BLW:** No **PH:** No **SS:** B pg 58; K pg 349 **BS:** 245

*HANGER, Peter; b 15 Apr 1792, Fort Defiance, d 3 Jul 1862, Waynesboro **RU:** Private, 32d or 93rd VMR, Capt Samuel Doak's Co, Augusta Co, attached to 5th VMR (McDowell) **CEM:** Zion Lutheran Church; Augusta; 297 Zion Church Rd **GS:** Yes **SP:** Mar (4 Apr 1822) Mary Allen (25 Dec 1797-27 Jan 1882), daug of Matthias Allen (1737-1815) and Anna Maria Christian. She recd pen **VI:** Son of Peter Hanger II (29 Jan 1761-23 Dec 1828) and Catherine Link (1767-1837). Recd BLW of 120 acres; was a Doctor **P:** Widow **BLW:** Yes **PH:** No **SS:** B pg 39; M pg 189; P BLW file; Fold3 **BS:** 245

*HARDING, Edward; b 9 Oct 1787, d 9 Nov 1873 **RU:** Private, USA Regular Army, enlisted 1815, obtained rank of 2d Lt while serving 3 yr tour in 1st Artillery Bn which included service at Fort McHenry, MD where retired in 1818 **CEM:** Salem Church; Loudoun; loc next to 14127 Harpers Ferry Rd; Hillsboro **GS:** Yes **SP:** Elizabeth (__) (1792-1882) **VI:** No further data **P:** Unk **BLW:** Unk **PH:** No **SS:** CI Army Register **BS:** 245

*HARDING, William; b 8 Oct 1795, d 19 Mar 1878 **RU:** Private, 45th VMR, Capt Elijah Harding's Co, Stafford County, attached to 1st VMR (Crutchfield) **CEM:** Wicomico United Methodist Church; Northumberland; Wicomico Church, Rt 200 **GS:** Yes **SP:** Sally Hudnall Cockrell (1796-12 Nov 1861) **VI:** Obtained rank of Capt after war **P:** No **BLW:** No **PH:** No **SS:** A rec 1832; B pg 190; **BS:** 245

*HARE, Jesse; b 11 Mar 1789, Uniontown, Dauphin, PA, d 14 Jan 1861 **RU:** Private, Captain James McKonkey's Co, MD Militia **CEM:** Spring Grove Hill; Lynchburg City; 300 Fort Ave **GS:** Unk **SP:** Catherine Welch, 11 Oct 1810, Baltimore, MD **VI:** No further data **P:** Widow **BLW:** Yes **PH:** No **SS:** BD pg 925; M pg 190 **BS:** 31

*HARPER, Joel Zane; b 16 May 1794, d 18 Oct 1864 **RU:** Private, DC Militia, Capt Craven T Peyton and Capt Charles McKnight's Cos **CEM:** Ivy Hill; Fauquier; Rt 50, across fr Buchanan Hall; Upperville **GS:** Yes **SP:** Mar (1820) Spotsylvania Co, Frances McCoull (1 May 1798-5 Dec 1878). She recd two pen **VI:** Recd two BLW 40 acres 1850 120 acres 1855 **P:** Widow **BLW:** Yes **PH:** No **SS:** M pg 192; O- pen file, Fold3 **BS:** 245

*HARRIS, Isaac; b 1790, d 7 Oct 1840 **RU:** Private, 56th VMR, Captain Thomas Gregg's Company, Loudoun Co., attached to 4th VMR (Beatty) **CEM:** Old Stone Church; Loudoun; Leesburg **GS:** Yes **SP:** No spouse info **VI:** No further data **P:** No **BLW:** No **PH:** No **SS:** A rec 1914; B pg 120 **BS:** 245

Key: *Additional veteran entry   **Corrected veteran entry   ***Deleted veteran entry
RU=Rank/Unit   CEM=Cemetery   GS=Gravestone   SP=Spousal Information   VI=Other Veteran Info   P=Pension
BLW=Bounty/Land Warrant   PH=Photo   SS=Service Source   BS=Burial Source   VMR= VA Military Regiment
LNR= Last Known Residence

*HARRIS, Edward; b 1782, d 1825 **RU**: Private, 38th VMR Capt Abraham Buford's Co, Goochland County, attached to 8th VMR (Wall) **CEM**: Tolersville Tavern; Louisa; 410 Old Tolersville Rd, Mineral **GS**: Yes **SP**: No spousal info **VI**: Permit for the tavern was issued in his name for several years **P**: No **BLW**: No **PH**: No **SS**: A rec 1746; B pg 239; K pg 82 **BS**: 245

*HARRIS (HARRISS), Moses; b unk, d 5 Nov 1873 **RU**: Private, 18th VMR Capt Abraham (or Aaron) Staple's Co, Patrick Co, attached to 6th VMR (Sharp and Coleman) **CEM**: Old Harris; Patrick; on ridge N Rt 691 nr Billy Martin Cem; **GS**: No **SP**: Mar (2 Oct 1866) Lucinda Clardy, prob 2nd wife. LNR 1878 Patrick C.H. Recd two pen **VI**: Recd two pen and two BLW. LNR Patrick Springs, 1871 **P**: Both **BLW**: Yes **PH**: No **SS**: B pg 157; M pg 194; O- pen file, Fold3 **BS**: 245

*HARRIS, William; b 25 Dec 1780, Buckingham, Co, d 29 Oct 1865, Bedford Co **RU**: Private, 100th VMR, wagon duty Capt William Moseley's Troop of Cavalry, Buckingham Co, attached to 1st VMR (Holcombe) **CEM**: Longwood; Bedford; Off Oakwood St; First R fr jct Rt 221; Bedford **GS**: Yes **SP**: Spouse not identified **VI**: No further data **P**: No **BLW**: No **PH**: No **SS**: L pg 602 **BS**: 245

HARRISON, Samuel Jordan; b 26 Mar 1769, Skinino, York Co; d 1846 **RU: Corporal, 4th VMR **CEM**: Old City Cemetery; Lynchburg; 401 Taylor St **GS**: Yes **SP**: Sarah H. Burton, daug of Capt Jesse Burton (__-1839, age 58) **VI**: Mayor of Lynchburg, 1808, 1814, 1817; was in mercantile and tobacco business, built the Franklin Hotel in 1818 **P**: No **BLW**: No **PH**: No **SS**: A rec 663 **BS**: 87 pg 50; 119

HARRISON, William; b 1781, MD, d 17 Jul 1859 **RU: Private, 20th VMR, Capt Richard L Lawson' Co, Princess Anne County **CEM**: Cedar Grove; Portsmouth; Effington St & Fort Ln **GS**: Yes **SP**: Johannah (__-23 Jul 1855, age 72) **VI**: Member of VA Soldiers of 1812 Society **P**: No **BLW**: No **PH**: No **SS**: A rec 781; B pg 165; L pg 539 **BS**: 65 pg 107

*HARRISON, William; b 1780, d 17 Jul 1859 **RU**: Commissary, 62nd VMR, Commissary Officer, Prince George Co **CEM**: Berkeley Plantation; Charles City; 12602 Harrison Landing Rd **GS**: Y **SP**: Mar Martha Ann Cocke (1798-1823) **VI**: Son of Carter Basset Harrison (1756-1808) and Mary Allen Howell, (1761-__) **P**: No **BLW**: No **PH**: No **SS**: A rec 796 **BS**: 245

*HARTER, Adam; b 12 Mar 1795, Franklin Co, d 22 May 1862 **RU**: Private, 75th VMR, Capt William Curring's Co, Montgomery County **CEM**: Weddie; Floyd; Shelder's Mill **GS**: Yes **SP**: Margaret Stigleman (30 Jun 1795 Augusta Co-4 Jan 1883 Floyd Co) **VI**: Son of Francis Harter, (1759-1832) and Catharine Kurtz, (1767-1836) **P**: Widow **BLW**: No **PH**: No **SS**: B pg 128; BD pg 940; M pg 194 **BS**: 245

*HARVEY, John; b 1780, d 19 Oct 1863 **RU**: Sergeant, 26th VMR, Captain Grief Barksdale's Company, Charlotte Co. attached to 4th VMR (Greenhill) **CEM**: Old City; Lynchburg City; 401 Taylor St **GS**: Yes **SP**: Anne (__) (1781-7 Dec 1852) **VI**: Farmer **P**: No **BLW**: No **PH**: No **SS**: A rec 2122. B pg 57 **BS**: 245

*HARWOOD, William Jr.; b unk, d Aug 1864 **RU**: Private 33rd VMR, Captain Francis Wicker's Co, Henrico County **CEM**: Hollywood; Richmond; 412 S Cherry St **GS**: No **SP**: No spouse info **VI**: No further data **P**: No **BLW**: No **PH**: No **SS**: K pg 829 **BS**: 245

HAWKINS, Samuel; b 1788; d (bur) 13 Feb 1839 **RU: Private, 19th VMR, Capt William Bryan's and Capt William Cole's Co, Richmond City **CEM**: Shockoe Hill; Richmond City; 100 Hospital St **GS**: Unk **SP**: Mar Dorothy (__) (2 Jun 1828) **VI**: No further data **P**: No **BLW**: No **PH**: No **SS**: A rec 498; L pgs, 181; 573 **BS**: 38 pg 17

*HAWKINS, Samuel; b 1776, d 1836 **RU**: Captain, Commanded a company in 13th VMR, Shenandoah Co, attached to 5th VMR (Mason-Preston) and 4th VMR (Boyd) **CEM**: Riverview; Shenandoah; across Rd 9746 fr Strasburg HS **GS**: Yes **SP**: No spousal info **VI**: Son of Joseph Hawkins. Obtained rank of Colonel after war period **P**: No **BLW**: No **PH**: No **SS**: B pg 184 **BS**: 245

*HAWLING, Isaac Wilcoxson; b 1786, d 24 Sep 1854 **RU**: Private 57th VMR Co under command of Lt Col Armistead Mason **CEM**: Union; Loudoun; 523 N King St, Leesburg **GS**: Yes **SP**: Mar (Loudoun Co, 16 Oct 1817) Frances Best (1793-12 Jun 1880) who recd a widow's pen **VI**: Recd BLW 1855 **P**: Widow **BLW**: Yes **PH**: No **SS**: B pg 119; M pg 196; O- pen file, Fold3 **BS**: 245

Key: *Additional veteran entry   **Corrected veteran entry   ***Deleted veteran entry
RU=Rank/Unit   CEM=Cemetery   GS=Gravestone   SP=Spousal Information   VI=Other Veteran Info   P=Pension
BLW=Bounty/Land Warrant PH=Photo   SS=Service Source   BS=Burial Source   VMR= VA Military Regiment
LNR= Last Known Residence

49

*HAYES, James; b 10 Jun 1785, d 31 May 1871 **RU:** Corporal, 68th VMR, Served in a company from James City or York Co **CEM:** Easters United Methodist Church; Mecklenburg; Boydton **GS:** Yes **SP:** Mar (25 Mar 1806) Martha Green **VI:** No further data **P:** No **BLW:** No **PH:** No **SS:** A rec 1316 **BS:** 245

*HEATH, James Ewell; b 8 Jul 1792, d 28 Jun 1862 **RU:** Lieutenant, 36th VMR, Capt John Linton's Troop of Calvary, Prince William Co **CEM:** Shockoe Hill; Richmond City; 100 Hospital St **GS:** Unk **SP:** Mar (Fairfield, Prince William Co, 21 Sep 1820) Elizabeth Ann Macon (1801-1860) **VI:** Son of John Heath and Sarah Ewell; editor of American Literary Messenger; author of book, *Edge Hill,* a novel about the American Revolution; was auditor of the Accounts of the State of Virginia **P:** No **BLW:** No **PH:** No **SS:** B pg 172; L pg 548 **BS:** 245

HEATH, Jesse; b 1765, d 1850 **RU: Private, 39th VMR, Capt Edward O Goodwin's Co, Petersburg **CEM:** Blandford; Petersburg; 111 Rochelle Ln **GS:** Yes **SP:** Mar Agnes Peebles, "his cousin" **VI:** Son of William Heath (1751-177x.) & Margaret Bonner. Memorialized in Blandford Cemetery with a plaque on Heartwell Peebles Heath's gravestone, as Jesse was buried in private cemetery near Petersburg **P:** No **BLW:** No **PH:** No **SS:** L pg 371 **BS:** 200

*HENDERSON, William, b 1773, d 1866 **RU:** 1st Sergeant, 69th or 84th VMR, Captains. George Wilson or John Wimbush Companies of Halifax Co, attached to 3rd VMR (Dickinson) **CEM:** Halifax Town; Halifax; GPS: 36.7639,-78.9269 **GS:** Unk **SP:** Sarah M Farmer **VI:** Son of Charles Henderson **P:** Yes **BLW:** No **PH:** No **SS:** A rec 564; B pg 91; M pg 198 **BS:** 245

*HENLEY, William D; b 1783, d 23 May 1838 **RU:** Private, Capt Edward James Company, Princess Anne Co, reassigned to Captain Cross's Co, 3rd VMR (Boykin's) **CEM:** Cedar Grove; Norfolk; 238 E Princess Anne Rd **GS:** Yes **SP:** Margaret E (__) (1797-Jun 1836) 2) Frances (__) **VI:** No further data **P:** Widow (W635385) **BLW:** No **PH:** No **SS:** B pg 165; BD pg 971; K pg 454 **BS:** 245

*HENRY, John H; b bef 1800, NJ, d 7 Jan 1850 **RU:** Private Captain Steven's Co, NJ Militia **CEM:** Cedar Grove; Norfolk; jct Salter St and E Princess Anne Rd **GS:** Yes **SP:** Mar Brunswick, NJ, Elizabeth Ryno, (__-Norfolk, 25 Apr 1881). Recd widows pen and BLW of 80 acres, 1855 **VI.** Grave loc Block 3rd A W, lot 21, space14. Recd BLW 80 acres, 1850. Titled "Capt" on GS **P:** Widow **BLW:** Both **PH:** No **SS:** CG pg 1606; M pg 198; O- pen file, Fold3 **BS:** 245

*HENRY, John III, b16 Feb 1796, d 7 Jan 1868 **RU:** Private, 26th VMR Capt John Richardson's Artillery Co, Charlotte County, attached to 4th VMR (Greenhill) **CEM:** Henry (AKA Red Hill); Charlotte; Red Hill Plantation, nr Brookneal **GS:** Yes **SP:** Mar (Lynchburg, 19 Oct 1826) Elvira Bruce McClelland (23 Apr 1808-2 Nov 1875) **VI:** Son of Patrick Henry (1736-1799) and Dorothea Spotswood Dandridge (1757-1831) **P:** No **BLW:** No **PH:** No **SS:** A rec 1528; B pg 57 **BS:** 245

*HERNDON, Joseph; b 1772, d 1832 **RU:** Paymaster, Staff Officer, 16th VMR Spotsylvania Co **CEM:** Gordon-Herndon; Spotsylvania; Post Oak **GS:** Yes **SP:** Lucy (__) **VI:** Son of Edward Herndon (1730-1799) and Mary Duerson (__-1806) **P:** No **BLW:** No **PH:** No **SS:** K pg 379 **BS:** 245

*HICKMAN, John T; b 1795, d 1884 **RU:** Corporal, 51st VMR, Capt Francis Ireland's Co, Frederick County, attached to 1st VMR (Taylor) **CEM:** Emanuel Lutheran Church; Shenandoah; Woodstock **GS:** No **SP:** No spouse info **VI:** No further data **P:** Yes **BLW:** No **PH:** No **SS:** A rec 2313; BD pg 981; M pg 200 **BS:** 245

*HIGGINBOTHAM, David; b 17 Oct 1775, Albemarle Co, d 20 Jan 1853 **RU:** Private, 47th or 88th VMR, Capt John Rothwell's, Co Albemarle County, attached to 7th VMR (Gray's) **CEM:** Hollywood; Richmond; 412 S Cherry St **GS:** Yes **SP:** Mary Emslie (1787-17 Jun 1872) **VI:** No further data **P:** No **BLW:** No **PH:** No **SS:** L pg 669; B pgs 35; 36 **BS:** 245

*HICKS, Benjamin A; b Jun 1798, Spotsylvania Co, d 22 Apr 1861 **RU:** Private 16th VMR, Capt Gabriel Long's Co, Spotsylvania County **CEM:** Shockoe Hill; Richmond City, 100 Hospital St **GS:** Yes **SP:** Mar (Caroline Co, 10 Nov 1835) Sarah B Sales (__-1891) Recd two widows pen **VI:** Son of Thomas H Hicks and Lucy Alsop. Recd BLW (1850) 40 acres, (1855) 120 acres **P:** Widow **BLW:** Yes **PH:** No **SS:** B pg 189; M pg 200; O- pen file, Fold3 **BS:** 245

Key: *Additional veteran entry   **Corrected veteran entry   ***Deleted veteran entry

RU=Rank/Unit	CEM=Cemetery GS=Gravestone SP=Spousal Information VI=Other Veteran Info P=Pension
BLW=Bounty/Land Warrant PH=Photo	SS=Service Source BS=Burial Source VMR= VA Military Regiment
LNR= Last Known Residence	

*HIGHT, Joel; b 27 Sep 1793, d 1866 **RU:** Corporal, 28th VMR, Capt William C Scott's Co, Nelson County, attached to 8th VMR (Cocke) **CEM:** Haines Chapel; Rockbridge; Vesuvius **GS:** Yes **SP:** Nancy Campbell (1 Sep 1794-__) **VI:** Son of Revolutionary War soldier, George Hight (1755-1837) and Lovia Agnes Lunsford (1760-1840) **P:** No **BLW:** No **PH:** No **SS:** K pg 104 **BS:** 245

*HILL, James; b 12 May 1774, d 3 Mar 1860 **RU:** Private, 56th or 57th VMR, Loudoun Co, attached to 5th VMR **CEM:** Ketoctin Baptist Church; Loudoun; Purcellville **GS:** Yes **SP:** No spouse info **VI:** No further data **P:** No **BLW:** No **PH:** No **SS:** A rec 1458 **BS:** 245

*HILL, John; b 1790 d 8 Jun 1860 **RU:** Private, 19th VMR, Richmond City **CEM:** Hollywood; Richmond; 412 S Cherry St **GS:** No **SP:** No spouse info **VI:** No further data **P:** No **BLW:** No **PH:** No **SS:** A rec 1614 **BS:** 245

*HILL, Robert F; b 1776, d 11 Jun 1851 **RU:** Sergeant, 87th VMR, Captain Blackwell Foster's Company, King William Co, attached to 2nd VMR (Sharp) **CEM:** Cedar Grove; Williamsburg; jct Rt 132, at Richmond Hill Ct **GS:** Yes **SP:** No spouse info **VI:** No further data **P:** No **BLW:** No **PH:** No **SS:** P pg 115; K pg 447 **BS:** 245

*HILL, Robert Garlick; b 8 Dec 1791, d 1 Oct 1816 **RU:** Paymaster, 30th VMR, officer in company from Caroline Co attached to 9th VMR (Boyd) **CEM:** Hill Family; Caroline; Mt Airy, Ruther Glen **GS:** Yes **SP:** No spouse info **VI:** Son of Humphrey Hill (1766-1841) and Mary Garlick (1771-1819) **P:** No **BLW:** No **PH:** No **SS:** A rec 1894; B pgs 55; 56 **BS:** 245

*HILL, William; b 17 Jun 1780, d 12 Apr 1830 **RU:** Captain, 87th VMR, commander, Artillery Company, King William Co, attached to Battalion of Artillery and 6th VMR (Reade) **CEM:** Springfield; King William; 2 mi S King William CH; R side Rt 621 **GS:** No **SP:** No spouse info **VI:** Son of James Hill (__-1802) and Mildred Clopton **P:** No **BLW:** No **PH:** No **SS:** B pg 115 **BS:** 245

*HILLMAN, Simeon; b 25 May 1789, d 16 Aug 1860 **RU:** Seaman, served 28 Jun 1813 to 28 Jun 1815 **CEM:** Mount Hebron; Winchester City; 305 E Boscawen St **GS:** Unk but plot old 60, grave 4 **SP:** Mar (Winchester, 11 Jun 1829) Charlotte Copenhaver (__-1 Dec 1892) who recd two pen. **VI:** Recd BLW 160 acres 1855 **P:** Widow **BLW:** Yes **PH:** No **SS:** M pg 202; O- pen files Fold3 **BS:** 245

*HINTON, John; b unk, d 28 Jan 1840 **RU:** Private, 39th VMR, Capt Cadwallader J Claiborne's Co, Petersburg, attached to 1st VMR (Byrne's) **CEM:** Blandford; Petersburg; 319 S Crater Rd **GS:** Yes **SP:** No spouse info **VI:** No further data **P:** No **BLW:** No **PH:** No **SS:** A rec 817; B pg 159 **BS:** 245

*HISER, Samuel; 29 Jan 1794, d 19 Jun 1878, Fort Defiance **RU:** Private 13th or 97th VMR, Capt Samuel Colville's (Colwell's) Co, Shenandoah County, attached to 6th VMR (Coleman) **CEM:** Salem Lutheran Church; Augusta: vic jct Rts 852 and Salem Ch Rd, Mount Sidney **GS:** Yes **SP:** Mar (19 Jul 1823) Catherine Showalter (21 Mar 1806-30 Sep 1881), recd two pen. LNR Fort Defiance, Augusta Co **VI:** Recd pen and two BLW; one 50 acres (1850); one (1855) 120 acres **P:** Both **BLW:** Yes **PH:** No **SS:** B pg 184; M pg 203; O- pen files Fold3 **BS:** 245

*HIX, Richard, J; b bef 1799, d unk **RU:** Private, 33rd VMR, Captain William Byrd Chamberlayne's Co, Henrico County, attached to 1st VMR (Taylor) **CEM:** Smith Family; Henrico; Tuckahoe **GS:** Yes **SP:** No spouse info **VI:** No further data **P:** No **BLW:** No **PH:** No **SS:** A rec 1481; B pg 99 **BS:** 245

HIX, Samuel; b 1795, James City Co, d 27 Apr 1870 **RU:** Private 52d VMR Capt William Taylor's Company, New Kent Co **CEM:** Shockoe Hill; Richmond City; 100 Hospital St **GS:** No **SP:** Mar (2 Sep 1819) Catherine Mullen (__-bef 18 Aug 1880). Recd pen, LNR Henrico Co **VI:** Son of Richard Hix and Drulla (__). He was one of 62 persons killed (27 Apr 1870) when upper floor VA State Capital collapsed. Was stonemason. Recd BLW **P:** Widow **BLW:** Yes **PH:** No **SS:** B pg 144; L pg 762; M pg 203; **BS:** 245

*HIX, William; b 1789, d Jul 1781 **RU:** Private, company of Bedford or Campbell Co, attached to 5th VMR **CEM:** Spring Hill, Lynchburg, 3000 Fort Ave **GS:** Yes **SP:** Ann R (__), (1800-Feb 1878) **VI:** No further data **P:** No **BLW:** No **PH:** No **SS:** A rec 1507; B pgs 43, 44, 53 **BS:** 245

Key: *Additional veteran entry   **Corrected veteran entry   ***Deleted veteran entry
RU=Rank/Unit   CEM=Cemetery   GS=Gravestone   SP=Spousal Information   VI=Other Veteran Info   P=Pension
BLW=Bounty/Land Warrant   PH=Photo   SS=Service Source   BS=Burial Source   VMR= VA Military Regiment
LNR= Last Known Residence

*HODGES, John; b 31 Dec 1786, d 31 Jul 1855, Portsmouth **RU:** Captain, commanded a company in 7th VMR, Norfolk County, attached to 3rd VMR (Veale) and in battle at Craney Island. Mustered into Reg Army, 26 Apr 1813 **CEM:** Walker; Chesapeake City; 2417 Water Mill Way **GS:** Yes **SP:** Jane Adelaide Gregory (11 Jan 1794-1875) **VI:** "Gen" rank listed on gravestone refers to his promotion to Brig Gen by President Tyler in Jan1826. Member General Assembly and county court, postmaster in Portsmouth. Operated ferry from his home **P:** No **BLW:** No **PH:** No **SS:** B pg 148; O Serv Rec Cards, Fold3 **BS:** 245

*HOFF, John; b c1788 (another source 1778), d 28 Jan 1859 **RU:** Private, 89th VMR, Prince William Co **CEM:** Mt Hebron; Winchester; 305 E Boscawen St **GS:** Yes **SP:** Maria (__) (9 Jun 1791-28 Jan 1863) **VI:** No further data **P:** No **BLW:** No **PH:** No **SS:** A rec 57 **BS:** 86 pg 53; 245

HOLBROOK, Josiah; b 1788, Derby, CT, d 20 Jun 1854 **RU: Private, MA Militia (Chamberlin), Capt J Howland's Co **CEM:** Old City; Lynchburg; 401 Taylor St **GS:** Yes **SP:** No spouse info **VI:** graduated Yale University; founder of Lyceum Movement; manufactured scientific apparatus for schools. Stone indicates he was born in MA and "of unbounded philanthropy, an ardent lover of science and a general benefactor and a most exemplary Christian. He came to his death suddenly in his geological and mineralogical pursuits in the vicinity of Lynchburg" **P:** No **BLW:** No **PH:** No **SS:** AO pg 30 **BS:** 207; 87 pg 53

*HOLLADAY, Joseph; b 1787, d 1855 **RU:** Captain, 59th VMR, company commander, Nansemond Co. attached to 3rd VMR (Boykin) **CEM:** Cedar Hill; Suffolk City; 105 Mahan St **GS:** Yes **SP:** No spouse info **VI:** Farmer and businessman **P:** No **BLW:** No **PH:** No **SS:** B pg 140 **BS:** 245

*HOLLIDAY, James; b, unk, d 1830 **RU:** Private. Frederick Co, company (probably Capt George Holliday) attached to 4th VMR (Beatty) **CEM:** Mt Hebron; Winchester; 305 E Boscawen St **GS:** Yes **SP:** No spouse info **VI:** Was one of the managers to build a new Presbyterian meeting house in 1778. A James Holliday, Jr. served in the same company **P:** No **BLW:** No **PH:** Yes **SS:** A rec 2313; B pgs 78; 79 **BS:** 245

*HOLMES, David; b 10 Mar 1769, York Co, PA, d 20 Aug 1832 **RU:** Governor; Mississippi Territory during war period, thus Commander of Militia **CEM:** Mt Hebron, Winchester; 305 E Boscawen St **GS:** Yes **SP:** No spouse info **VI:** Representative U.S. Congress from VA,1797-1809, Governor Mississippi Territory 1809-1817,Governor of state 1817; 1825, U.S. Senator from Mississippi **P:** No **BLW:** No **PH:** No **SS:** R pg 1325 **BS:** 245

HOLMES, Jeremiah; b 1788; d 1865 **RU: Corporal, 45th VMR, Capt Elijah Harding's Co, Stafford County **CEM:** Rockhill Baptist; Stafford; jct Rts 644 & 671 **GS:** Unk **SP:** Lydia McIntire Patton **VI:** No further data **P:** No **BLW:** No **PH:** No **SS:** A rec 237 **BS:** 26 pg 338

HOLMES, Joshua Rathborne; c25 Dec 1781, CT, d 23 May 1845 **RU: Private, 52nd VMR, Capt Robert Perkin's, Co, New Kent County **CEM:** Outside Old City Cemetery; Lynchburg; 401 Taylor St **GS:** Yes **SP:** Mar (Lynchburg, 1824) Martha Ann Bernard **VI:** Son of John Holmes (1756-1823) and Ann ; engaged in wool carding; member of city council 1829-1832, 1st Superintendent of Lynchburg Waterworks, a Mason **P:** No **BLW:** No **PH:** No **SS:** L pg 630 **BS:** 49; 87

*HOLMES, William; b 1793, d 8 Oct 1887 **RU:** Private, 57th VMR Loudoun Co **CEM:** Goose Creek Burying Ground; Loudoun; Rt 722, Lincoln **GS:** Yes **SP:** Eliza T (__) (__-1873) **VI:** No further data **P:** No **BLW:** No **PH:** No **SS:** A rec 483 **BS:** 245

*HOLT, William; b 9 Mar 1788, Charlotte Co, d 13 Dec 1857 **RU:** Private, 42nd or 105th VMR, Captain Thomas H Clark's Company, Pittsylvania Co., attached to 7th VMR (Saunders) **CEM:** Saint Johns Episcopal Church; Halifax; 197 Mountain Rd **GS:** Unk **SP:** Phoebe Fergurson **VI:** Son of Dudley Holt and Sarah Ann Jones, Clerk of Court for Halifax Co **P:** No **BLW:** Yes # 41398-40 acres-1850 **PH:** No **SS:** A rec 923; B pg 161; BD pg 1015 **BS:** 245

*HOOMES, John; b c1768, d 14 Mar1824 **RU:** 3rd Lieutenant, U.S. Army Dragoons, promoted 20 May 1813 **CEM:** Hoomes Family; Caroline; Bowling Green **GS:** Yes **SP:** No spouse info **VI:** Son of Rev War Ensign, John Hoomes (1749-1805) and Judith Churchill Allen (1748-1822); rank of major achieved after war period **P:** No **BLW:** No **PH:** No **SS:** AF **BS:** 245

Key: *Additional veteran entry   **Corrected veteran entry   ***Deleted veteran entry
RU=Rank/Unit   CEM=Cemetery   GS=Gravestone   SP=Spousal Information   VI=Other Veteran Info   P=Pension
BLW=Bounty/Land Warrant   PH=Photo   SS=Service Source   BS=Burial Source   VMR= VA Military Regiment
LNR= Last Known Residence

*HOOVER, Jacob; b 1789, d 1863 **RU**: Private, 13th VMR, Capt Samuel Hawkin's Co, Shenandoah Co, attached to 4th VMR (Boyd) & served under Capt Joseph Mauzy Company, Rockingham Co **CEM**: Massanutten; Shenandoah; Woodstock **GS**: No **SP**: No spouse info **VI**: No further data **P**: Yes **BLW**: Yes **PH**: No **SS**: A rec 2135; B pg 184; BD pg 1019; M pg 206 **BS**: 245

*HOOVER, John; b 1777, d 11 Sep 1838 **RU**: Private 13th VMR, Capt George Shrum's Company, Shenandoah Co, attached to 1st VMR (Truehart) and secondly 2nd VMR, Maj Nathaniel Perkin's Cmd **CEM**: Mt Hebron; Winchester City; 305 E Boscawen St **GS**: Yes **SP**: Barbara (__) **VI**: No further data **P**: Widow applied **BLW**: Yes **PH**: Yes **SS**: A rec 2173; B pgs 183, 242; M pg 206 **BS**: 31

HOPKINS, Charles; b 1789; d 8 Aug 1839 **RU: Private, 115th VMR, York Co **CEM**: Hopkins / Cook / Smith; York; Tabb **GS**: Unk **SP**: Mary Elizabeth Gibbs (1793-1871) **VI**: No further data **P**: No **BLW**: No **PH**: No **SS**: A rec 60 **BS**: 49

*HOPKINS, James; b 9 Mar 1794, d 30 Apr 1852 **RU**: Sergeant, 58th or 116th VMR Co of Rockingham County attached to Flying Camp (McDowell's) or 6th VMR (Coleman) **CEM**: Cook Creek Presbyterian Church; Rockingham; Harrisonburg **GS**: Yes **SP**: No spouse info **VI**: No further data **P**: No **BLW**: No **PH**: No **SS**: A rec 166 or 179; B pgs 181-182 **BS**: 245

*HOPKINS, Price Wiley; b 1777, d 1845 **RU**: Private, 10th VMR, Captain Walter Otey's Troop of Cavalry, Bedford Co, attached to Major Woodford's Cavalry Squadron **CEM**: Bedford County; Bedford City; 1131 Park St **GS**: Unk **SP**: Frances G Claytor **VI**: Son of Francis Hopkins (New Kent Co, 27 Feb 1737-1804) and Jane Cox (1737-1815) **P**: No **BLW**: No **PH**: No **SS**: A rec 320; B pg 43 **BS**: 245

HORSLEY, John, Jr; b 1785, d 19 Dec 1827 **RU: 1st Sergeant, 17th VMR, Capt John Miller's Co, Cumberland County, attached to McDowell's Flying Camp **CEM**: Horsley/Yancey; Buckingham; "Travelers Rest," Rt 604 **GS**: Yes **SP**: Mar (25 Sep 1817) Mary Chambers Yancey (22 Jan 1792-May 1873), daug of Major Charles Yancey of Louisa Co & Anne Spencer (30 May 1769-29 May 1795) **VI**: First attorney in Nelson Co 1808 **P**: No **BLW**: No **PH**: No **SS**: L pg 30 **BS**: 209; 245

*HOSE, Philip; b 1780, PA, d Unk **RU**: Private, Frederick County unit attached to 1st VMR (Taylor) **CEM**: Old Chapel; Clarke; Millwood **GS**: No **SP**: Mar (Hagerstown, MD, Oct 1780) Margaret Fluriet **VI**: No further data **P**: No **BLW**: No **PH**: No **SS**: A rec 1539; B pg 78; **BS**: 86 pg 30

HOWARD, Beal; b 1760; d 11 Nov 1820 **RU: Private, DC Militia, Capt Charles McKnight Co **CEM**: Christ Church Episcopal; Alexandria; Wilkes & Hamilton **GS**: Yes **SP**: Elizabeth Fillmore **VI**: A butcher in Fairfax **P**: Yes **BLW**: No **PH**: No **SS**: A rec 278; BD pg 1030 **BS**: 34 pg 100

*HOWARD, George; b 1792, d 27 Oct 1860 **RU**: Sergeant, 19th VMR, Richmond City, attached to 2nd VMR (Ballowe) **CEM**: Shockoe Hill; Richmond; 100 Hospital St **GS**: Yes **SP**: Mar (2 Apr 1838) Sarah Roberts (__-28 Jan 1897) **VI**: No further data **P**: Yes & widow **BLW**: Yes **PH**: Yes **SS**: A rec 369; M pg 208 **BS**: 31

*HOWARD, John: b 1774. Bowling Green, Caroline Co, d 11 May 1868 **RU**: Corporal, 19th VMR, Capt Andrew Stevenson's Artillery Co, Richmond City, attached to 2d VMR (Ballowe) **CEM**: Shockoe Hill; Richmond City; 100 Hospital St **GS** Yes **SP**: Mar (Henrico Co, 9 Mar 1810) Nancy Sizer Dickinson (22 Oct 1790-22 Jun 1884), daug of Jonathan Dickerson and Crocia Sizer. Recd two pen **VI**: Recd three BLW, two of 40 acres ea, one of 80 acres **P**: Widow **BLW**: Yes **PH**: No **SS**: B pg 175; M pg 208; O- pen files, fold3 **BS**: 245

*HOWELL, Daniel; b 11 Aug 1798, d 30 Mar 1886 **RU**: Private, 11th & 119th VMR, Capt George J Davidsons Artillery Co, Harrison County (now WVA) attached to Battalion of Artillery **CEM**: Old Topeco; Floyd; Rt 221 **GS**: Yes **SP**: No spouse info **VI**: No further data **P**: No **BLW**: No **PH**: No **SS**: A rec 1226; B pg 97 **BS**: 245

Key: *Additional veteran entry   **Corrected veteran entry   ***Deleted veteran entry
RU=Rank/Unit   CEM=Cemetery   GS=Gravestone   SP=Spousal Information   VI=Other Veteran Info   P=Pension
BLW=Bounty/Land Warrant   PH=Photo   SS=Service Source   BS=Burial Source   VMR= VA Military Regiment
LNR= Last Known Residence

HUFFMAN, Andrew; b 9 Feb 1787; d 9 Sep 1883, **RU: Private, 48th VMR, Capt Edmond Sherman's Co, Botetourt County, attached to 6th VMR **CEM:** Huffman Family; Craig; Broad Run **GS:** Yes **SP:** Mar (Craig Co, 1 Nov 1808) Mary Polly Paxton (20 Nov 1787-1875), daug of Thomas Paxton and Mary Margaret Barclay **VI:** Tombstone cites service in War of 1812 **P:** Yes **BLW:** No **PH:** No **SS:** G; BD pg 1042; B pg 46 **BS:** 106 pg 1; 245

*HUFFMAN, Jacob Trolinger; b 1789, d 28 Dec 1863 **RU:** Private, 121st VMR, Capt Thomas Burwell's Co, Botetourt County, attached to 5th VMR (McDowell) **CEM:** Fairview; Craig; on McClover Hill Lane on L past first stream fr jct with Rt 42 **GS:** Yes **SP:** Mar (12 Oct 1809) Pamela Lake (1787-14 Sep 1876), LNR P O Simondsville, Craig Co. Recd two pen as widow **VI:** Son of John Huffman and Mary Molly Trolinger. Recd two BLW, one 40 acres 1850 and 120 acres 1855 **P:** Widow **BLW:** Yes **PH:** No **SS:** B pg 241; M pg 210; O- pen files, fold3 **BS:** 245

*HUFFMAN, John; b 21 Jan 1791, d 13 Oct 1863 **RU:** Private, 116th VMR, Capt William Harrison's Co, Rockingham County, attached to 1st VMR (Trueheart) & 6th VMR (Coleman's) **CEM:** Pleasant Grove Methodist Church; Rockingham; Summit Church Rd, Mt Crawford **GS:** Yes **SP:** Susanna Swope, (21 Jun 1791-12 Dec 1868) **VI:** No further info **P:** No further data **BLW:** No **PH:** No **SS:** K pgs 51, 52 (Note: service record might be applied to another John Huffman born in Rockingham Co, 4 Dec 1796, d May 1843, buried at Friedens United Methodist Church; Friedens, Rockingham Co) **BS:** 245

*HUGHES, Thomas; b 3 Oct 1782, d 14 Mar 1843 **RU:** Private, Company from Frederick County attached to Battalion of Artillery **CEM:** Goose Creek Burying Ground; Loudoun; Rt 722 Lincoln **GS:** Yes **SP:** Martha Canby (27 Oct 1791-28 Sep 1851) **VI:** No further data **P:** No **BLW:** No **PH:** No **SS:** Fold3 Gen Index Card **BS:** 245

*HULL, John C; b 1886, d 10 Apr 1841 **RU:** Sergeant, 49th VMR, MD Militia, Private, Capt John Sample's Co, MD Militia **CEM:** Saint George Episcopal Burial Ground; Fredericksburg **GS:** Yes **SP:** No spouse info **VI:** No further data **P:** Drew pension (#SO-25563) for service as private in MD Militia **BLW:** No **PH:** No **SS:** A rec 323; BD pg 1047 **BS:** 245

*HUME, Joseph; b 1 Sep 1775, d 5 May 1851 **RU:** Captain, 82nd VMR, company commander, Madison Co, attached to 1st VMR (Crutchfield) **CEM:** Hume Cemetery; Madison; Madison town **GS:** Unk **SP:** Mar (12 Feb 1816) Elizabeth Jones (9 Jul 1785-15 Aug 1862), daug of Ambrose Jones (1745-24 Dec 1812) and Mary Waggoner **VI:** No further data **P:** No **BLW:** No **PH:** No **SS:** B pg 126 **BS:** 245

***HUNGERFORD, Thomas: Deleted as not identified as being buried in that cemetery

*HUNTER, Benjamin; b 1770, d 23 Jun 1845 **RU:** Major, 117th VMR, staff officer, Campbell Co **CEM:** Hunter-Marshall Family; Appomattox; in town of Appomattox **GS:** Yes **SP:** No spouse info **VI:** No further data **P:** No **BLW:** No **PH:** No **SS:** B pg 53 **BS:** 245

HUNTER, George Washington; b 1794; d 25 Mar 1856 **RU: Acting Major, 60th VMR (Minor), Fairfax Co **CEM:** Pohick Episcopal Church; Fairfax; 9301 Richmond Hwy; Lorton **GS:** Yes **SP:** 1) Sarah Ann Tyler, 2) Angeline French Moore (1772-1856) **VI:** Styled Colonel; Son of John Hunter and Jane Broadwater, reinterred from "Summer Hill" in 1940 to Pohick Church in Lorton **P:** No **BLW:** No **PH:** No **SS:** A rec 2118 **BS:** 34 pg 47

*HUPP, George Franklin; b 14 Aug 1792, d 23 Dec 1885 **RU:** 13th VMR Paymaster, Shenandoah Co **CEM:** Mount Zion United Methodist Church (AKA Old Strasburg); Shenandoah; vic jct W Queen and High St, Strasburg **GS:** Unk **SP:** Catherine Spengler **VI:** Recd pen; appl for BLW **P:** Yes **BLW:** Applied **PH:** No **SS:** B pg 184; M pg 212; O- pen files **BS:** 245

*HURST, William; b 1785, d 14 Oct 1857 **RU:** Private, 61st VMR Capt Gabriel Hughes Co and four other company commanders, Mathew Co **CEM:** Friendship; Mathews; jct Rts 14 and 697 **GS:** Yes **SP:** Mar (Mathews Co, 17 Nov 1808) Sarah Hall (__-aft 1781) Recd pen, LNR Mathews Ct House, 1871 **VI:** Recd BLW 160 acres 1850. Titled "Captain" after war period **P:** Widow **BLW:** Yes **PH:** No **SS:** B pg 128; K pg 295; M pg 213; O- pen files, fold3 **BS:** 245

Key: *Additional veteran entry   **Corrected veteran entry   ***Deleted veteran entry
RU=Rank/Unit   CEM=Cemetery   GS=Gravestone   SP=Spousal Information   VI=Other Veteran Info   P=Pension
BLW=Bounty/Land Warrant   PH=Photo   SS=Service Source   BS=Burial Source   VMR= VA Military Regiment
LNR= Last Known Residence

54

*HURT, William; b 22 Mar 1778, d 12 Jul1852 **RU**: Lieutenant; 10th and 91st VMR, Capt John Gray's Co, Bedford Co, attached to 2d VMR (Ambler-Brown) **CEM**: Finch Family; Bedford; Moneta **GS**: Yes **SP**: No spousal info **VI**: No further data **P**: No **BLW**: No **PH**: No **SS**: B pg 42; K pg 72; O- serv index card, fold3 **BS**: 245

*HUTCHINSON, Franklin S; b unk, d 19 May 1838 **RU**: Major, Infantry, US Army **CEM**: Arlington National; Arlington **GS**: Yes **SP**: No spouse info **VI**: No further data **P**: No **BLW**: No **PH**: No **BS**: 245

HUTCHISON, John; b 21 Feb 1768, d 18 Feb 1843 **RU: Sergeant, 60th VMR (Minor), Fairfax Co, and 57th VMR, Loudoun Co **CEM**: Prospect Hill; Prince William; Rt 624, 6 mi fr Haymarket **GS**: Yes **SP**: Nancy (__) (1777-1864) **VI**: No further data **P**: No **BLW**: No **PH**: N **SS**: A rec 1745; 1746; B pg 119-122 **BS**: 248; pg 150

*IRONMONGER, Charles B; b 1799, Accomack Co, d 1887 **RU**: Private, 2d VMR Capt William Outten's Co, Accomack Co **CEM**: Forest Lawn; Norfolk; 8100 Granby St **GS**: Yes **SP**: Mar 1) Mary Ann Brittingham (1813-1846) 2) Sarah McCready (19 Jan 1821-3 Dec 1905) **VI**: Applied for pen but unit comdr had no rec at NARA, thus declined **P**: No **BLW**: No **PH**: No **SS**: B pg 33; M pg 214; O- pen files, Fold3 **BS**: 245

*ISAACS, Samuel, b 1794, d 13 Mar 1878 **RU**: Sergeant In a company in Lt Col Lynn's 1st Regt DC Militia **CEM**: Methodist Protestant; Alexandria; 1500 Wilkes St **GS**: Yes **SP**: No spouse info **VI**: No further data **P**: No **BLW**: No **PH**: No **SS**: A rec 1722; O- Serv index card, Fold3 **BS**: 245

*ISBELL, William; b 26 Jul 1777, Cumberland Co, d 1853 **RU**: Ensign, 24th VMR, Capt Walter L Fontaine's Artillery Co, Buckingham County, or Capt John Gannaway's Co, attached to 8th VMR (Wall's) **CEM**: Rose Hill Plantation; Buckingham; Andersonville **GS**: No **SP**: Ann Murray Walton (23 Jun 1779 Cumberland Co,- Cumberland Co, 26 Nov 1863) **VI**: Son of Lewis Isbell (1751-1796) & Ann Anderson (1760-unk); a Reverend **P**: No **BLW**: No **PH**: No **SS**: A rec 1780; B pg 50 **BS**: 245

*JACKSON, Daniel Franklin; b c1792, d 9 Apr 1872 **RU**: Private, 13th or 97th VMR probably Daniel Strickler's Co, Shenandoah County, attached to 6th VMR (Coleman) **CEM**: Richard Jackson Family; Rappahannock; I mi fr Washington at Rosemont **GS**: Unk **SP**: Mary Corder, daug of John, applied for pension but was rejected for lack of identification of unit served **VI**: Son of Richard Jackson (c 1750-1820) and Phebe Updike (__-25 Mar 1833). Butcher was his trade **P**: No **BLW**: No **PH**: No **SS**: B pg 185; M pg 214; O- fold3 pen files **BS**: 245

*JAMES, Cyrus Basye; b 7 Aug 1786, Gloucester Co, d 30 Dec 1865 **RU**: Private, 61st VMR, Capt Francis Jarvis and Capt Thomas Tabb's companies, Mathews Co **CEM**: Ironmonger Family; York; 611A Shirley Rd, Grafton **GS**: Name on plaque **SP**: Mary White (13 Sep 1786-Jan 1873), daug of Edward White (1757-1809) and Pembroke Singleton (1762-1788); recd two pen **VI**: Son of Thomas James (1759-1828) and Betsy Davis (1763-__); recd BLW 80 acres 1855 **P**: Widow **BLW**: Yes **PH**: No **SS**: B pg 129; M pg 215; O- pen files, fold3 **BS**: 245

JAMES, Edward; b 21 Jun 1787, d 8 Oct 1814 **RU: Sergeant, 20th VMR, Capt Jonathan Woodhouse's Co, Princess Anne County, attached to 2d VMR (Sharp) **CEM**: Red Mill Farm; Virginia Beach; left of 2181 Hedgelawn Way **GS**: Yes **SP**: Jennet Henley Shepard (1785-27 Nov 1835) **VI**: No further data **P**: No **BLW**: No **PH**: No **SS**: A rec 1725; B pg 165; L pg 843 **BS**: 245

JAMESON, William; b 20 Sep 1791; d 6 Oct 1873 **RU: Midshipman, USS *United States* **CEM**: Cedar Grove; Norfolk City; 238 E Princess Anne Rd **GS**: Yes **SP**: Mar (16 Oct 1816) by Rev Low; Catherine Moore Rose (1790-__), eldest daug of Mrs Mary Rose of Norfolk. Marriage notice from the *Richmond Compiler*, 21 Oct 1819, pg 3. Her middle name is from her tombstone **VI**: Enlisted as a midshipman 1811. Aboard the USS *United States* which captured the *Macedonian* in the harbor of Vera Cruz in 1818 "during the absence of Com. M. C. Perry." He was styled "Lt William Jameson of U.S. Navy" in his marriage notice. Commissioned Lieutenant on 5 March 1817, Master Commandant on 9 Feb 1837, Captain on 4 Jun 1844. Reserved List on 13 Sep 1855, and as Commodore, Retired List on 4 Apr 1867 **P**: No **BLW**: No **PH**: No **SS**: AQ **BS**: 49

Key: *Additional veteran entry   **Corrected veteran entry   ***Deleted veteran entry
RU=Rank/Unit   CEM=Cemetery   GS=Gravestone   SP=Spousal Information   VI=Other Veteran Info   P=Pension
BLW=Bounty/Land Warrant   PH=Photo   SS=Service Source   BS=Burial Source   VMR= VA Military Regiment
LNR= Last Known Residence

JARVIS, William, Sr; b 1770, d 2 Jan 1831, (Will) **RU: Captain, 27th VMR, company commander, Northampton Co **CEM:** Piney Forest (Old Jarvis Place); Northampton; Rt 645, 0.1 mi S of Rt 644 **GS:** Yes **SP:** 1) Frances Goffigon (10 Nov 1776-21 Oct 1796), daug of Nathaniel Goffigon and Frances Dunton, mar 2) Ann Wilkins (24 Jul 1781-6 Dec 1805), daug of William Wilkins Sr and Elizabeth Goffigon, mar 3) Elizabeth Upshur Robins, (1788-1 Jul 1844), daug of John Robins and Susanna Teackle **VI:** No further data **P:** No **BLW:** No **PH:** No **SS:** K pg 113 **BS:** 20 pg 44

*JARVIS, William; b 1783, Henrico Co, d 14 Oct 1861, Henrico Co **RU:** Ensign, 74th VMR, Captain Hezekiah Henley's Company, Hanover Co, attached to 1st VMR (Clarke) **CEM:** Shockoe Hill, Richmond; 100 Hospital St **GS:** Yes **SP:** Mar (Henrico Co, 11 Feb 1806) Frances Jones (__-1871) **VI:** No further data **P:** No **BLW:** No **PH:** No **SS:** K pg 483 B pg 94 **BS:** 245

*JENKINS, John Cole; b unk, d 1835 **RU:** Corporal, 54th VMR, Norfolk Borough **CEM:** Cedar Hill; Suffolk City; Hill St **GS:** Yes **SP:** Elizabeth Madden, (1796-1865) **VI:** Grave loc block D, section 54 **P:** No **BLW:** No **PH:** No **SS:** A rec 1324 **BS:** 245

*JENKINS, Stephen; b 10 Jul 1792, d 27Jul 1861, Winchester **RU:** Private; 3rd VMR, Capt Richard Proctor's Co, Orange County, attached to 5th VMR (Mason & Preston) **CEM:** Hebron; Winchester; GPS 39.18170, -78.1572, 305 E Boscawen St **GS:** Yes **SP:** Mary (__) (1797-1875) **VI:** Grave loc section 386. Received two BLW of 60 acres each **P:** No **BLW:** Yes **PH:** No **SS:** A rec 1849; B pg 156; BD pg 1078; M pg 216 **BS:** 245

*JENKINS, William K; b c1789, d 26 Jun 1852, **RU:** Private; Frederick Co company attached to 4th VMR (Beatty) or 1st VMR (Taylor) **CEM:** Green Hill; Clarke; Berryville **GS:** Yes **SP:** Mar (Frederick Co, 6 Dec 1808) Mary Wilson **VI:** No further data **P:** No **BLW:** No **PH:** No **SS:** A recs 1550; 1551 or 1556; B pgs 78; 79 **BS:** 245

*JENNELLE, Lewis M; b 1795, d 1885 **RU:** Private 75th VMR, Capt James Hoge's Co, Montgomery County, attached to 4th VMR (Koontz) **CEM:** Jennelle Family; Montgomery; SW jct Rt 613 and Old Yellow Sulphur Rd **GS:** Yes **SP:** Malvinah Turner (1810-1885) **VI:** Recd pen and BLW of 80 acres **P:** Yes **BLW:** Yes **PH:** No **SS:** B pg 138; M pg 216; O- pen files, fold3 **BS:** 245

JENNINGS, Thomas; b 21 Nov 1778, Pittsylvania Co, d 31 Dec 1849 **RU: Private, 4th VMR **CEM:** Jennings Family; Carroll; jct Rts 767 & 753 **GS:** Yes **SP:** Sarah Cocke Clifton (1781- 7 Jan 1855) **VI:** Son of William Jennings & Elizabeth Ogle **P:** No **BLW:** No **PH:** No **SS:** A rec 1892 **BS:** 90 pg 127

*JESUP, Thomas Sidney; b 1761, Berkeley Co, d 1860, Washington, DC **RU:** Colonel (Brevet); U.S. Army 25th Inf at battles on Niagara frontier and at Lundy Lane, Jul 1814 **CEM:** Arlington National, Arlington, off George Washington Parkway **GS:** Unk **SP:** No spouse info **VI:** Entered service 2d LT (3 May 1808) in 7th Inf, promoted 1/LT Dec 1809, promoted Brig. Major and served as Adjutant General to BG Hull in 1812, promoted to Captain (Jan 1813) and Major (6 Apr 1813) while serving in the 19th Inf, then transferred to the 25th Inf and promoted to Brevet LT Colonel (5 Jun 1814) for distinguished service at the Battle of Chippewa, promoted (25 Jul 1814) to Brevet Colonel for gallantry and distinguished skill in the Battle of Niagara; Where he was seriously wounded in the Battle of Lundy Lane. After the war, he continued with further service and was promoted to Brevet Major General (8 May 1828) and served as the Army's Commander in Florida in the Seminole War and was wounded (24 Jan 1838) near Jupiter's Inlet. Then he served in the Mexican War and lastly served as the Army's QM General to his death in 1860 **P:** No **BLW:** No **PH:** No **SS:** F pg 251 **BS:** 53 pg 19

*JETER, Ira; b 8 Mar 1796, d 3 Jan 1876 **RU:** Sergeant, Capt Laurence T Dade's Co of Artillery, Orange Co, attached to 1st VMR (Yancey) **CEM:** East Hill; Salem City; jct Rt 460 and Lynchburg Turnpike **GS:** Yes **SP:** No spousal info **VI:** No further data **P:** Yes **BLW:** No **PH:** No **SS:** B pg 156; M pg 216; O-pen files, fold3 **BS:** 245

*JETER, John; b c1786, d aft 1814, **RU:** 1st VMR, Captain; 1st VMR, company commander. Amelia Co, attached to Battalion of Artillery (Freeman) **CEM:** Jeter Family; Amelia; Rodophil **GS:** Unk **SP:** No spouse info **VI:** Son of Rodophil Jeter (1765-1845) and Lucy Gills (1765-__); Nickname "Black Jack" **P:** No **BLW:** No **PH:** No **SS:** A rec 2273; B pg 37 **BS:** 245

Key: *Additional veteran entry   **Corrected veteran entry   ***Deleted veteran entry
RU=Rank/Unit   CEM=Cemetery   GS=Gravestone   SP=Spousal Information   VI=Other Veteran Info   P=Pension
BLW=Bounty/Land Warrant   PH=Photo   SS=Service Source   BS=Burial Source   VMR= VA Military Regiment
LNR= Last Known Residence

56

*JETER, Tilmon (Tilman); b c1780, d 1847, **RU:** Captain; 1st VMR; Troop of Calvary commander, Amelia County, attached to Major Woodford's Squadron **CEM:** Jeter Family; Amelia; 3 mi N of Jetersville **GS:** Unk **SP:** No spouse info **VI:** His home where buried was a Masonic lodge and tavern in 1823 and in 1945 became the home of the family **P:** No **BLW:** No **PH:** No **SS:** A rec 2283; B pg 37; K pg 145 **BS:** 245

*JOHNSON, Gerard E; b 4 Jun 1769, d 16 Jun 1850 **RU:** Private, Lieutenant Sneed's Co, 2nd US Dragoons **CEM:** South River Meeting House Graveyard; Lynchburg City; 5810 Fort Ave **GS:** Yes **SP:** No spouse info **VI:** Pension indicates disabled in service **P:** Yes **BLW:** Yes **PH:** No **SS:** BD pg 1084 **BS:** 245

*JOHNSON, John; b 18 Oct 1789, d 1 Jan 1875 **RU:** Sergeant, Captain Tom Wilkinson's Co, Loudoun County attached to 6th VMR (Reade) **CEM:** New Valley Baptist; Loudoun: Luckeets **GS:** Unk **SP:** Mar (28 Oct 1814) Catherine Fry (1789-1879) **VI:** No further data **P:** Yes, widow also **BLW:** Yes **PH:** No **SS:** B pg 122; M pg 217 **BS:** 245

*JOHNSON, Robert; b 13 Sep 1790, Frederick Co, MD, d 16 Feb 1873 **RU:** Ensign, Capt Barton Hackney's 3rd Regt (Stembel), MD Militia **CEM:** Old Presbyterian; Loudoun; So Church St, behind Providence Primitive Baptist Church, Lovettsville **GS:** Yes **SP:** Malinda Thrasher (1789-1859) **VI:** Recd pen and BLW of 160 acres in 1855 **P:** Yes **BLW:** Yes **PH:** No **SS:** O- pen files, fold3 **BS:** 245

*JOHNSON, William; b 1780, d 27 Aug 1848 **RU:** Private, 114th VMR, Lt Hook's Co, Hampshire Co, now in West Virginia **CEM:** Union; Loudoun; Leesburg **GS:** Yes **SP:** Margaret Ish (__-3 Mar 1861) **VI:** No further data **P:** Widow applied (#26652) **BLW:** No **PH:** No **SS:** BD pg 1090; M pg 217 **BS:** 245

*JOHNSTON, Joseph; b 4 Dec 1785, d 6 Jan 1854 **RU:** Corporal, 77th VMR, Capt Isaac Heiskell's Co, Hampshire County (now WVA), attached to 1st VMR (Connell), part of Gen Joel Leftwich Brigade **CEM:** Johnson Graveyard; Frederick; nr Trone Methodist Church on Timber Ridge **GS:** Unk **SP:** Mar (28 Feb 1805) Ann McGinnis (18 May 1782-20 Aug 1864) **VI:** Son of William Johnston and Elizabeth Hancher **P:** No **BLW:** No **PH:** No **SS:** A rec 1448; B pgs 21; 221 **BS:** 56 pg 214

*JONAS, Daniel; b 1797, Kanawha Co, WVA, d 18 Jan 1869 **RU:** Private, 105th VMR Capt St John Dixon's Co, Washington County, attached to 5th VMR (Mason-Preston) **CEM:** St Peters Lutheran Church; Wythe; Rt 619, Cripple Creek **GS:** Yes **SP:** Mar 14 Feb 1814, Rockbridge Co, Elizabeth Newcomer (Wythe Co, 1796-29 Aug 1886) Recd two pen **VI:** Recd two BLW one 40 acres, 1850 and one 120 acres, 1855 **P:** Widow **BLW:** Yes **PH:** No **SS:** B pg 218; M pg 218; O- pen files, fold3 **BS:** 245

*JONES, Abraham; b 13 Apr 1774, Callaway, d 1839, Callaway **RU:** Private, 42 VMR or 101st VMR, Pittsylvania Co attached to 6th VMR **CEM:** Pigg River Primitive Baptist Church; Franklin; Swenfield Rd **GS:** No **SP:** Mar (26 Jul 1813) Sally Turnbull Hale (widow of Joseph Hale) **VI:** Son of Thomas Jones & Joanna Hill **P:** No **BLW:** No **PH:** No **SS:** A rec 26; B pgs 101,102 **BS:** 245

*JONES, Benjamin; b 22 Feb 1786, Sussex Co, d 6 Nov 1856 **RU:** Private, 33rd VMR, served in a company of Dinwiddie Co **CEM:** Blandford; Petersburg City; 319 So Crater Rd **GS:** No but in plot: Ward D (old Ground), Sec 16, Square 3, E 1/2 corner **SP:** No spousal info **VI:** Son of Robert Jones and Mary (__). Committed suicide **P:** Yes **BLW:** Yes **PH:** No **SS:** B pg 65; M pg 218; O- Serv index card fold3 **BS:** 245

JONES, Catesby; b 5 Mar 1788; d 18 Jul 1845 **RU: Captain, 21st VMR, commanded company, Gloucester Co **CEM:** Ware Episcopal Church; Gloucester; 7825 John Clayton Memorial Rd, Gloucester **GS:** Yes **SP:** 1) Mary Brook, (25 Oct 1796-9 Apr 1836), 2), (24 Jan 1837) Mary Anne Brook Pollard (3 Oct 1808-24 Nov 1886). She mar 2) T Montague in 1855. She buried Newington Baptist Church near Gloucester C.H. **VI:** No further data **P:** No **BLW:** No **PH:** No **SS:** B pg 83 **BS:** 82 pg 50; 76

*JONES, Gabriel; b 1768, d 10 Jul 1835 **RU:** Private, 21st VMR Capt John C Pryor's Troop of Cavalry, Gloucester Co, attached to 4th VMR **CEM:** Kinloch; Culpeper; E side Rt 15, 2 mi S of Culpeper **GS:** Yes **SP:** Mary P (__) (__-aft 1855) Recd pen and BLW **VI:** No further data **P:** Widow **BLW:** Widow **PH:** No **SS:** B pg 83; M pg 218; O- pen files fold3 **BS:** 245

Key: *Additional veteran entry   **Corrected veteran entry   ***Deleted veteran entry
RU=Rank/Unit   CEM=Cemetery   GS=Gravestone   SP=Spousal Information   VI=Other Veteran Info   P=Pension
BLW=Bounty/Land Warrant   PH=Photo   SS=Service Source   BS=Burial Source   VMR= VA Military Regiment
LNR= Last Known Residence

57

*JONES, Richard; b 17 Mar 1789, d 2 Mar 1870 **RU:** Private, Captain George Town's Co of Artillery, Pittsylvania Co, attached to 3rd VMR (Dickinson) **CEM:** Jones (Richard); Pittsylvania; Sheva **GS:** Yes **SP:** No spouse info **VI:** No further data **P:** Yes **BLW:** Unk **PH:** No **SS:** B pg 162; Mpg 219 **BS:** 245

*JONES, Thomas; b 1773, d 1812 **RU:** Captain, 74th VMR company commander, Hanover Co., attached to 8th VMR (Magnien) and attached to Colonel William Trueheart's 1st VMR **CEM:** Saint Johns Church; Hampton; 100 W Queens Way **GS:** Yes **SP:** No spouse info **VI:** No further data **P:** No **BLW:** No **PH:** No **SS:** B pg 95; L pg 505 **BS:** 245

*JONES, Stephen; b 2 Feb 1787, d 18 Aug 1818 **RU:** Private, 72 VMR, Capt George W Camp's Co, Russell County **CEM:** Jones #293; Lee; on Alt US 58, Dryden **GS:** Unk **SP:** Mary "Polly" Parsons **VI:** Son of Samuel Jones & Selah (or Sarah) (__) **P:** No **BLW:** No **PH:** No **SS:** A rec 2480; B pg 183 **BS:** 245

*JONES, Thomas; b 18 Aug 1781, d 9 Nov 1860 Chesterfield Co **RU:** Private, Capt George Town's Co of Artillery, attached to 3rd VMR (Dickerson's), Pittsylvania Co **CEM:** Blandford; Petersburg; 319 S Crater Rd **GS:** Yes **SP:** Mary Lee (1790-1848) **VI:** Son of Joseph Jones & Jane Atkinson **P:** No **BLW:** No **PH:** No **SS:** B pg 161; BD pg 1100 **BS:** 245

*JONES, Thomas; b 1790, Shenandoah Co, d 4 Apr 1815, Shenandoah Co **RU:** Lieutenant, 97th VMR, Shenandoah Co **CEM:** Green Hill; Page; Rt 340 vic jct Rt 7; Berryville **GS:** Yes **SP:** Mar (12 Aug 1810) Nancy Wood (1790-1850) **VI:** Son of George Jones & Margaret Morgan **P:** No **BLW:** No **PH:** No **SS:** A rec 2689; B pgs 184-185 **BS:** 245

*JONES, Thomas Sr; b 17 Oct 1793, Washington Co, d 14 Apr 1860 **RU:** Private, 35th VMR, Capt Samuel Graham's Co, Wythe County, attached to 4th VMR (McDowell's, Koontz, Chilton) **CEM:** Jones; Scott; GPS 36.61704,-82.68571 **GS:** Unk **SP:** Frances Klepper (Hawkins Co, TN, 1801-7 Apr 1887, Scott Co) **VI:** Son of Wells Jones (Washington Co, aft 1770-Hawkins Co, TN, 1835) and Nancy Hannah Lewis (Scott Co, 1772- 1850) **P:** Yes # 98-18148 **BLW:** No **PH:** No **SS:** B pg 204; BD pg 1100 **BS:** 245

*JONES, William; b 6 Mar 1784, d 17 Apr 1859 **RU:** Private, 22nd VMR Capt Stephan Pool's Co, Mecklenburg Co, attached to 6th VMR (Sharp's) **CEM:** Zion Methodist Church; Mecklenburg; Busy Bee Rd, South Hill **GS:** Yes **SP:** Mary B Jackson, (1829-1904) **VI:** No further data **P:** No **BLW: PH:** No **SS:** A rec 2961; B pgs 130-131 **BS:** 245

*JONES, Wood; b 1787, d 1878 **RU:** Private, 49th VMR, Capt Samuel Jeter's Artillery Co, Nottoway County, attached to 7th VMR (Gray) **CEM:** Lakeview; Nottoway; jct 8th St and Memorial Dr, Blackstone **GS:** Yes **SP:** No spousal info **VI:** Disinterment fr family cem, (1941-1942) on expansion of Fort Pickett. Residing 1870 in Bellefonte, Nottoway Co. Applied for pen which was rejected for name not found on muster roll, but neighbors signed depositions indicated he had 1812 service **P:** Applied **BLW:** No **PH:** No **SS:** B pg 154; M pg 218; O- pen files fold3 **BS:** 245

*JORDAN, Merit; b 2 Jul 1792, d 18 May 1872, Norfolk City **RU:** Private, 29th VMR, Capt Joseph Atkinson's Troop of Calvary, Isle of Wight Co **CEM:** Jordan Family; Isle of Wight; Farm Field, on Moonlight Rd Rt 627 **GS:** Yes **SP:** Paulina (__) (Nov 1801-aft 1878) Recd two pen LNR Corpus Christi, TX **VI:** Recd two pen and BLW of 80 acres, 1855 **P:** Both **BLW:** Yes **PH:** No **SS:** B pg 102; M pg 219; O-pen files fold3 **BS:** 245

KAUFFMAN, Daniel; b 20 Dec 1781; d Unk **RU: Ensign, 6th VMR **CEM:** Kauffman Family; Page; Kauffman Mill Camp **GS:** Yes **SP:** No spouse info **VI:** No further data **P:** No **BLW:** No **PH:** No **SS:** A rec 2275 **BS:** 115 pg 173

*KEELING, Henry; b 1792, d 5 Jan 1839 **RU:** Private 20th VMR , Capt Lemuel Cornick's Co and Capt William Stone's Co, Princess Anne County **CEM:** Elmwood; Norfolk City; 238 E, Princess Anne Rd **GS:** Yes at Elm ext; Block 13; Lot 11, space 5-9 N **SP:** Frances Wilson (1795-1872) applied for pen **VI:** Obituary in American Beacon Daily Advertiser **P:** Widow applied **BLW:** No **PH:** No **SS:** B pgs 164,166; K pg 252; L pg 471; M pg 220 **BS:** 245

Key: *Additional veteran entry   **Corrected veteran entry   ***Deleted veteran entry
RU=Rank/Unit   CEM=Cemetery   GS=Gravestone   SP=Spousal Information   VI=Other Veteran Info   P=Pension
BLW=Bounty/Land Warrant   PH=Photo   SS=Service Source   BS=Burial Source   VMR= VA Military Regiment
LNR= Last Known Residence

*KEELING, Henry; b 26 Dec 1794, Norfolk City, d 19 Nov 1870 **RU**: Private, 20th VMR, Capt John Painter's (Paynter's) Co, Princess Anne County **CEM**: Shockoe Hill; Richmond City; 100 Hospital St **GS**: No unmarked in Keeling plot **SP**: Jane Catherine Charlton (1787-1860) **VI**: Founding father University of VA, and President of the Virginia Baptist Seminary which was chartered as Richmond College in 1840; Name listed on pastor's plaque at the 3rd Baptist church **P**: Yes **BLW**: No **PH**: No **SS**: B pg 165; O- pen files fold3 **BS**: 245

*KEELING, William S or T; b1793, Mar 1828 **RU**: Private, 20th VMR, Captain Lemuel Cornick's Company, Princess Anne County **CEM**: Cedar Grove, Norfolk, 238 E Princess Anne St **GS**: Yes **SP**: No spouse info **VI**: No further data **P**: No **BLW**: No **PH**: No **SS**: A rec 246 B pg 164 **BS**: 245

*KEEN, William Sr; b 1780, Bedford Co, d unk **RU**: Private, 72th VMR, Captain William Cromwell's or Captain George Kindrick's Cos, Russell County attached to Major Bradley's Regt **CEM**: Keen William; Russell; Belfast **GS**: Yes **SP**: Rebecca Jackson, (Russell Co, 7 Jun 1785- Russell Co, 22 Nov 1839) **VI**: No further data **P**: No **BLW**: No **PH**: No **SS**: A rec 408; M pg 183 **BS**: 245

*KEESEE, Jacob Burton; b 1792, d 1 Jan 1872 **RU**: Corporal, 16th or 96th VMR, Capt William Palmer's Co, Brunswick County, attached to 5th VMW (Mason-Preston) **CEM**; Hollywood; Richmond; 412 S Cherry St **GS**: Yes **SP**: No spousal info **VI**: Recd two pen and BLW 80 acres 1855 **P**: Yes **BLW**: Yes **PH**: No **SS**: B pg 49; M pg 220; O- pen files fold3 **BS**: 245

*KEESLING, Peter; b 19 Aug 1792, d 26 Jul 1857 **RU**: Private, 35th VMR, Captain Samuel Graham's or Captain Joseph Steffey's Company, Wythe Co. **CEM**: Keesling; Wythe; Rural Retreat **GS**: Yes **SP**: Catherine (__) (1794-1867) **VI**: Son of George Kisling (1759-1840) and Catherine (__) (1766-1808) **P**: No **BLW**: No **PH**: No **SS**: A rec 656 **BS**: 245

*KEEZEL, George; b 1795, d 1 Dec 1862 **RU**: Private, 116th VMR, Capt William McMahon's Troop of Calvary, Rockingham Co, attached to Major Woodford's Squadron **CEM**: Cross Keys; Rockingham; on Rt 679; S of jct with Rt 276 **GS**: Yes **SP**: Mar (13 Jan 1852) Amanda M Fitz Allen Peale (1815-23 Jun 1870). Recd pen 1878 **VI**: No further data **P**: Widow **BLW**: No **PH**: No **SS**: B pg 182; M pg 220; O- pen files fold3 **BS**: 245

*KELLER, George; b 13 Sep 1770, d 7 Mar 1859 **RU**: Private/Sergeant, 13th VMR/97th VMR, served in one of these companies of Shenandoah Co attached to 6th VMR (Coleman's) **CEM**: Walnut Springs Christ Church; Shenandoah; Oranda **GS**: Yes **SP**: Frances Feckley (Shenandoah County, 7 Sep 1758 - Shenandoah Co, 1 Apr 1847) **VI**: Son of Joseph Keller (1740-1825) and Leah (____) (1744-1840) **P**: No **BLW**: No **PH**: No **SS**: A rec 128435, B pgs 184-185 **BS**: 245

*KELLER, George; b 1789, d 16 Apr 1856 **RU**: Private, 32nd & 93rd VMR, in a company, Augusta Co, attached to McDowell's Flying Camp **CEM**: Keller on Dr Knopp's Farm; Augusta; Churchville **GS**: Unk **SP**: 1) Sarah Catherine Shafer (16 Mar 1810, 2) (8 Dec 1853) Jane Paxton (1814-14 Jun 1855), 3) (2 Aug 1855) Winny Woods **VI**: Son of George Keller and Sophie Mowry **P**: No **BLW**: No **PH**: No **SS**: A rec 1287; B pgs 39-40 **BS**: 245

*KELLER, Jacob B; b 13 Apr 1785, d 28 Aug 1846 **RU**: Private, 13th or 97th VMR, Shenandoah Co, attached to 4th VMR **CEM**: Zion Lutheran Church, Shenandoah, Hamburg **GS**: Yes **SP**: Mar (19 Aug 1819) Rebecca Coffman **VI**: No further data **P**: No **BLW**: No **PH**: No **SS**: A rec 1304 j; B pg 185-5 **BS**: 245

*KELLER, Jacob; b 26 Oct 1774, d 17 May 1836 **RU**: Ensign, 13th or 97th VMR, Shenandoah Co, attached to 6th VMR (Coleman) **CEM**: Keller Family, Shenandoah, Rt 853, 2mi W of Thom's Brook, Carmel **GS**: Yes **SP**: Susanna Summers, (1779-1858) **VI**: Son of George Keller (1731-1788) and Barbara Zimmerman (unk-1798); A doctor. **P**: No **BLW**: No **PH**: No **SS**: A rec 1305 **BS**: 245

Key: *Additional veteran entry   **Corrected veteran entry   ***Deleted veteran entry
RU=Rank/Unit   CEM=Cemetery   GS=Gravestone   SP=Spousal Information   VI=Other Veteran Info   P=Pension
BLW=Bounty/Land Warrant   PH=Photo   SS=Service Source   BS=Burial Source   VMR= VA Military Regiment
LNR= Last Known Residence

59

*KELLER, Lewis; b 3 Oct 1792, d 8 May 1866 **RU:** Private, 32d VMR, Capt Hugh Young, Augusta County Militia, attached to 2d VMR (Ballowe) **CEM:** Keller Family; Augusta; Dr Knopps farm **GS:** Yes but not legible **SP:** Mar 1) Catherine Fall, 2) (21 Aug 1846 Augusta Co) Elizabeth Palmer (__-aft 1878). Recd two pen and BLW 40 acres (1850) and 120 acres (1855) **VI:** Son of George Keller (1758-1844) and Sophie Mowry (1766-1859) **P:** Widow **BLW:** Widow **PH:** No **SS:** B pg 40; M pg 220; O- pen files fold3 **BS:** 245

KELLO, Samuel; b unk; d 1814 **RU: 1st Lieutenant, 65th VMR Capt John Critchlow's Artillery Company Southampton Co, attached to 3rd VMR (Boykin) **CEM:** Probably Millfield Plantation; Southampton; Millfield Rd (Rt. 605) **GS:** No **SP:** Martha Claiborne Simmons who pre-deceased her husband **VI:** Son of Samuel Kello (__-1802) and Margaret Belches. Samuel the father was the 2nd Clerk of Court of Southampton Co. Samuel Kello II was the 3rd clerk of court. Southampton Co, *Court Minute Book (1830-1835)* p. 216, 19 Nov 1832: "It appears to the Court from evidence introduced that Samuel Kello is the only child and heir of Samuel Kello, deceased, who died intestate in the service during the late war, leaving no widow" and *Order Book (1814-1816)*, p. 21, 19 Dec 1814 James Rochelle (Deputy Clerk) qualified as Administrator of the estate of Samuel Kello, deceased. There is no stone for Samuel Kello II, but he is presumed to be buried here. His son, Samuel Kello III (1807-1875), became sheriff of Southampton Co, Clerk of Southampton Co. Died during the war **P:** No **BLW:** No **PH:** No **SS:** A rec 1821; B pg 186 **BS:** 245

*KELLY, John; b 31 Aug 1781, d 13 Dec 1833 **RU:** 2nd VMR, Capt William Custis's, Company, Accomack Co **CEM:** Kelly Grave; Accomack; Keller **GS:** Yes **SP:** No spouse info **VI:** No further data **P:** No **BLW:** No **PH:** No **SS:** B pg 32; L pg 263 **BS:** 245

KENDALL, Joshua; b 30 Mar 1796; d 23 Jun 1852 **RU: Private, 45th VMR Capt Elijah Harding's Co, Stafford County **CEM:** Kendall Family; Stafford; Garrisonville, Vista Woods Subdivision **GS:** Yes **SP:** Eleanor (__), (29 Aug 1788-12 Jan 1859). Age 61 on 1850 Census, but dates come fr tombstone **VI:** Age 51 on 1850 census, but dates come fr tombstone, Son of Henry Kendall (1749-1810) **P:** No **BLW:** No **PH:** No **SS:** A rec 473 **BS:** 26 pg 262

KENDALL, William; b 27 Jan 1792; d 2 Jan 1854 **RU: Private, 45th VMR, Capt John C. Edrington's Co of Drafts, Stafford County, attached to 6th VMR (Coleman) **CEM:** Kendall Family; Stafford; Garrisonville, Vista Woods Subdivision **GS:** Yes **SP** Lucy T (__) (3 Feb 1795-9 Mar 1859) **VI:** Probably the son of Henry Kendall (1749-1810) as buried in cemetery lot with parents and brother **P:** No **BLW:** No **PH:** No **SS:** L pg 521 **BS:** 26 pg 262; 31

*KENT, Charles; b 1784, MA d 17 Mar 1871 **RU:** Captain, 39th VMR, Commanded a company in Petersburg City, attached to 6th VMR (Dickinson, Scott, Coleman) **CEM:** Blandford; Petersburg City; 111 Rochelle Ln **GS:** Yes **SP:** No spouse info **VI:** A native of MA that moved to Petersburg in 1807 **P:** No **BLW:** No **PH:** No **SS:** B pg 160 **BS:** 245

*KENT, James Randal; b 23 Oct 1792, d 29 May 1867 **RU:** Private; 75th VMR, Capt James Hoge's Co, Montgomery County, attached to 4th VMR (Huston-Wooding) **CEM:** Kent Family; Montgomery; GPS 37.19400,-80.58269; nr end Rt 623 nr New River, McCoy **GS:** Yes **SP:** Mary Cloyd (15 Feb 1800-5 Feb 1858), daug of Maj Gen Gordon Cloyd (1771-1833) and Elizabeth McGavock (1776-1830) **VI:** Son of Joseph Kent (7 Nov 1765-20 Oct 1843) and Margaret McGavock (1769-1827). Was Sheriff and Justice, Montgomery Co, (1822-1823), and founder of Preston and Olin Institute, a Methodist school in Blacksburg, that became VA Tech **P:** No **BLW:** No **PH:** No **SS:** A rec 1769; B pg 138 **BS:** 245

*KERLIN, David; b 4 May 1784, d 10 Jan 1862 **RU:** Private, 58th or 116th VMR, Capt William McMahon's Troop of Cavalry, Rockingham Co, attached to Major Woodford's Squadron **CEM:** Solomon's Lutheran Church; Shenandoah; Solomon's church Rd, Rt 727 **GS:** Yes **SP:** Mar (17 Feb 1815) Barbara Hudson (28 Feb 1790-15 Apr 1874). Recd two pensions. LNR Moore's Store, Shenandoah Co **VI:** No further data **P:** Widow **BLW:** No **PH:** No **SS:** B pg 182; M pg 221; O- pen files fold3 **BS:** 245

Key: *Additional veteran entry   **Corrected veteran entry   ***Deleted veteran entry
RU=Rank/Unit   CEM=Cemetery   GS=Gravestone   SP=Spousal Information   VI=Other Veteran Info   P=Pension
BLW=Bounty/Land Warrant   PH=Photo   SS=Service Source   BS=Burial Source   VMR= VA Military Regiment
LNR= Last Known Residence

*KERR, James b 1796 DC, d 10 Oct 1859, DC **RU**: Private, 1st DC Regt commanded by Lt Col Adam Lynn **CEM**: St Marys Catholic Church; Alexandria City; jct GW Memorial Parkway and Church St **GS**: Unk **SP**: No spousal info **VI**: Buried 12 Oct 1859 **P**: No **BLW**: No **PH**: No **SS**: O- Serv index Card **BS**: 245

*KERR, Thomas; b 1 Dec 1797, d 21 Dec 1875 **RU**: Private, 94th VMR Capt Jeremiah H Neill (Neil)'s Co, Lee County, attached to 7th VMR (Saunders) **CEM**: New Monmouth Presbyterian; Lexington City; 2348 West Midland Rd **GS**: Yes Plot Old Section, Row P **SP**: Nancy McClung Patton (14 Mar 1815-17 Dec 1893) **VI**: Recd two pen **P**: Yes **BLW**: No **PH**: No **SS**: O- pen Files, Fold3 **BS**: 245

KESLER, Henry; b 1793 Botetourt Co; d 22 Jan 1850 **RU: Private, 121st VMR, Capt Washington West's Company, Botetourt Co **CEM**: Wingo Family; Franklin; Rt 662 nr Smith Mountain Lake in Woods top of knoll **GS**: Yes **SP**: Mar (Franklin Co, 14 Jan 1818) (bond), to Elizabeth Howsman, Peter Howsman surety (Albemarle Co, 28 Feb 1796-10 Jan 1887) **VI**: Son of Jacob B Kesslar (PA, 22 Sep 1756 -20 Jul 1824) and Elizabeth Shearer (20 Mar 1769-13 Aug 1844)**P**: Widow **BLW**: No **PH**: No **SS**: BD pg 1121; B pg 46; M pg 221 **BS**: 118 pg 203; 245

*KESSLER, John; b 23 Nov 1788 (pen), Botetourt Co, d 1864 **RU**: Private, 121St VMR Capt Washington West's Company, Botetourt Co, attached to 4th VMR (Boyd) **CEM**: Ross; Craig; off Rt 42 N betw jct with Rts 629 and 641 **GS**: Yes **SP**: Mar (7 Aug 1813) Elizabeth Housman (1796-aft 1878), Recd pen Simmonsville, Craig Co 1878 **VI**: Son of Jacob B Kessler of Montgomery Co, PA (22 Sep 1756-20 Jul 1824) and Elizabeth Shearer (20 Jul 1769-13 Aug 1844) **P**: Widow **BLW**: No **PH**: No **SS**: B pg 46; M pg 221;O- pen files, Fold3 **BS**: 245

*KEYSER, Fleming; b 26 Jan 1794, Bath Co, d 6 Apr 1864 **RU**: Private, 12th Inf Regt, US Army, Capt Thomas Post's Co. **CEM**: Emory United Methodist; Alleghany; GPS: 37.9227,- 79.9625 **GS**: No **SP**: Margaret C (____) **VI**: Son of William Keyser (1755-1837) and Keziah Snead (1761-1849); Enlisted (Staunton, VA 12 Jun 1813); discharged, (Buffalo, NY, 31 May 1815) **P**: Wife **BLW**: No **PH**: No **SS**: C pg 104; BD pg 1122 **BS**: 245

*KIBLER, Jacob, b 19 Feb 1796, Shenandoah Co, d 19 Jun 1876 **RU**: Private 3rd VMR, Capt Richard Proctor's Co, Orange County, attached to 4th VMR (Beatty) **CEM**: Kibbler Family; Page; Big Spring **GS**: Unk **SP**: Mar (Shenandoah Co, 16 Mar 1818) Nancy V Dearing (Fauquier Co,11 Mar 1802-Page Co, 1872) **VI**: Pen orig filed under Jacob Kibbler, Jr **P**: Yes **BLW**: No **PH**: No **SS**: B pg 156; M pg 221; O- pen files, Fold3 **BS**: 245

*KILGORE, Hiram; b 25 Mar 1792, Russell Co, d 16 Dec 1870 **RU**: Lieutenant 87th VMR, Capt William Hill's Co of Artillery, King William County, attached to Battalion of Arty **CEM**: Rollins Family; Scott; Clinchport **GS**: Yes **SP**: Lucy Ann Pennington (9 Jul 1827-16 Sep 1914), daug of John Dees Pennington (1802-1856) and Rachel Zion (1792-1870). Recd two pen LNR 1878 Estillville, Scott Co **VI**: Son of William Kilgore (5 Jun 1769-1857) and Virginia Jane Osburn (1774-1851). Styled Colonel on Gr St. Received two BLW of 80 acres each **P**: Widow **BLW**: Yes **PH**: No **SS**: B pg 115; M pg 222; O- pen files, Fold3 **BS**: 245

*KING, Augustine; b 7 Apr 1789, d 2 May 1835 **RU**: Private, Wagon master 115th VMR, Capt Henry Howard's Co, York County **CEM**: Prospect Hill; Warren; 200 West Prospect St, Front Royal **GS**: Yes **SP**: Mar (Shenandoah Co, 12 Sep 1811) Verlinda Northcraft (1789-aft 1881). Recd two pen and BLW, 160 acres, 1855 **VI**: No further data **P**: Widow **BLW**: Widow **PH**: No **SS**: B pg 67; M pg 222; O- pen files, Fold3 **BS**: 245

*KING, John; b 24 Sep 1773 d 21 Aug 1852 **RU**: Corporal 35th VMR, Capt Jacob Fishback's Co of Artillery, Wythe County, attached to Brigade of Artillery **CEM**: Kings Grove UMC; Wythe; Crockett **GS**: Yes **SP**: Sarah Nifong, (27 Oct 1772-27 Mar 1855) **VI**: No further data **P**: No **BLW**: No **PH**: No **SS**: A rec 807, B pg 204 **BS**: 245

Key: *Additional veteran entry   **Corrected veteran entry   ***Deleted veteran entry
RU=Rank/Unit   CEM=Cemetery   GS=Gravestone   SP=Spousal Information   VI=Other Veteran Info   P=Pension
BLW=Bounty/Land Warrant   PH=Photo   SS=Service Source   BS=Burial Source   VMR= VA Military Regiment
LNR= Last Known Residence

61

*KING, John "Jack", Jr; b 11 Jan 1788, d 13 Nov 1862 **RU:** Private, 64th VMR company, Henry Co **CEM:** King; Henry; Leatherwood **GS:** Unk **SP:** Mary Wills (__-1874) **VI:** Son of John King (1758-1821) and Mary Elizabeth Seward; served in House of Delegates, (1842-1843) **P:** No **BLW:** No **PH:** No **SS:** A rec 926; B pg 101 **BS:** 245

*KING, Miles Jr; b 9 Jun 1786, d 8 Dec 1849 **RU:** Captain, 54th VMR, Company commander of Norfolk Light Artillery Blues, Norfolk **CEM:** Saint Pauls Episcopal Churchyard; Norfolk; 201 St Pauls Blvd **GS:** Unk **SP:** 1) Rebecca Calvert (10 Mar 1788-28 Feb 1827), daug of Jonathan Calvert and Elizabeth Newton, 2) Mary Little Fisher (1813-1893) **VI:** Son of Miles King (1746-1814) and Martha Kirby (1765-1849). Mayor of Norfolk, member of Virginia Legislature, Naval agent for Port of Norfolk **P:** 2nd wife applied **BLW:** No **PH:** No **SS:** B pg 145, M pg 200 **BS:** 245

*KIGER, Isaac; b 27 Mar 1790, d 17 Aug 1851 **RU:** Sergeant, VMR, Capt Charles Brent's Company, Frederick Co, attached to 4th VMR (Beatty) **CEM:** Hebron; Winchester; GPS 39.18170, -78.1572, 305 E Boscawen St **GS:** Yes **SP:** Mar (Frederick Co, 31 Jan 1814) Lydia Rutter **VI:** No further data **P:** No **BLW:** Yes **PH:** Yes **SS:** B pg 78; BD pg 1123; M pg 222 **BS:** 31; 81

*KINDER, George; b 11 Oct 1779, d 22 Mar 1851 **RU:** Private, 35th VMR Capt Christopher Brown's Co, Wythe County **CEM:** Saint Pauls Lutheran Church; Wythe; 330 St Pauls Church Rd, Rural Retreat **GS:** Yes **SP:** No spouse info **VI:** Son of John Peter Kinder, Sr (d 1807) who was a Rev War soldier and Margaret (___) (__-1845) **P:** No **BLW:** No **PH:** No **SS:** A rec 224; B pg 204 **BS:** 245

*KIRK, William: b 8 Jun 1773, d 24 Jan 1844 **RU:** Private, 98th VMR Capt Ralph Hubbard's Co, and Capt Green Blanton's Co Mecklenburg County, attached to 1st VMR (Byrne) and 5th VMR **CEM:** Brushy Hill; Tazewell; Thompson Valley **GS:** No **SP:** Margaret Brooks (1779-1853) **VI:** Reportedly killed fr fall fr horse **P:** Yes **BLW:** No **PH:** No **SS:** B pg 130; M pg 223;O- pen files, Fold3 **BS:** 245

*KISNER, Samuel; b 1785, d unk **RU:** Rank E-7, West VA Regular Army Inf **CEM:** Arlington National, Arlington; Jefferson Davis Hwy, Rt 110 **SP:** No spouse info **VI:** No further data **P:** Yes **BLW:** No **PH:** No **SS:** Fold3- Pension Numerical Index **BS:** 245

*KLEIN, Jacob; b 1777, d 1833 **RU:** Private 54th VMR Capt John West's Co, Norfolk Borough **CEM:** Cedar Grove; Norfolk; Jct Salter St and E Princess Anne Rd **GS:** Yes **SP:** Esther Colgate (1791-18 Jul 1882), daug of Robert Colgate (1758-1826) and Sarah Ellen Bowles (__-1846). Applied for pen **VI:** No further data **P:** Widow applied **BLW:** No **PH:** No **SS:** B pg 146; M pg 223; A-*Ancestry.com* Index to Serv Rec **BS:** 245

*KLEIN, Lewis Allen; b 13 Mar 1783, d 14 Apr 1837 **RU:** Private, 57th VMR, Captain George W Blincoe's Co, Loudoun County **CEM:** Waterford Union of Churches; Loudoun; Waterford **GS:** Yes **SP:** No spouse info **VI:** No further data **P:** Yes **BLW:** No **PH:** No **SS:** B pg 119; M pg 1136 **BS:** 245

*KOOGLER, Jacob; b 28 Jul 1776, d 27 Feb 1854 **RU:** Private, 1st Regt (Gano's) Ohio Volunteers & Militia **CEM:** Cooks Creek Presbyterian Church; Harrisonburg City; 4222 Mt Clinton Pike **GS:** Yes **SP:** Mary (___) (8 May 1780-5 Dec 1839) **VI:** No further data **P:** No **BLW:** PH: No **SS:** A rec 5100 **BS:** 262

*KOONTZ, John; b 5 Oct 1793, d 25 Jun 1869 **RU:** Musician, 13th VMR, Captain Samuel Hawkins or Moses Walton's Companies, Shenandoah Co. attached to 5th VMR (Mason and Preston's) **CEM:** Union Church; Shenandoah; Mount Jackson **GS:** Yes **SP:** Mary Johnston Thompson (30 May 1797-28 Jan 1882), daug of John Thompson (1766-1826) and Anna Gillespie (1753-1818) **VI:** No further data **P:** Widow **BLW:** No **PH:** No **SS:** B pgs 184-185; M pg 224 **BS:** 245

*KNIGHT, Bailey Alexander; b before 1794, d aft 7 Aug 1820 **RU:** Private, 45th VMR Capt John Edrington's Company, Stafford Co **CEM:** Knight-English Family; Stafford; Rt 709 **GS:** Yes, with other names **SP:** Frances (__) **VI:** No further data **P:** Yes, Wife **BLW:** No **PH:** No **SS:** L pg 301 **BS:** 26 pg 264

*KNISELEY(KNICELY)(KNISLEY), John; b 12 Aug 1790, d 20 Oct 1871, Fort Valley, **RU:** Private 97th VMR Capt Jesse Allen's Trp of Cav, Shenandoah Co **CEM:** Knisley Family; Shenandoah; Detrick **GS:** Yes **SP:** Mar (29 Jan 1816) (Bond) Barbara (__) who recd pen **VI:** Recd BLW 40 acres (1853) **P:** Widow **BLW:** Yes **PH:** No **SS:** B pg 184; M pg 223; O- pen rec Fold3 **BS:** 245

Key: *Additional veteran entry   **Corrected veteran entry   ***Deleted veteran entry
RU=Rank/Unit   CEM=Cemetery   GS=Gravestone   SP=Spousal Information   VI=Other Veteran Info   P=Pension
BLW=Bounty/Land Warrant   PH=Photo   SS=Service Source   BS=Burial Source   VMR= VA Military Regiment
LNR= Last Known Residence

*KNOX, James; b 1788, d 12 Apr 1872 **RU:** Private, 39th VMR, Captain Edward Pescud's Co Petersburg City, attached to 6th VMR (Dickinson, Scott, Coleman) **CEM:** Blandford, Petersburg City; 319 S Grater Rd **GS:** No **SP:** Mar (4 Feb 1836) Jane Donnan, daug of David Donnan (1778-1857) and Mary Stewart (1787-1850) **VI:** Railroad Contractor **P:** No **BLW:** No **PH:** No **SS:** A rec 4814; B pg 160 **BS:** 245

*KREMER, George, Sr; b 1796, d 14 Sep 1871, Winchester **RU:** Private:31st VMR, Capt St. George Tucker's, Troop of Calvary, Frederick Co, attached 5th VMR **CEM:** Hebron; Winchester; GPS 39.18170, -78.1572, 305 E Boscawen St **GS:** Yes **SP:** Susan Sonner (1796-1868) **VI:** Son of Conrad and Catherine Helphinstine. Constable of Winchester **P:** Applied **BLW:** No **PH:** No **SS:** A rec 5389; B pg 80; BD pg 1142 **BS:** 245; 86 pg 49

*KUNKLE, Jacob; b Sep 1781, d 8 Jul 1866, **RU:** Private 93rd VMR Capt Jesse Dold's Troop of Cav, Augusta Co, attached to Maj Woodford's Squadron **CEM:** Bethel Presbyterian Church; Augusta; 563 Bethel Green Rd **GS:** Yes **SP:** Mar (Augusta Co, 6 May 1808) Mary Magdalene Bumgardner (25 Feb 1792- 6 May 1875), daug of Jacob Bumgardner (1767-1857) and Mary M Waddell (1765-1849), Recd two widows pen. LNR P.O. Elizabeth Furnace, Augusta Co **VI:** Recd BLW 40 acres (1850) and 160 acres (1855) **P:** Widow **BLW:** Yes **PH:** No **SS:** B pg 39; M pg 224; O- pen rec Fold3 **BS:** 245

*KURTZ, Isaac; b 5 Jul 1790, d 3 Feb 1861 **RU:** Private; 31st VMR, Capt Thomas Robert's Company, Frederick Co, attached to 4th VMR (Beatty) **CEM:** Hebron; Winchester; GPS 39.18170, -78.1572, 305 E Boscawen St **GS:** Yes **SP:** Frances T Fitzhugh (10 Feb 1808-12 Feb 1874) **VI:** No further data **P:** No **BLW:** No **PH:** No **SS:** L pg 70

*LACK, John; b Mar 1786; d 1849 (Will) **RU:** Private, 47th VMR, Capt John Field's Company, Albemarle Co, attached to 8th VMR (Walls) **CEM:** Black Family; Fluvanna; vic Rts 619 & 660 **GS:** Yes **SP:** Mar (Fluvanna Co, 6 Mar 1813) by John Goodman to Elizabeth Irvin **VI:** No further data **P:** No **BLW:** No **PH:** No **SS:** K pg 91 **BS:** 93 pg 6

LACY, Fleming; b unk; d unk **RU: Private, 19th VMR, Capt William Richardson's Co, Richmond City, attached to 1st Corps d'Elite (Randolph) **CEM:** Lacy Family; Goochland; off Rt 615 nr Old Forest Grove Church **GS:** Unk **SP:** Mar (Goochland Co, 5 Oct 1816) Elizabeth H Richards, daug of John Richards, Sr. **VI:** No further data **P:** No **BLW:** No **PH:** No **SS:** K pg 263 **BS:** 78 pg 176

LAKE, Timothy; b 18 May 1786; d 10 Oct 1869 **RU: Private, 48th VMR, Capt James Casmill's Company, Botetourt Co **CEM:** Mill Creek; Botetourt; 6 mi N of Troutville **GS:** Yes **SP:** Mary Magdalena Coffman (Kauffman) **VI:** No further data **P:** No **BLW:** No **PH:** No **SS:** K pg 7 **BS:** 155 pg 19

*LAND, Peter; b 1763, Princess Anne Co, d Aug 1845 **RU:** Captain, 20th VMR, commanded a company, Princess Anne Co, attached to 8th VMR (Wall) **CEM:** Smith Family; Chesapeake City; Mount Pleasant Farm **GS:** Yes and also on family GS **SP:** Elizabeth Keeling (1764-1832) **VI:** No further data **P:** No **BLW:** No **PH:** No **SS:** L pg 530 **BS:** 245

*LANDES, Daniel; b 1780, York Co, PA, d 3 Mar 1866 **RU:** Private, 5th or 34th VMR, Capt Reuben Moore's Co, Culpeper County, attached to 6th VMR (Coleman) **CEM:** Landes Family; Augusta; 1002 Fadley Rd, Rt 646, Weyers Cave **GS:** BS 245 submitter indicates could be buried at nearby Salem Church **SP:** Catherine Woodington **VI:** Son of Christian Landes (Switzerland, 17 Jan 1728- York Co, PA, 1782) and Mariah Bixler. Recd two pen and two BLW- 40 acres (1850) and 120 acres (1855) **P:** Yes **BLW:** Yes **PH:** No **SS:** B pg 62; M pg 224; O- pen rec Fold3 **BS:** 245

*LANDES/LANDIS, Samuel; b 20 Apr 1788, York Co. PA, d 20 Dec 1849 **RU:** Sergeant, Shanks Detachment, Pennsylvania Militia **CEM:** Landes; Augusta; Weyers Cave **GS:** Yes **SP:** Elizabeth Jane Welch (8 Mar 1804-27 Aug 1882) **VI:** Son of John Landes (1752-1819) and Catherine Miller (1749-1834) **P:** No **BLW:** No **PH:** No **SS:** A rec 6973 **BS:** 245

*LARY, John; b 1797, d 30 Jun 1848 **RU:** Private in Capt William McMahon's Troop of Cav, attached to Maj Woolford's Squadron, Rockingham Co **CEM:** Isaiah Clem; Shenandoah; Camp Roosevelt Rd; GPS: 38.44331, -78.30399 **GS:** Yes **SP:** No spouse info **VI:** No further data **P:** No **BLW:** No **PH:** No **SS:** K pg 188 **BS:** 245

Key: *Additional veteran entry   **Corrected veteran entry   ***Deleted veteran entry
RU=Rank/Unit   CEM=Cemetery   GS=Gravestone   SP=Spousal Information   VI=Other Veteran Info   P=Pension
BLW=Bounty/Land Warrant   PH=Photo   SS=Service Source   BS=Burial Source   VMR= VA Military Regiment
LNR= Last Known Residence

63

*LAREW, Jacob; b 21 Feb 1793, Augusta County, d 3 Feb 1875 Brownsburg(h) **RU:** Sergeant 93rd VMR, Capt Samuel Doak's Co, Augusta Co, attached to 5th VMR (McDowell) **CEM:** New Providence Presbyterian Church; Rockbridge; 1208 New Providence Rd **GS:** Yes **SP:** 1) Ann Scott (1793-11 Jun 1861), 2) Margaret G Warwick (__-aft 1878) who recd two pen **VI:** Son of Jacob LaRue (Hunterdon Co, NJ 1741-1816) and Mary Isabella Fortiner (1746-1798). Recd two pen **P:** Both **BLW:** Yes **PH:** No **SS:** B pg 39; M pg 225; O- pen rec Fold3 **BS:** 245

*LARUE, Jabez; b 1 Nov 1768, d 12 Sep 1828 **RU:** Sergeant in Rangers, US Volunteers **CEM:** Green Hill; Clarke; Berryville **GS:** Yes **SP:** No spouse info **VI:** Son of Isaac LaRue (1712-1795) and Phebe Carmen (1725-1804) **P:** No **BLW:** No **PH:** No **SS:** A rec 9315 **BS:** 245

*LAVENDER, William H; b 1792, d 1860 **RU:** Private, 90th VMR Capt Robert L Coleman's Company, Amherst Co, attached to Gen Cock's. 8th VMR **CEM:** Lavender Family; Amherst; Stone Wall Creek Rd, Gidsville **GS:** Unmarked but probably his **SP:** Sarah Pamplin (1803-aft 1860) Recd widow's pen **VI:** No further data **P:** Both **BLW:** Unk **PH:** No **SS:** B pgs 27, 38; M pg 225; O- pen rec Fold3 **BS:** 245

*LAW, Nathaniel; b 5 Jun 1792, d 1859 **RU:** Private 110th VMR, Capt John Pinkard's Company, Franklin Co **CEM:** Law Family; Franklin; off McNeil Rd Rt 718, on private rd .2 mi S in field **GS:** No **SP:** Sally (__-aft 1859) recd pen **VI:** Son of Danial B Law (1762, Halifax Co-Feb 1852) **P:** Widow **BLW:** No **PH:** No **SS:** B pg 77; M pg 225; O- pen files Fold3 **BS:** 245

*LAWRENCE, William; b 1774 d 22 Apr 1849 **RU:** Private, 32nd or 93rd VMR, Capt Christopher Morris's Company, Augusta Co, attached to 4th VMR (Huston and Wooding) **CEM:** Western State Hospital; Staunton City; 103 Valley Center Drive **GS:** No **SP:** No spouse info **VI:** No further data **P:** No **BLW:** No **PH:** No **SS:** A rec 11454; B pg 40 **BS:** 245

*LAWSON, John; b 10 Nov 1769, d 21 Apr 1860 **RU:** Sergeant 58th or 116th VMR, Capt Robert McGill's Co, Rockingham County, attached to McDowell's Flying Camp **CEM:** Thomas Q Wyant; Rockingham; Beldor **GS:** Yes **SP:** Eve Harness (1 Mar 1786-23 Mar 1877) Recd two pen **VI:** Recd two BLW of 80 Acres ea in 1850 and 1855 **P:** Widow **BLW:** Yes **PH:** No **SS:** B pg 182; M pg 226; O- pen files and Serv Index cards, Fold3 **BS:** 245

*LAYTON, Charles Grimes; b 1782, d 11 Nov 1856 **RU:** 2d Lieutenant, 109th VMR, Capt Carter Berkeley's Artillery Co, Middlesex County **CEM:** Hundley Family; Essex; Rose Hill, Center Cross **GS:** No **SP:** Mar (Middlesex Co, 13 Jan 1816) Elizabeth Hundley (2 Apr 1799-23 Mar 1816) **VI:** Son of John Layton and Mildred Sibley. Obtained rank of Captain (BS indicates during War of 1812 but SS not identified) **P:** No **BLW:** No **PH:** No **SS:** B pg 132; L pg 133 **BS:** 245

*LEAVELL, Burwell: b 11 Apr 1775, d 28 Dec 1847 **RU:** Private, 16th VMR, company in Spotsylvania Co **CEM:** Leavell Family; Spotsylvania; 6107 Sweetbrier Dr, Fox Point Sub Division, GPS: 38.22262, -72.54920 **GS:** Yes **SP:** Mar 1) Mary Purvis (1775-10 Sep 1801), 2) Anne Goulder Spindle (1785-1849) **VI:** No further data **P:** No **BLW:** No **PH:** No **SS:** A rec 12830 **BS:** 245

*LEE; John; b 1774, d 6 Oct 1831 **RU:** Sergeant, 53rd VMR, Capt James Dunningham's Artillery Co, Campbell County, attached to Gen Cocke's Detachment **CEM:** Lee Family; Campbell; Lynch Station Quadrant on dirt road 200 yds S of Rt 626 and 75 yds W of Railroad **GS:** Yes **SP:** Tabitha Arnold, (1775-9 Nov 1857) **VI:** No further data **P:** No **BLW:** No **PH:** No **SS:** B pg 53 **BS:** 245 quotes obituary from "The Lynchburg Virginian"

*LEE, John Hite; b 30 Jul 1796, d 30 Jun 1832 **RU:** Lieutenant, US Navy **CEM:** Cedar Grove; Norfolk; GPS: 36.8586, -76.2831 **GS:** Yes **SP:** Mar 1) Elizabeth Prosser (1825), daug of William Prosser, 2) Elizabeth Dabney (Widow) **VI:** Son of Theodoric Lee (1766-1867) **P:** No **BLW:** No **PH:** No **SS:** BD pg 1168 **BS:** 245

*LEE, John R; b 1799, d Feb 1880 **RU:** Private, 90th VMR Capt Cornelius Sale's or Capt Isaac Tinsley's Cos, Amherst County, attached to 8th VMR (Wall) **CEM:** Presbyterian; Lynchburg; 2020 Grace St **GS:** Yes **SP:** Elizabeth (__) (__-aft Feb 1880) **VI:** Recd BLW **P:** No **BLW:** Yes **PH:** No **SS:** B pg 38; M pg 227 **BS:** 245

Key: *Additional veteran entry   **Corrected veteran entry   ***Deleted veteran entry
RU=Rank/Unit   CEM=Cemetery   GS=Gravestone   SP=Spousal Information   VI=Other Veteran Info   P=Pension
BLW=Bounty/Land Warrant   PH=Photo   SS=Service Source   BS=Burial Source   VMR= VA Military Regiment
LNR= Last Known Residence

64

LEE, Robert H; b 1770, Warwick Co; d 1841, Warwick Co **RU: Corporal, 115th VMR, Capt Humphrey H. Wynne's Company, Warwick Co **CEM:** Lee/Davis/Young; James City; Nr Lee Hall on 51 Curtis Dr **GS:** Yes **SP:** 1) Jane Shepherd, 2) Sarah Kirby **VI:** Son of Henry (Harry) Lee. Father of Richard Decatur Lee who built Lee Hall **P:** No **BLW:** No **PH:** No **SS:** B pg 67-8; L pg 450 **BS:** 31; 49

*LEFTWICH, James; b 1797, d 2 Jul 1852 **RU:** Trumpeter, 10th VMR, Capt Walter Otey's Co, Bedford County, attached to Woodford's Squadron **CEM:** Leftwich Family; Bedford; nr Bunker Hill **GS:** Yes **SP:** No spouse info **VI:** A Reverend **P:** No **BLW:** No **PH:** No **SS:** A rec 14646; B pg 43; K pg 189 **BS:** 245

LEFTWICH, John "Jack"; b 1783; d 1833 **RU: Private, 10th VMR, Capt Mark Anthony's Co, Bedford County, attached to 4th VMR **CEM:** Clark Family; Campbell; jct Lawyers & Missionary Manor Rds **GS:** Yes **SP:** Sarah Walton (1784-1829) **VI:** An extensive planter and tobacco broker. Only child of Major General Joel Leftwich and Nancy Turner (1761-1822) **P:** Spouse **BLW:** No **PH:** No **SS:** BD pg 1170; B pg 42 **BS:** 49; 245

*LEFTWICH, Thomas Lumpkin; b 22 Oct 1792, d 13 Jan 1872, **RU:** QM Sergeant, 3rd VMR commanded by Lt Col William Dickinson at Ellicott Mills, MD, 1814 **CEM:** Longwood; Bedford; loc next Rd; NW of jct Rts 221 and Park St; Bedford city **GS:** Yes **SP:** Mar (Bedford City, 22 Dec 1829) Mildred Otey Turner (1804-1889). Recd pen and BLW of 160 acres **VI:** Son of William Leftwich, Jr (1768-1848) of Amherst Co and Frances Otey (1772-1825) **P:** Widow **BLW:** Widow **PH:** No **SS:** B pg 43; M pg 227; O pen files and serv index card, Fold3 **BS:** 245

*LEIGH, William; b c1783, d 19 Jul 1781; **RU:** Captain; 84th VMR, Company commander, Halifax Co, attached to 2nd Corps d'Elite (Green); **CEM:** St. John's Episcopal Church; Halifax; Rt 360 E of jct Rt 501 **GS:** Yes **SP:** Mar (15 Dec 1807) Rebecca Watkins **VI:** Son of Reverend William Leigh and Elizabeth Cary Watkins; Judge in Halifax County. Portrait hangs in County Court room **P:** No **BLW:** No **PH:** No **SS:** A rec 14967; B pg 90; K pg 217 **BS:** 49

LESTER, John; b 1793; d 26 Aug 1870 **RU: Private, Captain Anthony Turner's Co, Richmond City **CEM:** St John's Church; Richmond City; 24th & Broad, Church Hill **GS:** Unk **SP:** Mar 1) (Richmond Co, 21 Nov 1820) by Rev William H Hart to Jane Miller, daug of Peter Miller. Marriage notice in the *Richmond Compiler*, (28 Nov 1820) pg 3, 2) Annie (___) who drew pension **VI:** No further data **P:** Spouse **BLW:** Yes **PH:** No **SS:** B pg 175; K pg 361 **BS:** 63 pg 345; 465

*LESTER, John, Jr: b 7 Mar 1776, Montgomery Co, d 21 Sep 1851, Floyd County **RU:** Sergeant 101st VMR, Capt Tunstall Shelton's Co, Pittsylvania County, attached to 2nd Corps D'Elite **CEM:** Lester Family; Floyd; Rt 738 **GS:** Unk **SP:** Mar (Montgomery Co, 5 Oct 1802) Mary Ann Terry **VI:** Son of John Lester and Catherine Plickenstalver; was Sheriff (1844), and Justice of the Peace (1831), Floyd Co **P:** No **BLW:** No **PH:** No **SS:** B pg 162; K pg 225 **BS:** 245

*LEONARD, Frederick, Jr; b 1 Jan 1776, d 5 Apr 1864 **RU:** Private, 10th or 92d VMR, Capt Willie Jones's Co, Bedford County, attached to 5th VMR (Mason-Preston) **CEM:** Malone; Washington; GPS:36.6325, -82.2242 **GS:** Unk **SP:** Mar (Sullivan Co, TN 1813) Anne Wright (1795-1874) **VI:** Son of Frederick Leonard, Sr. (1761-1845) and Anna Marie Braun/Brown (1762-1855) **P:** Spouse **BLW:** No **PH:** No **SS:** A rec 15937; B pg 198-9; BD pg 1173 **BS:** 245

*LEWARK, Joseph; b 12 Feb 1796, Greensboro, Guilford Co, NC, d 25 Mar 1882 **RU:** Private 70th or 105th VMR, Capt Abram Fulkerson's Co, Washington County, attached to 5th VMR and 7th VMR (Saunders) **CEM:** Sinking Spring; Abington; 136 East Main St **GS:** Yes **SP:** Mar (Guilford Co, 22 Nov 1822) NC Jemma Hutton (1803-1869), daug of Arnold Hutton and Isabell (___) **VI:** Son of John Lewark and Mary E Hutton. Recd BLW 80 acres 1855, Recd pension but later discontinued for disloyalty during and after Confederate War **P:** Yes **BLW:** Yes **PH:** No **SS:** M pg 228; O- pen files, Fold3 **BS:** 245

*LEWIS, William. b 1798, d 15 May 1847 **RU:** Private 83rd VMR, Capt William Cousin's Co, Dinwiddie County, attached to 1st VMR (Byrne) **CEM:** Blandford; Petersburg City; 111 Rochelle Ln **GS:** Unk **SP:** No spousal info **VI:** Son of David Lewis and Mary (___) **P:** No **BLW:** No **PH:** No **SS:** B pg 66; K pg 140 **BS:** 245

Key: *Additional veteran entry   **Corrected veteran entry   ***Deleted veteran entry
RU=Rank/Unit   CEM=Cemetery   GS=Gravestone   SP=Spousal Information   VI=Other Veteran Info   P=Pension
BLW=Bounty/Land Warrant   PH=Photo   SS=Service Source   BS=Burial Source   VMR= VA Military Regiment
LNR= Last Known Residence

*LEWIS, William; b 3 Nov 1772, d 29 Dec 1828 **RU**: Private Capt Ward's Co, U.S. Army Infantry, 29th Regt **CEM**: Central; Rockingham; SE side Main St; bef Rt 1607; Port Republic **GS**: Yes **SP**: Martha Palmer (24 Sep 1788-16 Mar 1860) **VI**: BLW # 124 320 12 **P**: No **BLW**: Yes **PH**: No **SS**: BD pg 1180 **BS**: 245

*LICHLITER, Adam; b 5 Nov 1792, d 29 Nov 1873 **RU**: Private 97th VMR, Capt Walter Hambaugh's Co, Shenandoah County, attached to Flying Camp (McDowell) **CEM**: Lichliter Family; Shenandoah; E side, Fort Valley Rd on Knoll by Blue Silo, S of Detrick **GS**: Yes **SP**: Mar (31 Oct 1816) Catherine Keller (25 Dec 1799-19 Mar 1888), recd two pen 1878 **VI**: Recd two BLW of 80 acres ea, 1850, 1855 **P**: Both **BLW**: Yes **PH**: No **SS**: B pg 184; M pg 228; O- pen files Fold3 **BS**: 245

*LINCOLN, David; b 28 Jun 1781, d 26 Apr 1849 (BLW files) **RU**: 58th VMR, Capt Adam Harnsberger's Co, Rockingham County, attached to 1st VMR (Trueheart) **CEM**: Lacey Springs; Rockingham; Rt 806; Lacey Springs Rd; SE jct with Rt 805 **GS**: Yes **SP**: Mar (6 Dec 1809) Catherine Bright (6 Feb 1783-15 May 1873). Recd BLW of 160 acres (1856) and pen in 1871 **VI**: Son of Jacob Lincoln (Birdsboro, PA, 18 Nov 1751-Linville, Rockingham Co, 20 Feb 1822) and Dorcas Robinson (15 Mar 1763-25 Jun 1840) **P**: Widow **BLW**: Widow **PH**: No **SS**: B pg 181; M pg 229; O- pen files, Fold3 **BS**: 245

*LINEBAUGH, William; b 14 Nov 1796, Frederick Co, MD, d 30 Oct 1881 **RU**: Served in MD Militia **CEM**: Wise Family; Rockingham; 7 mi N of Keezletown on Dancing Bear Lane loc betw Armentrout Path and Mountain Valley Rd **GS**: Yes **SP**: Mar 1) Elizabeth (Betsy) Wise (1808-1850), 2) (Melrose, Rockingham Co, 2 Jan 1852) Fanny E Moore (__-Hagerstown, MD, 13 Apr 1885). Appl for pen rejected **VI**: Recd pen and BLW, LNR Harrisonburg **P**: Yes **BLW**: Yes **PH**: No **SS**: M pg 229; O- pen files Fold3 **BS**: 245

*LINGAN, James McCubban; b 13 May 1751, d 18 Jul 1812, Baltimore, MD **RU**: Brigadier General, MD Militia **CEM**: Arlington National; Arlington; off George Washington Parkway **GS**: Yes **SP**: Janet Henderson (1765- 1832) **VI**: He was killed in a mob riot at the printing office of the *Federalist* where he had taken refuge for being one of those that was supporting the editor of the *Federalist* newspaper's right of free speech in criticizing the US involvement in the war. He was first buried at the Harlem Cem in DC, then reinterred to the Arlington National Cem, section 4, burial site 89A. (Internment Record); (Another source indicates Section 1, lot 2481 WS). He was a Captain in the Revolutionary War, was wounded and was prisoner on a British ship for 3.5 years and received bounty land of 300 acres for his service. In the Revolution he served at Long Island, York Island, and at Fort Washington where he was taken prisoner. After the revolution he was the Collector for the Port of Georgetown and given the title of General **P**: No **BLW**: Yes (Rev War) **PH**: No **SS**: Al pg 9 **BS**: 53 pg 19; 245

*LINDSAY. Henry; b 24 Oct 1789, Spotsylvania Co, d 21 May 1874 Albemarle, Co **RU**: Corporal, 47th or 88th VMR, Capt John Rothwell's Co, Albemarle County, attached to 7th VMR (Gray) **CEM**: Oakwood (AKA Charlottesville); Charlottesville City; jct Elliott Ave & S 1st St **GS**: Yes **SP**: Frances D (__) (__-18 Dec 1893) Recd two pen LNR Charlottesville **VI**: Recd BLW 80 acres, 1855 **P**: Both **BLW**: yes **PH**: No **SS**: B pg 36; M pg 229; O- pen files, Fold3 **BS**: 245

*LIPSCOMB, James; b 1775, d 14 Apr 1875 **RU**: Private, 107th VMR, Capt Jonathan Wamsley's Co, Isaac Booth's Regt, Randolph County, attached to 5th VMR (Mason-Preston) **CEM**: Fairview; Giles; betw Riverside Ave and Fletcher St **GS**: Yes War of 1812 service inscribed **SP**: Mar (Oct 1814) Anna M (__) (__-aft Apr 1875) Recd two pen **VI**: Recd two BLW of 80 acres ea, 1850, 1855 **P**: Both **BLW**: yes **PH**: No **SS**: B pg 173; M pg 229; O- pen files, Fold3 **BS**: 245

*LIPSCOMB, Sterling; b 1 Mar 1788, d 8 Dec 1867 **RU**: Sergeant 87th VMR, Capt Charles H Braxton's Troop of Cavalry, King William County, attached to 1st VMR (Holcombe) **CEM**: Sweet Hall; King William; GPS 37.57004, -76.90371; spur of Rt 624 nr Pamunkey River **GS**: Yes **SP**: Probably Louisa (__) (15 Oct 1816-20 Jun 1885) as adjacent stone in cem **VI**: Obtained rank of captain after war period. Applied for BLW; however not approved probably because not enough service time during war period **P**: No **BLW**: Applied **PH**: No **SS**: B pg 115; Fold3 application for BLW file **BS**: 245

Key: *Additional veteran entry   **Corrected veteran entry   ***Deleted veteran entry
RU=Rank/Unit   CEM=Cemetery   GS=Gravestone   SP=Spousal Information   VI=Other Veteran Info   P=Pension
BLW=Bounty/Land Warrant   PH=Photo   SS=Service Source   BS=Burial Source   VMR= VA Military Regiment
LNR= Last Known Residence

*LITTON, John Whitley Sr; b 11 Nov 1775, Russell Co, d 21 Jan 1853 **RU:** Private, 72nd VMR, Captain William Cromwell's or Captain George Kindrick's Co, Russell County, attached to Major Bradley's Regt **CEM:** Litton Hill; Russell; Elk Garden **GS:** Govt gravestone cites service. **SP:** Sarah Ann Pearl Fullen (17 Jul 1771-16 Oct 1843), daug of James Fullen (1751-1817) and Sarah Whitley (1753-1820) **VI:** Son of Solomon Caleb Litton (1751-1843) and Martha Duncan (1756-1821). Established Litton Hill in Elk Garden **P:** No **BLW:** No **PH:** No **SS:** B pg 183 **BS:** 245

LIVINGSTON, Samuel; b 1772; d 3 Apr 1829 **RU: Sergeant, 7th and 95th VMR, Captain Emmerson's "Portsmouth Light Artillery Blues" Co attached to 6th VMR **CEM:** Trinity Episcopal; Portsmouth; 500 Court St **GS:** Unk **SP:** Margaret (__) (1779-1828) **VI:** No further data **P:** No **BLW:** No **PH:** Yes **SS:** A rec 22144 **BS:** 245

*LOHR, John; b 11 Nov1792, d 19 Apr 1881 **RU:** Private, 93rd VMR Capt Jesse Dold's Troop of Calvary, Augusta County, attached to Major Woodford's Squadron **CEM:** Mount Zion; Augusta; 918 Free Mason Run Rd; Rt 747; Stokesville **GS:** Unk **SP:** Mar (Rockingham Co, 14 Jan 1822) Margaret Martin (__-aft Jun 1882) Recd two widow pen **VI:** Recd two pen and 40 acres BLW **P:** Both **BLW:** Yes **PH:** No **SS:** B pg 39; M pg 230; O- pen files, Fold3 **BS:** 245

*LONG, Frederick; b 1780, d 1826 **RU:** Private, 116th VMR, Captain William Harrison's Company, Rockingham Co, attached to 6th VMR (Coleman) **CEM:** Radar Lutheran Church; Rockingham; Timberville **GS:** Yes **SP:** No spouse info **VI:** No further data **P:** No **BLW:** No **PH:** No **SS:** B pg 181; K pg 51 **BS:** 31; 245

*LONG, William: b 1777, d 1861 **RU:** Private, **RU:** Private, 78th VMR, Capt Timothy Dalton's or Capt Lewis Hail's (Hale's) Co, Grayson County, attached to 4th VMR **CEM:** Long-Phipps Family; Grayson; GPS: 36,57641, -81.20306; West of Rt US 21 and S of Peach Bottom Creek, Long Gap **GS:** Unk **SP:** Elizabeth "Betsey" Howell (1780-Jan 1870) **VI:** Son of William Long (1755-1821) and Catherine "Caty" (__) (__-1828) **P:** No **BLW:** No **PH:** No **SS:** A rec 24163; B pg 86 **BS:** 245

*LOVE, Chappell; b 1783, d 1843 **RU:** Private, 73rd VMR, Captain Levi Clay's Co, Lunenburg County, attached to 6th VMR (Sharp) **CEM:** Love-Young Family; Mecklenburg; Baskerville **GS:** Unk **SP:** No spouse info **VI:** No further data **P:** Yes **BLW:** No **PH:** No **SS:** B pg 125 **BS:** 245

*LOVE, Henry Hicks; b 4 Jun 1787, d 5 Jul 1871 **RU:** Private, 54th VMR, Captain James R. Nimmo's Co, Norfolk Borough, attached to 5th VMR (Mason-Preston) **CEM:** Love Family; Lunenburg; Victoria **GS:** Unk **SP:** Mary G. (__) (1802-1857) **VI:** No further data **P:** Spouse **BLW:** No **PH:** No **SS:** A rec 25801; B pg 145; BD pg 1198 **BS:** 245

*LOVE, Richard H; b 1786, d 1832 **RU:** Cornet, 60th VMR, Captain George Graham's Troop of Cavalry "Fairfax Light Dragoons", Fairfax County, attached to 1st Corps DeElite (Randolph) **CEM:** Union; Loudoun; Leesburg **GS:** Unk **SP:** No spouse info **VI:** Cem plat A; Lot 128; Site 3 **P:** No **BLW:** No **PH:** No **SS:** A rec 25876; B pg 71; 240 **BS:** 245

*LOVETT/LOVITT, Reuben, b 18 Sep 1765, d 11 Aug 1819 **RU:** Lieutenant, 20th VMR, Captain James Harrison's, Company, Princess Anne Co **CEM:** Lovitt Family; VA Beach; Court House 3445 Princess Anne Rd **GS:** Unk **SP:** Elizabeth (__) (12 Jan 1773-7 Sep 1846) **VI:** No further data **P:** No **BLW:** No **PH:** No **SS:** B pg 164; K pg 403 **BS:** 245

*LUNDY, Amos; b 1792, d 2 Jan 1859 **RU:** Private, 78th VMR, Captain James Anderson's Company. Grayson Co **CEM:** Nuckolls; Grayson; GPS: 36.63551,-80.95944 **GS:** No **SP:** Mary Bedsaul (1801-1885), daug of Elisha Bedsaul and Margaret Edwards **VI:** Son of John Lundy (1751-1831) and Rebecca Silverthorn (1754-1839) **P:** Yes, widow Mary **BLW:** Yes **PH:** No **SS:** A rec 28656; B pg 86; BD pg 1206; M pg 232 **BS:** 245

*LYNCH (LINCH), James, b 1782, d aft 1825 **RU:** Private 63rd VMR, Capt Josiah Penick's Co, Prince Edward County, attached to 7th VMR (Gray) **CEM:** Lynch Family: Bedford; High Point Rd Subdivision NW side, the older of two cem there **GS:** Unk **SP:** Mar (Bedford city, 1 Dec 1806) Elizabeth Ashwell (1786-1870), daug of John Ashwell and Eleanor Stump **VI:** No further info **P:** Yes **BLW:** No **PH:** No **SS:** K pg 346; M pg 229; O- pen files Fold3 **BS:** 245

Key: *Additional veteran entry   **Corrected veteran entry   ***Deleted veteran entry
RU=Rank/Unit   CEM=Cemetery   GS=Gravestone   SP=Spousal Information   VI=Other Veteran Info   P=Pension
BLW=Bounty/Land Warrant   PH=Photo   SS=Service Source   BS=Burial Source   VMR= VA Military Regiment
LNR= Last Known Residence

67

*LYNCH, James; b 1780, d 13 Aug 1825 (bur) **RU**: Private, 23rd VMR, Capt Benjamin Graves's Co, Chesterfield County, attached to 4th VMR (McDowell, Koontz, Chilton) **CEM**: Shockoe Hill; Richmond; 100 Hospital St **GS**: Unk **SP**: Mar (nr Petersburg, 28 Mar 1809) Jane Anthony **VI**: A person this name (perhaps him) also served in Chesterfield Co in Capt David Weisiger's Co and Capt Alexander Gibb's Co **P**: Yes **BLW**: No **PH**: No **SS**: B pgs 60, 61; L pgs 354; 823; O- pen files, Fold3 **BS**: 245

*LYON, Elisha; b 1773, d 4 Apr 1842 **RU**: Private 75th VMR, Capt William Pepper's Co, Montgomery County, attached to 4th VMR (McDowell, Koontz, Chilton) **CEM**: Whitlow Family; Patrick; 4 mi NE Woolmine off Charity Rd, Charity **GS**: Yes **SP**: Mar (2 Aug 1792) Rhoda Hatcher (__-aft 1855). Recd BLW 80 acres 1850 and 80 acres 1855 **VI**: No further data **P**: No **BLW**: Widow- two **PH**: No **SS**: B pg 138; M pg 232; O- pen files; Fold3 **BS**: 245

*MAIDEN, William; b 1 Oct 1795, d 29 Dec 1875 **RU**: Private 58th or 116th VMR, Capt Joseph Larew's Co, Rockingham County, attached to 6th VMR (Dickinson, Scott, Coleman) **CEM**: Maiden Family; Rockingham; E of jcts Rts 628 & 626 nr Swift Run Gap on Rt 33 **GS**: Yes-crude stone and inscription **SP**: Mar (Rockingham Co, 12 Mar 1816) Sarah Harris Gardner (Albemarle Co, 24 Apr 1798-16 Oct 1880). Recd two widow pen, daug of John Gardner & Anna Epperson **VI**: Son of James Maiden, Sr (__-1795) and Theodosia Lee (1753-1847) Recd two pen and applied for BLW **P**: Both **BLW**: Applied **PH**: No **SS**: B pg 27; O- pen files, Fold3 **BS**: 245

MANKIN, Charles; b 1778; d 10 Nov 1840 **RU: Sergeant, 1st DC Regiment of Militia **CEM**: Methodist Protestant; Alexandria; Wilkes St **GS**: Yes **SP**: Mar (Alexandria, 29 Jul 1819) Elizabeth Merlerouz **VI**: No further data **P**: Widow applied for pen, indicating he was a Lt, but was denied due to lack of proof of service, probably because she indicated he was in a VA unit and not a DC unit **BLW**: No **PH**: No **SS**: M pg 237 **BS**: 31; 245

*MANKIN, David; b bef 1780, d 4 Dec 1824 **RU**: Captain, commanded a company in the 1st DC Regiment commanded by Lt Col Adam Lynn **CEM**: Methodist Protestant; Alexandria; Wilkes St **GS**: No **SP**: Ann (__) (__-c1818) **VI**: No further data **P**: No **BLW**: No **PH**: No **SS**: O- pen rec for Jacob Mellon a soldier in his company; fold3 **BS**: 31

*MARSHALL, Henry; b 2 Mar 1795, d 19 Aug 1877 **RU**: Private, 116th VMR Capt Thomas Hopkin's Co, Rockingham County, attached to 6th VMR (Coleman) **CEM**: Miller Marshall; Shenandoah; nr Narrow Passage, 3 mi SW Woodstock **GS**: Unk **SP**: Julia Lambert (1 Nov 1794-17 Sep 1883). Recd two pen **VI**: Recd two pen and BLW, 8 May 1851 **P**: Both **BLW**: Yes **PH**: No **SS**: B pg 182; O- pen files, Fold3 **BS**: 245

*MARSHALL, James Douglas; b 12 Feb 1786, d 16 Jun 1877 **RU**: Private 64th VMR, Capt Joseph Larew's Co, Henry County, attached to 6th VMR (Dickinson, Scott, Coleman) **CEM**: Horsepasture Christian Church; Henry; 1146 Horsepasture Price Rd (Rt 692) **GS**: Yes **SP**: Mar (Henry Co, 22 May 1820) Susanna C. Weaver (1790-1883) Recd pen **VI**: Son of Reuben Marshall of Scotland and Mollie Douglas of Ireland. Applied for pen but rejected **P**: Widow **BLW**: No **PH**: No **SS**: B pg 27; O- pen files, Fold3 **BS**: 245

*MARTIN, Alexander A; b 1788, d 8 Jan 1859 **RU**: Sergeant, 54th VMR, in a company, Norfolk Borough **CEM**: Cedar Grove; Portsmouth; Effington St & Fort Lane **GS**: Yes **SP**: No spouse info **VI**: Gravestone located block 4th AE, lot 22 **P**: No **BLW**: No **PH**: No **SS**: A rec 233; B pg 145 **BS**: 245

*MARTIN, Joseph; b 22 Sep 1785, d 3 Nov 1859 **RU**: Lt Col, 64th VMR, Entered serv as a private in 64th from Henry Co (Note: *Https//commons.wikimedia.org* fr their sources indicate he served as colonel in command of a regiment of militia during war period, however service not found in Butler's "Guide" (SS: B), county records or in NARA records) **CEM**: Martin Family; Henry; nr jct Rts 57 & 710; Leatherwood **GS**: Yes **SP**: Sarah Hughes (1792-1883), daug of Achelous Hughes of Patrick Co **VI**: Served VA House of Delegates and Senate; and member Constitutional Convention (1829-1830) **P**: No **BLW**: No **PH**: No **SS**: A rec 1053; *https//commons.wikimedia.org* **BS**: 245

Key: *Additional veteran entry   **Corrected veteran entry   ***Deleted veteran entry
RU=Rank/Unit   CEM=Cemetery   GS=Gravestone   SP=Spousal Information   VI=Other Veteran Info   P=Pension
BLW=Bounty/Land Warrant   PH=Photo   SS=Service Source   BS=Burial Source   VMR= VA Military Regiment
LNR= Last Known Residence

*MARTIN, Peter; b 1779, d 9 Jun 1847 **RU**: Private, 39th VMR, Captain Alexander Taylor's "Petersburg Republican" Light Infantry company, Petersburg City, attached to 2d VMR (Sharp) **CEM**: Blandford; Petersburg City; 319 So Crater Rd **GS**: Unk **SP**: No spousal info **VI**: Son of Peter Martin **P**: No **BLW**: No **PH**: No **SS**: A rec 1207; B pg 160 **BS**: 245

*MARTIN, William C; b 1765, d 1809 **RU**: Private 42d or 101st VMR, Capt Nathaniel Terry's Co, Pittsylvania County, attached to 4th VMR (Greenhill) **CEM**: Old Hickey-Martin; Henry; GPS: 36.81719,-79.97507; Oak Level **GS**: No **SP**: Sarah Dodd (Amelia Co, 30 May 1767-21 Feb 1851). She applied for widows pension **VI**: Probably son of William Martin and Rachel (__) a Rev War soldier **P**: Widow applied **BLW**: No **PH**: No **SS**: B pg 162; O Fold3 Pen files; M pg 239 **BS**: 245

*MASON, Daniel; b 8 Aug 1792, Rappahannock Co, d by 20 Apr 1845 (burial) **RU**: Private; 121st VMR, Capt Andrew Lewis's Co, Botetourt County, attached to 4th VMR (McDowell, Koontz and Chilton) **CEM**: Presbyterian; Lynchburg City; 2020 Grace St **GS**: Yes **SP**: Sarah Porter (29 Jul 1793-7 Jan 1869) **VI**: GS loc Grave 3; Lot 5; Range 5. He was a doctor **P**: No **BLW**: No **PH**: No **SS**: A rec 2142; B pg 46; K pg 24 **BS**: 245

*MASON, George; b 1797, d 1870 **RU**: Ensign, 68th VMR. Capt John H. Smith's Co, James City County **CEM**: Mason Family (Gunston Hall); Fairfax; 10709 Gunston Hall Rd; Lorton **GS**: Yes **SP**: Sally E. (__) **VI**: No further data **P**: Spouse **BLW**: No **SS**: BD pg 1255; M pg 239 **BS**: 245

*MASON, John; b 4 Apr 1766, Clermont Woods, Fairfax Co, d 19 Mar 1849 **RU**: Brigadier Gen DC Militia. Was appointed Commissioner of Prisoners War of 1812 **CEM**: Christ Church Episcopal; Alexandria; jct Wilkes & Hamilton **GS**: Yes **SP**: Mar (Annapolis, MD, 10 Feb 1796) Anna Maria Murray, daug of James Murray and Sarah Ennalls Maynadier **VI**: Merchant in Georgetown and large landowner **P**: No **BLW**: No **PH**: No **SS**: *https//en.wikimedia.org* **BS**: 245

MASON, Jonathan; b 20 Jan 1793, d 17 Nov 1860 **RU: Private, 31st VMR, Frederick Co **CEM**: Hironimus; Frederick; Old Mill Ln, Whitacre **GS**: Yes **SP**: Mar (Frederick Co, 20 Mar 1822); (returned by Joseph Dalby), Helen Braithwaite **VI**: No further data **P**: No **BLW**: No **PH**: No **SS**: A rec 2354 **BS**: 56 pg 264

*MASSIE, Josiah; b 1783, Fauquier Co, d 1 May 1852 **RU**: Lieutenant, VMR Company of Frederick Co, attached to 4th VMR (Boyd) **CEM**: Mt Hebron; Winchester; 305 E Boscawen St **GS**: Yes **SP**: Elizabeth Ball, daug of William and Elizabeth (Riley) Ball **VI**: Rank of Major on GS; bur Centenary Reformed UCC portion of Mt Hebron Cem **P**: No **BLW**: No **PH**: Yes **SS**: A rec 2806 **BS**: 245

*MASSIE, William Wright; b 3 Mar 1795, Frederick Co, d 29 Jul 1862 **RU**: Private 90th VMR, Capt William Coleman's Co, Nelson County, attached to Cocke's Detachment **CEM**: Level Green Estate; Nelson; 524 Level Green Rd, Rt 679, Roseland **GS**: Yes **SP**: Mar 1) Sarah Tate Steptoe (1796-1828), 2) (29 Oct 1832), her sister Frances M Steptoe (__-14 May 1895) who recd widows pen **VI**: Recd BLW and pen **P**: Both **BLW**: Yes **PH**: No **SS**: B pg 38; M pg 239; O- pen files, Fold3 **BS**: 245

*MATHEWS, John P: b 8 Aug 1790, d 4 Dec 1850 **RU**: Private, 76th, or 104th, or 118th VMR Capt James Hurry's Co, Monongalia County, attached to 6th VMR (Dickinson, Scott, Coleman) **CEM**: East End; Wythe; E Goodwin Lane, Wytheville **GS**: Yes **SP**: No spousal info **VI**: Recd pen **P**: Yes **BLW**: No **PH**: No **SS**: B pg 136; M pg 239; O- pen files, Fold3 **BS**: 245

*MATHEWS, William; b 21Dec 1786, Norfolk Co, d 18 May 1880 **RU**: Private, 95th VMR, Capt Arthur Emmerson's "Portsmouth's Light Infantry Blues" Artillery Co, Norfolk County attached to 3rd VMR (Boykin) **CEM**: Walker's Presbyterian Church; Appomattox; Hixburg **GS**: Yes **SP**: Mar (28 Jul 1814) Margaret Susan Owens (__-11 May 1879) who recd pen **VI**: Gr st shows rank of Captain obtained after war. Recd BLW 40 acres **P**: Widow **BLW**: Yes **PH**: No **SS**: B pg 148; M pg 239; O- pen files, Fold3 **BS**: 245

*MAURICE, James; b 1770 DL, d 2 Jan 1831 **RU**: Major, 54th VMR, Staff Officer of Regiment, Norfolk Borough **CEM**: Basilica of Saint Mary Churchyard; Norfolk; 232 Chapel St **GS**: Yes **SP**: No spouse info **VI**: No further data **P**: No **BLW**: No **PH**: No **SS**: B pg 145 **BS**: 245

Key: *Additional veteran entry   **Corrected veteran entry   ***Deleted veteran entry
RU=Rank/Unit   CEM=Cemetery   GS=Gravestone   SP=Spousal Information   VI=Other Veteran Info   P=Pension
BLW=Bounty/Land Warrant   PH=Photo   SS=Service Source   BS=Burial Source   VMR= VA Military Regiment
LNR= Last Known Residence

69

*MAUZY, Joseph; b 14 Aug 1779, Fauquier Co, d 20 Dec 1863 **RU**: Captain, 58th VMR, commanded a company in Rockingham County, attached to 4th VMR (McDowell, Koontz, Chilton) and Maj Washington's Command **CEM**: Mount Olivet; Rockingham; Rt 843, Cemetery Rd; McGaheysville **GS**: Yes **SP**: Mar (12 Sep 1805) Christina Kisling (5 Jun 1783-3 Jul 1874), daug of Jacob Kisling and Anna Barbara Baer. Recd pen Feb 1781 **VI**: County supervisor, (1835-1849); and county surveyor; editor of Rockingham Register, recd title of Colonel; recd BLW 120 acres 1855 **P**: Widow **BLW**: Yes **PH**: No **SS**: B pg 182; M pg 240; O-pen files, Fold3 **BS**: 245

MAY, Andrew (or James Andrew); b 1794, probably Shenandoah Co, d 1830 **RU: Private, Captain John W Bayliss's Artillery Co, Shenandoah County, attached to Battalion of Artillery **CEM**: Smith; Rockingham; behind Bennett's Run School House; Bergton **GS**: Unk **SP**: Mar (18 Jul 1825) Margaret (Peggy) Smith (21 Apr 1799-aft Apr 1878), daug of Lorenzo F Smith and Christina Agatha Sonifrank. Recd two pen and BLW, (1855) of 80 acres **VI**: Son of Johannes George May (Lancaster Co, PA, 1 Oct 1758-1815) and Martha Magdalene Houghman (1755-1816) **P**: Widow **BLW**: Widow **PH**: No **SS**: B pg 184; M pg 240; O- pen files, Fold3 **BS**: 245

*MAY, James; b 7 Nov 1785, d 20 Oct 1863 **RU**: Private, 58th or 116th VMR, a company of Rockingham County, attached to 6th VMR Coleman **CEM**: May; Rockingham; May Creek Lane, Criders, Bergton **GS**: Unk **SP**: Mar 1) Mary Tusing (1790-1837), 2) Sarah Shaver (1812-1878) **VI**: Son of Johannes George May (Lancaster Co, PA, 1 Oct 1758-1815) and Martha Magdalene Houghman (1755-1816) **P**: No **BLW**: No **PH**: No **SS**: A rec 5617; B pg 181 **BS**: 245

*MAYO, Philip; b 2 Nov 1793, d 21 Mar 1857 RU: Private, 19th VMR, Capt William Murphy, "Light Infantry Blues" Co, Richmond City, attached to 1st Corps De 'Lite (Randolph) **CEM**: Shockoe Hill; Richmond City; 100 Hospital St **GS**: Yes **SP**: Mar (Warren Co, NC, 6 Feb 1823) Caroline Elizabeth Atkinson (21 Jun 1806-15 Feb 1896), daug of Roger Atkinson & Agnes Poythress. Recd pen **VI**: Recd BLW 120 acres. Brother to Joseph Mayor of Richmond. Was Court Clerk, Henrico Co **P**: Widow **BLW**: Yes **PH**: No **SS**: B pg 175; M pg 240; O- pen files, Fold3 **BS**: 245

*McBRIDE, Isaiah; b 12 May 1777, d 30 Jun 1830 **RU**: Captain, 8th VMR, commanded a company in Rockbridge Co, attached to 5th VMR (McDowell) **CEM**: New Providence Presbyterian Church; Rockbridge; 1208 New Providence Rd; Rapine **GS**: Yes **SP**: Mar (17 Aug 1816) Annie J McChesney (9 Feb 1799-12 Dec 1872), daug of Robert McChesney & Elizabeth Johnston **VI**: Son of John McBride. An author of a book of Scottish poems **P**: No **BLW**: No **PH**: No **SS**: B pg 179 **BS**: 245

*McCANDLISH, Thomas C; b 1773, d 11 Aug 1832 **RU**: Captain, 54th, Company Commander of Troop of Cavalry, Norfolk Borough **CEM**: Cedar Grove; Norfolk; 238 E Princess Anne Rd **GS**: Yes **SP**: No spouse info **VI**: No further data **P**: No **BLW**: No **PH**: No **SS**: B pg 145 **BS**: 245

*McCANN, James; b 20 Aug 1790, d 13 May 1848 **RU**: Private; 31st VMR, Capt William Morris's Company, Frederick Co, attached to Battalion of Artillery **CEM**: Union Church; Shenandoah; Mt Jackson **GS**: Yes **SP**: Sarah S (__) (1802-1876) **VI**: No further data **P**: No **BLW**: No **PH**: No **SS**: A rec 7880; B pg 79; S pg 106 **BS**: 245

*McCARTY, John; b 1795, d 13 Sep 1852 **RU**: Private 87th VMR, Major Thomas Hill's Detachment, King William Co. **CEM**: Hollywood Cemetery; Richmond; 412 S Cherry St **GS**: No **SP**: Ann Lucinda (__) (25 Jul 1798-9 Jul 1854) **VI**: No further data **P**: No **BLW**: No **PH**: Yes **SS**: B pg 115; K pg 48 **BS**: 31

*McCARTY, William Mason; b 1792, d 1863 **RU**: Sergeant, 41st VMR (Branham's), in a company of Richmond Co **CEM**: Shockoe Hill; Richmond City; 100 Hospital St **GS**: Yes **SP**: Mar 1) Emily Rutger Mason (1796-1835), 2) (19 Jun 1838) Mary B Burwell **VI**: Son of Dennis McCarty and Felix(__). Governor of Florida Territory, 1827 and US Representative (1840-1841) **P**: Widow **BLW**: No **PH**: Yes **SS**: A rec 8343; M pg 232; O-pen files, Fold3 **BS**: 31; 245

McCLUNG, Henry; b 22 Apr 1773, d 3 Apr 1846 **RU**: Captain, 8th VMR, commanded an artillery Company, Rockbridge Co, attached to a Battalion of Artillery **CEM**: Thornrose; Staunton City; 1041 West Beverley St **GS**: Yes **SP**: Elizabeth (_) (17 May 1782-9 Mar 1869) **VI**: No further data **P**: No **BLW**: No **PH**: **SS**: B pg 180 **BS**: 245

Key: *Additional veteran entry **Corrected veteran entry ***Deleted veteran entry
RU=Rank/Unit   CEM=Cemetery   GS=Gravestone   SP=Spousal Information   VI=Other Veteran Info   P=Pension
BLW=Bounty/Land Warrant   PH=Photo   SS=Service Source   BS=Burial Source   VMR= VA Military Regiment
LNR= Last Known Residence

70

*McCLUNG, James; b 7 Dec 1767, d 14 Jan 1817 **RU:** Captain, 79th VMR, Company Commander, Greenbrier Co, attached to 4th VMR (Boyd's) **CEM:** Timber Ridge Presbyterian Churchyard; Rockbridge; Timber Ridge **GS:** Yes **SP:** Elizabeth McPheeters, (30 Jun 1770-13 Apr 1813), daug of John McPheeters and Elizabeth Campbell **VI:** Son of Henry McClung (1739-1818) and Esther Caruthers (1744-1818) **P:** No **BLW:** No **PH:** No **SS:** B pg 87 **BS:** 245

*McCOMB, William; b 26 May 1794, d 22 Jul 1886 **RU:** Private, Capt Jesse Dold's Troop of Cavalry, 93rd VMR attached to Major Woodford's Squadron **CEM:** Tinkling Spring Presbyterian Church; Augusta; Fishersville **GS:** Yes **SP:** Sarah Lewis Hughes (29 Oct 1812-15 Mar 1892) **VI:** Son of James McComb (1765-1846) and Susannah Henderson McComb (1769-1848) **P:** Self and Wife (#WC-34567 #SC-20458) **BLW:** No **PH:** No **SS:** BD pg 1221; B pg 39; M pg 233 **BS:** 245

*McCUE, John; b 17 Feb 1793, d 18 May 1862 **RU:** Private in 5th VMR (McDowell's) and in many Augusta Co companies attached to regt **CEM:** Tinkling Spring Presbyterian Church; Augusta; Fishersville, 11 mi N of Staunton **GS:** Yes **SP:** No spouse info **VI:** No further data **P:** No **BLW:** No **PH:** No **SS:** A rec 12232; B pg 40-41 **BS:** 245

*McCUE, Moses; b 23 Dec 1769, d 28 Apr 1847 **RU:** Major, Staff Officer, Caroline Co **CEM:** Tinkling Springs Presbyterian Church; Augusta; Fishersville; 11 mi N of Staunton **GS:** Yes **SP:** Sarah (__) (5 Nov 1773-6 Apr 1856) **VI:** No further data **P:** No **BLW:** No **PH:** No **SS:** B pg 39 **BS:** 245

*McCLURE, Hugh; b 29 Oct 1796, d 7 Oct 1876 **RU:** Private, Capt John Link's Co, attached to 2nd Corps D'Elite (Green's) **CEM:** Tinkling Spring Presbyterian Church; Augusta; Fishersville 11 mi N of Staunton **GS:** Yes **SP:** No spouse info **VI:** No further data **P:** No **BLW:** No **PH:** No **SS:** B pg 4 **BS:** 245

*McCUNE, James; b 1791, d 22 Apr 1845 **RU:** Ensign, 32nd and 93rd VMR, Capt Christopher Morris's Co, Augusta County, attached to 4th VMR (Huston-Wooding) **CEM:** Tinkling Spring Presbyterian Church; Augusta; Fishersville 11 mi N of Staunton **GS:** Yes **SP:** No spouse info **VI:** No further data **P:** No **BLW:** No **PH:** No **SS:** A rec 12698; B pg 40 **BS:** 245

*McDOWELL, John; b 1 Aug 1770, d 20 Jan 1849 **RU:** Private, 32nd or 93rd VMR, Augusta Co, attached to 5th VMR (McDowell's) **CEM:** Trinity Episcopal Churchyard; Staunton City; 214 W Beverly St **GS:** Yes **SP:** No spouse info **VI:** No further data **P:** No **BLW:** No **PH:** No **SS:** A rec 14284 **BS:** 245

*McDOWELL, William A; b 19 Nov 1798, d 15 Sep 1822 **RU:** Private, 32nd or 93rd VMR, in a company, Augusta County, attached to 5th VMR (McDowell) **CEM:** Trinity Episcopal Churchyard; Staunton City; 214 W Beverly St **GS:** Yes **SP:** No spouse info **VI:** No further data **P:** No **BLW:** No **PH:** No **SS:** A rec 14358 **BS:** 245

*McFARLANE, Alexander; b 11 Oct 1790, d 14 Jan 1874 **RU:** Sergeant; 72d VMR, Capt John Hamon's Co, Russell Co, attached to 5th VMR (Mason-Preston) **CEM:** McFarlane; Russell Co; GPS 36.96330,-81.95170, W in field at end of McFarlane Ln **GS:** Unk **SP:** No spousal info **VI:** Son of James McFarlane (Augusta Co, 1767-24 Mar 1830) and Jane Price (1768-1820) **P:** No **BLW:** No **PH:** No **SS:** A rec 15083; B pg 183 **BS:** 245

*McFARLANE, James; b 1767, Augusta Co, d 24 Mar 1830 **RU:** Lt Col commissioned 1 Jan 1807, commanded 72d VMR Russell Co **CEM:** McFarlane; Russell; GPS: 36.96331, -81.95170 loc west of the end of McFarlane Ln vic field, Lebanon **GS:** No **SP:** Jane Price (1768-29 Sep 1820) **VI:** Son of William McFarland (Augusta Co, 1732-1801 Tazewell Co) and Elizabeth Gibson (1735- 1782) **P:** No **BLW:** No **PH:** No **SS:** B pg 72 **BS:** 245

McGRUDER, Sublet; b 10 Oct 1781; d 3 Jul 1853 **RU: 1st Sergeant, 19th VMR (Ambler), Capt George Booker's Co, Richmond City **CEM:** Hollywood; Richmond City; 412 S Cherry St, Sec K, lot 44 **GS:** Yes **SP:** Mar 1) (Richmond City, 20 Nov 1817) Mary M Woolfolk (12 Jul 1799-Richmond Co-18 Feb 1833), 2) Ann Hite. From Bible of Sublet McGruder and Bible of Edwin Wrotham, Lib of VA **VI:** Partner in firm of McGruder & Wrotham, dry goods & groceries in Richmond **P:** No **BLW:** Yes **PH:** Yes **SS:** A rec 16850; BD pg 1233; B pg 174 **BS:** 237; 49

Key: *Additional veteran entry   **Corrected veteran entry   ***Deleted veteran entry
RU=Rank/Unit   CEM=Cemetery   GS=Gravestone   SP=Spousal Information   VI=Other Veteran Info   P=Pension
BLW=Bounty/Land Warrant   PH=Photo   SS=Service Source   BS=Burial Source   VMR= VA Military Regiment
LNR= Last Known Residence

71

*McINTIRE, Charles; b 7 Oct 1783, d 3 May 1858 **RU:** Private, Frederick Co company attached to 4th VMR (Beatty) **CEM:** McIntire Family; Frederick; 4 mi fr Hinkle nr Morgan Co, WVA line **GS:** Yes **SP:** Catherine (__) (21 May 1783-19 May 1857) **VI:** No further data **P:** No **BLW:** No **PH:** No **SS:** A rec 17412 **BS:** 56 pg 255

*McINTIRE, William; b 27 Sep 1788, d 15 Jan 1863 **RU:** Private, 119th VMR Harrison Co (now WVA) **CEM:** McIntire Family; Frederick; 4 mi fr Hinkle; nr Morgan County, WVA line **GS:** Yes **SP:** No spouse info **VI:** No further data **P:** No **BLW:** No **PH:** No **SS:** A rec 17519 **BS:** 56 pg 255

*McKEE, John Telford; b 14 Apr 1783, d 30 Apr 1857 **RU:** 8th VMR, Capt Daniel Hoffman's Co, Rockbridge County, attached to McDowell's Flying Camp **CEM:** McKee Family; Rockbridge; Rt 631 **GS:** Yes **SP:** Agnes Hanna (1779-1847) **VI:** Son of James McKee (1752-1832) and Jane Telford, (1754-1800) **P:** No **BLW:** No **PH:** No **SS:** B pg 179; K pg 15 **BS:** 245

*McKEE, Joseph; b 1777, d 1852 **RU:** Private, Frederick Co company attached to 1st VMR (Taylor) **CEM:** Bethel Church; Frederick; Bethel Church Rd (Rt 610), Gore **GS:** Yes **SP:** Mar (Frederick Co, 5 Feb 1799) Elizabeth Reid (1780-(__), daug of Jeremiah Reid and Elizabeth McMahon **VI:** Son of Robert and Elizabeth (Ferguson) McKee **P:** No **BLW:** No **PH:** No **SS:** A rec 18060 **BS:** 56 pg 257; 245

*McKINNEY, James; b 1787, NC, d 7 May 1859 **RU:** Private, 70th, 72th, 94th or 105th VMR, in a company from Russell, Washington or Lee Counties. Note; Scott Co. evolved from those counties in 1814, attached to 5th VMR (Mason, Preston) **CEM:** McKinney-Carter Family; Scott; Duffield **GS:** Yes **SP:** Nancy Martin (1792-__) **VI:** Son of John McKinney and Sarah Harris **P:** No **BLW:** No **PH:** No **SS:** B pgs 118; 183; 198; 199 **BS:** 245

*McLAUGHLIN, Jno (John or Jonathan), b 31 Dec 1770, d 21 Nov 1833 **RU:** Lieutenant, 116th VMR, Capt Thomas Hopkin's Co, Rockingham County, attached to McDowell's Flying Camp **CEM:** Topping Castle; Caroline; GPS: 37.92332,-77.54283; nr S end unmarked Rd, E of jct Rts 700 and 658 **GS:** Yes **SP:** Mary Overton Minor (1775-1848) **VI:** Son of John McLaughlin and Sarah Mackie of Scotland **P:** No **BLW:** No **PH:** No **SS:** A rec 20075; B pg 182; K pg 19 **BS:** 245

MEREDITH, Reuben; b 24 May 1791, d 20 Aug 1840 **RU: Private, 19th VMR, Capt Wilson Bryant's Co, Richmond City (Note: BS 245 indicates, he was a Surgeon but not in NARA records) **CEM:** Hollywood; Richmond City; 412 S Cherry St **GS:** Yes **SP:** Mar 1) Elizabeth Anderson, 2) Mary Letitia Clarkson (1801-1880) who recd two widow pen **VI:** Son of Elisha Meredith and Anne Clopton; graduate Univ PA Medical School; served in VA Legislature **P:** Widow **BLW:** No **PH:** No **SS:** B pgs 100, 174; O pen files Fold3 **BS:** 237; 245

*MESMER, Jacob; b 3 Oct 1787, d 9 Jan 1855 **RU:** Sergeant, 31st VMR, Capt Michael Coyle's Co, Frederick Co, attached to 1st VMR (Taylor) **CEM:** Mt Hebron; Winchester City; 305 E Boscawen St **GS:** Unk, buried old lot 272, Grave 4 **SP:** Mar (Frederick Co, 29 Apr 1819) Ann Smith Jackson; recd widow's pen **VI:** No further data **P:** Widow **BLW:** No **PH:** No **SS:** B pg 79; M pg 241; O- pen files Fold3 **BS:** 245

*MEYERS, Martin, Sr; b 13 Jan 1771, MD or GER, d 23 Jan 1851 **RU:** Private, MD Militia, Capt James F. Houston Co **CEM:** Middleton; Frederick; 5 mi W of Middleton on a hill **GS:** Yes **SP:** Catherine (__) (8 Aug 1799-15 Aug 1856 **VI:** Minister German Baptist Church for 30 years **P:** Yes **BLW:** No **PH:** No **SS:** BD pg 1337 **BS:** 151; 277

*MIDDLETON, John S; b 4 Jul 1788, d 15 Dec 1867 **RU:** Captain 37th VMR, Commanded company in Northumberland Co Militia **CEM:** Elmwood; Norfolk City; 238 E. Princess Anne Rd **GS:** Yes **SP:** No spousal info **VI:** Styled "Captain" on grave stone **P:** No **BLW:** No **PH:** No **SS:** B pg 153; M pg 241 **BS:** 245

MILLAN, John; b 11 Sep 1783, Fairfax Co; d 18 Feb 1858 **RU: Captain, 60th VMR (Minor), company commander, Fairfax Co **CEM:** Millan Family; Fairfax; 4600 West Ox Rd, Fairfax **GS:** Yes **SP:** Elizabeth Reid (23 Sep 1781-11 Jul 187) **VI:** Son of Thomas Millan (1750-1828) **P:** No **BLW:** No **PH:** No **SS:** A rec 29400 **BS:** 93

---

*MILLER, Abraham; b 16 Jan 1788, Shenandoah Co, d 11 Apr 1847 **RU**: Private 13th or 97th VMR, served in either Capt Samuel Hopewell's, Capt Jacob Fry's, or Capt John Sloan's Co, Shenandoah County, attached to 4th VMR (Huston & Wooding) **CEM**: Sangerville Church; Augusta; 26 Vance Rd (Rt 755) **GS**: Yes **SP**: Mar 1) unk 2) (1814) Christina Arnold (Burlington, Mineral Co, WVA Nov 1793-11 Apr 1873) **VI**: Son of Jacob Miller (Washington Co, MD, 2 Oct 1748-11 Jul 1815) & Anna Martha Wine (1753-1795) **P**: No **BLW**: No **PH**: No **SS**: B pg 185; O- Serv Index Cards, Fold3 **BS**: 245

*MILLER, Anderson: b 9 Jan 1776, New Kent Co, d 12 Aug 1850 **RU**: Captain, 19th VMR, commanded a company, Richmond City **CEM**: The Grove, AKA Locust Grove and Miller Hill; Nottoway; nr Burkeville at top of hill **GS**: Unk **SP**: Sarah Thweat (Halifax Co, 16 Nov 1783-26 Jul 1851), daug of Giles Thweat and Sarah Barksdale **VI**: No further data **P**: No **BLW**: No **PH**: No **SS**: B pg 175 **BS**: 245

*MILLER, Christian; b Feb 1785, d 12 Mar 1852 **RU**: Private 116th VMR Capt Henry Welch's Co, Rockingham County, attached to 6th VMR (Dickinson, Scott, Coleman) **CEM**: Oak Lawn; Rockingham; jct Rts 704 & 257, Bridgewater **GS**: Yes **SP**: Susannah Flory (1792-23 Sep1871), daug of John Flory (1766-1845) and Catherine Garber (1771-1835) **VI**: No further data **P**: No **BLW**: No **PH**: No **SS**: B pg 182; O- Serv Index Cards, Fold3 **BS**: 245

*MILLER, George; b 1768, d 11 May 1845 **RU**: Private, 116th VMR, Captain William Harrison's Company, Rockingham Co, attached to 6th VMR (Coleman) **CEM**: Radar Lutheran Church; Rockingham; Timberville **GS**: Yes **SP**: No spouse info **VI**: No further data **P**: No **BLW**: No **PH**: No **SS**: B pg 181; K pg 51 **BS**: 31; 245

*MILLER, George; b 3 May 1774, d 23 Apr 1848 **RU**: Corporal, 56th VMR, Captain Thomas Gregg's Company, Loudoun Co. attached to 4th VMR (Beatty) **CEM**: Saint Pauls Lutheran Church; Loudoun; Neersville **GS**: Yes **SP**: Mary (__) **VI**: No further data **P**: No **BLW**: No **PH**: No **SS**: A rec 30141; B pg 120 **BS**: 245

*MILLER, Jacob; b 1790, d 1842 **RU**: Ensign, 56th or 57th VMR, company of Loudoun Co. attached to 5th VMR **CEM**: Saint Pauls Lutheran Church; Loudoun Co; Neersville **GS**: No **SP**: No spouse info **VI**: No further data **P**: Unk **BLW**: No **PH**: No **SS**: B pgs 119-121 **BS**: 245

*MILLER, Jacob; b 1769, d 5 Aug 1861 **RU**: 58th or 116th VMR, Rockingham Co company attached to 6th VMR (Coleman) **CEM**: Elk Run Rockingham; Rockingham & Spotswood Ave; Elkton **GS**: Yes **SP**: No spouse info **VI**: No further data **P**: No **BLW**: No **PH**: No **SS**: A rec 30474 **BS**: 245

*MILLER, Jacob; b 25 Dec 1780, d 20 Sep 1818 **RU**: 58th VMR, Capt Adam Harnsberger's Co, Rockingham County, attached to 1st VMR (Trueheart) **CEM**: Rader Lutheran Church; Rockingham; Timberville **GS**: Unk **SP**: No spouse info **VI**: No further data **P**: No **BLW**: No **PH**: No **SS**: A rec 30428 **BS**: 245

*MILLER, John; b 21 Mar 1775, d 1 Oct 1815 **RU**: Private; 31st VMR, Capt Thomas Robert's Co, Mt Hebron; Frederick Co, attached to 4th VMR (Beatty) **CEM**: Mt Hebron; Winchester; 305 E Boscawen St **GS**: Yes **SP**: Margaret Sperry **VI**: Son of Godfrey and Anna Maria (Kurtz) Miller. Was a store keeper in Winchester **P**: No **BLW**: No **PH**: Yes **SS**: B pg 79; S pg 106 **BS**: 31; 245

*MILLER, John; b unk, d Oct 1856 **RU**: Private 33rd VMR, Captain Samuel Brown's Company, Henrico Co. **CEM**: Hollywood Cemetery; Richmond City; 412 S Cherry St **GS**: No **SP**: No spouse info **VI**: No further data **P**: No **BLW**: No **PH**: No **SS**: L pg 177 **BS**: 245

*MILLER, Martin; b 19 Nov 1792, d 31 Mar 1874 **RU**: Private, 32nd or 93rd VMR, Augusta Co, attached to 5th VMR (McDowell's) **CEM**: Mount Tabor Lutheran Church; Augusta; Middlebrook **GS**: Yes **SP**: Mary Ann Mizer (1 Aug 1797-1 Aug 1874), daug of Elizabeth Kiplinger (1763-1848) **VI**: No further data **P**: No **BLW**: No **PH**: No **SS**: A rec 1205; B pgs 39-40 **BS**: 245

*MILLER, Mathias, b unk, d 28 Dec 1834 **RU**: Private, 13th or 97th VMR, a company in Shenandoah Co attached to 6th VMR (Coleman's) **CEM**: Mill Creek; Page; Luray **GS**: No **SP**: No spouse info **VI**: No further data **P**: No **BLW**: No **PH**: No **SS**: A rec 31219

Key: *Additional veteran entry   **Corrected veteran entry   ***Deleted veteran entry
RU=Rank/Unit   CEM=Cemetery   GS=Gravestone   SP=Spousal Information   VI=Other Veteran Info   P=Pension
BLW=Bounty/Land Warrant   PH=Photo   SS=Service Source   BS=Burial Source   VMR= VA Military Regiment
LNR= Last Known Residence

73

*MILLER, Peter; b 1781, Germany, d 19 Mar 1865 **RU:** Private, Shenandoah Co attached to Flying Camp McDowell's **CEM:** Atwood Family; Page; Rileyville **GS:** Yes **SP:** Mary Delaney **VI:** Was a miller and chair maker **P:** No **BLW:** No **PH:** No **SS:** A rec 31338 **BS:** 245

*MILLER, Peter; b 21 Aug 1792, d 25 Mar 1853 **RU:** Private, Frederick Co VMR, company attached to 1st VMR (Taylors) **CEM:** Mt Hebron; Winchester; 305 E Boscawen St **GS:** Yes **SP:** Mar (Frederick Co, 29 Apr 1806) Priscilla Watson **VI:** No further data **P:** No **BLW:** No **PH:** Yes **SS:** A rec 31354 **BS:** 31

*MILLER, Samuel; b 18 Feb 1782, d 1 Jan 1862 **RU:** Private, 13th or 97th VMR, Capt Jacob Fry's Co, Shenandoah County, attached to 4th VMR (Huston & Wooding) **CEM:** New Monmouth Presbyterian; Lexington City; 2348 W Midland Trail **GS:** Yes **SP:** No spouse info **VI:** No further data **P:** Yes **BLW:** No **PH:** No **SS:** pg 1294; B pg 185; M pg 242 **BS:** 245

*MILLS, William; b 1797, d 19 Sep 1830 **RU:** Private, 36th Infantry, Captain James Spence's Co **CEM:** Trinity United Methodist Church; Alexandria; GPS: 38.8023, -77.05746; 2911 Cameron Mills Rd **GS:** Yes **SP:** No spouse info **VI:** Son of Margery (__) (1766-1831). Tombstone inscription, Son of Margery and Capt William, enlisted Georgetown, Washington DC, 13 Aug 1814; discharged, Washington DC, 20 Mar 1815 **P:** 3 SC-12067 **BLW:** No **PH:** No **SS:** BD pg 197; C pg 127 **BS:** 245

*MILLS, William; b unk, d Oct 1815 **RU:** Corporal, 56th VMR a company in Loudoun Co **CEM:** Fairfax Friends; Loudoun; Waterford **GS:** Yes **SP:** No spouse info **VI:** No further data **P:** No **BLW:** No **PH:** No **SS:** A rec 32679 **BS:** 245

*MILLS, William Nelson; b 3 Mar 1783, d 4 Oct 1852 **RU:** Private, Major King's Detachment DC Militia **CEM:** Presbyterian Church; Alexandria; GPS: 38.80015, -77.05791; Wilkes St & Hamilton Ln **GS:** Yes **SP:** Mar (27 Feb 1806) Ann Leap (1789-1831), daug of Jacob Leap and Ann (__) **VI:** Son of William Mills and Ann (__) **P:** No **BLW:** No **PH:** No **SS:** A rec 32654 **BS:** 245

MILSTEAD, Samuel; b unk (stone chipped) Charles City Co, MD; d 10 Oct 1830 **RU: Private, 45th VMR, Captain Elijah Harding's Company, Stafford Co, attached to 1st VMR (Crutchfield) **CEM:** Sunnyside; Prince William; off Rt 642; 6.5 mi N of Dumfries **GS:** Yes **SP:** No spouse info **VI:** No further data **P:** No **BLW:** No **PH:** No **SS:** B pgs 190; 247 **BS:** 31

MINOR, George; b 1774 (1850 Census); d aft 1850 **RU: Lt Colonel, 60th VMR, Company commander, Fairfax Co **CEM:** Birch/Payne Family; Arlington; N Sycamore & 28th **GS:** No **SP:** Ann Birch **VI:** Commission date of 24 Apr 1809 **P:** No **BLW:** No **PH:** No **SS:** B pg 71 **BS:** 245

*MINOR, Hubbard; b 1 Aug 1795, d 14 Oct 1875 **RU:** Sergeant, 30th VMR, Capt Thomas Royston's Co, Caroline County, attached to 41st VMR, Richmond County serving under command of General Hungerford's Brigade **CEM:** Fredericksburg City; Fredericksburg City; Washington Ave **GS:** Yes **SP:** Mar (Spring Forest, Spotsylvania Co, 12 Jul 1826) Malvina Crutchfield (1801-1866) **VI:** Was a doctor. Recd two BLW, one of 40 acres Act of 1850, one 120 acres Act of 1855 **P:** No **BLW:** Yes **PH:** No **SS:** B pg 56; M pg 243; O- pen files, Fold3 **BS:** 245

*MITCHELL, William Jr.; b 24 Feb 1797, d 3 Sep 1852 **RU:** Private 19th VMR, Captain Samuel Adams's Co, Richmond City **CEM:** Hollywood Cemetery; Richmond; 412 S Cherry St **GS:** No **SP:** No spouse info **VI:** No further data **P:** No **BLW:** **PH:** Yes **SS:** B pg 174; L pg 79 **BS:** 31

*MOHLER, John; b 1789, MD, d 2 Jul 1871 **RU:** Private, 93rd VMR, Capt Archibald Stuart's Co, Augusta County, attached to Flying Camp (McDowell) **CEM:** Wilson Springs; Rockbridge; on Rt 39 off Rt 623, Tucker Farm, site of Bethesda Church **GS:** Unk **SP:** Mar (Rockbridge Co, 4 Dec 1810) Elizabeth Amick (1787-31 Jan 1867) **VI:** Recd pen and two BLW; One 40 acres (1850), other 120 acres (1855) **P:** Yes **BLW:** Yes **PH:** No **SS:** B pg 40; M pg 244; O- pen files, Fold3 **BS:** 245

*MOHLER, John; b 15 Apr 1772, Lancaster Co, PA, d 17 Apr 1835 **RU:** Ensign, 8th VMR, Capt Walter Stuart's Detachment of Mounted Riflemen, Rockbridge Co, attached to McDowell's Flying Camp **CEM:** Falling Springs Presbyterian Church; Rockbridge; 410 Falling Springs Rd, Glasgow **GS:** Unk **SP:** Mar (PA, 1794) Magdalene Rhinehart (1778-1853) **VI:** Recd pen, no further data **P:** Yes **BLW:** No **PH:** No **SS:** B pg 180; BD pg 1303; O- pen files fold3 **BS:** 245

Key: *Additional veteran entry   **Corrected veteran entry   ***Deleted veteran entry
RU=Rank/Unit   CEM=Cemetery   GS=Gravestone   SP=Spousal Information   VI=Other Veteran Info   P=Pension
BLW=Bounty/Land Warrant   PH=Photo   SS=Service Source   BS=Burial Source   VMR= VA Military Regiment
LNR= Last Known Residence

74

MONCURE, Edwin Conway; b 1780; d 19 Aug 1815 **RU: Private, 45th VMR, Capt John Edrington's Co, Stafford County, attached to 6th VMR (Coleman), **CEM:** Aquia Episcopal; Stafford; jct routes 1 and 610 **GS:** Yes **SP:** Eleanor Glascock, daug of George Glascock, (__-Sep 1859, age 70) **VI:** Reinterred from Clermont Cem as part of Quantico Marine Base expansion **P:** No **BLW:** No **PH:** No **SS:** L pg 302 **BS:** 26 pg 120; 31

MONCURE, John; b 23 Nov 1779; d 24 May 1864 **RU: 1st Lieutenant, 45th VMR, Capt John Edrington's Co, Stafford County, attached to 6th VMR (Coleman) **CEM:** Glencairn; Stafford; US Rt 1, Glencairn Estate **GS:** Yes **SP:** Mar (29 Dec 1802) Catherine Storke Peyton (20 Jul 1786-10 Apr 1865) **VI:** No further data **P:** Applied, but no record of approval **BLW:** No **PH:** No **SS:** BD pg 1304; B pg 190 **BS:** 26 pg 216

MONCURE, John, Jr; b 24 Dec 1793, "Somerset"; d 3 Aug 1876 **RU: Private, 45th VMR, Capt John Edrington's Co, Stafford County, attached to 6th VMR (Coleman) **CEM:** Aquia Episcopal; Stafford; jct routes 1 and 610 **GS:** Yes **SP:** 1) Esther J Vowles (1771-13 Aug 1822), 2) Frances (Fanny) Daniel (__-11 Sep 1874, age 74) **VI:** Son of John Moncure (1772-1822) and Alice Peachy Gaskins (1774-1860), served as agent for Mutual Insurance Society **P:** No **BLW:** No **PH:** No **SS:** L pg 302 **BS:** 26 pg 122

MONCURE, John, Sr; b 1771, "Somerset"; d 13 Aug 1822 **RU: Paymaster, 45th VMR, Capt John Edrington's Co, Stafford County, attached to 6th VMR (Coleman) **CEM:** Aquia Episcopal; Stafford; jct of routes 1 and 610 **GS:** Unk **SP:** Alice Peachy Gaskins (__-9 May 1860, age 86) consort of John Moncure **VI:** Member of Virginia Legislature fr Stafford Co **P:** Yes **BLW:** No **PH:** No **SS:** A rec 1899; BD pg 1304; B pg 90; M pg 244 **BS:** 26 pg 122

*MONROE, James; b 28 Apr 1758, Westmoreland Co, d 4 Jul 1831 **RU:** U S Secretary of State and Secretary of War under President James Madison **CEM:** Hollywood; Richmond City; 412 S Cherry St **GS:** Yes **SP:** Elizabeth Kortright (1768-1830), daug of Laurence Kortright and Hannah Aspinwall **VI:** Son of Spence Monroe and Elizabeth Jones of Westmoreland Co. Fifth President of the United States of America (1817-1825) **P:** No **BLW:** No **PH:** Yes **SS:** Historical fact **BS:** 31

*MONTAGUE, William Valentine; b 3 Sep 1797, d 10 Sep 1865, Norfolk **RU:** Private, 19th VMR, Company of Richmond City **CEM:** Cedar Grove; Norfolk; 238 Princess Anne Rd **GS:** Yes **SP:** Mary Ann Barrack, (15 Jan 1801-30 Aug 1840) **VI:** Son of William Montague and Elizabeth Valentine **P:** No **BLW:** **PH:** No **SS:** A rec 2376 **BS:** 245

*MONTGOMERY, James; b 31 Dec 1786, d 21 Jan 1866 **RU:** Private, 8th VMR, Company of Rockbridge, Co, attached to 5th VMR (McDowell's) **CEM:** Oxford Presbyterian; Lexington City; GPS: 37.45175,-79.33689; 18 Churchview Lane **GS:** Yes **SP:** Mar (5 Oct 1807) Martha Hall (8 Jan 1783-4 Apr 1854), daug of James Hall (1745-1816) and Martha Gilmore **VI:** No further data **P:** No **BLW:** No **PH:** No **SS:** A rec 256; B pg 179 **BS:** 245

*MONTGOMERY, John; b 1788, d 9 Apr 1829 **RU:** Corporal, 30th VMR, Capt William Gray's Co, Caroline County, attached to 16th VMR Spotsylvania Co and 9th VMR, King and Queen Co **CEM:** Bell Family; Rockbridge; nr Cameron Hall across from Iron Bridge on left end of rd, Goshen **GS:** Unk **SP:** Elizabeth Nelson (1 Sep 1796-9 Jan 1858), daug of Alexander Nelson (1749-1834) and Nancy Elizabeth Ann Mathews **VI:** Son of John Montgomery (1792-1818) and Agnes Hughart (1762-1824) **P:** Yes **BLW:** No **PH:** No **SS:** A rec 2376; B pgs 55,188; M pg 244; O- pen files Fold 3 **BS:** 245

*MONTGOMERY, Robert; b 1 Feb 1773, d 1 Oct 1859; **RU:** Private, 78th VMR, Capt Timothy Dalton's Co, Carroll County, attached to 4th VMR (McDowell, Koontz, Chilton) **CEM:** Quaker-Nester; Carroll; Rt 624, Nester School Rd, .07 mi fr jct with Rt 638 **GS:** Yes **SP:** Mar (Martinsville, NC, 18 Jun 1802) Sarah Hiatt (1786-23 Jun 1873), daug of Joseph Haitt and Keziah Mills. As widow applied for pen **VI:** Son of William Montgomery (1752-1811) and Rebecca Erwin (1755-1806) **P:** Widow **BLW:** No **PH:** No **SS:** B pg 86; M pg 244; O- pen files, Fold3 **BS:** 245

Key: *Additional veteran entry **Corrected veteran entry ***Deleted veteran entry
RU=Rank/Unit   CEM=Cemetery   GS=Gravestone   SP=Spousal Information   VI=Other Veteran Info   P=Pension
BLW=Bounty/Land Warrant PH=Photo   SS=Service Source   BS=Burial Source   VMR= VA Military Regiment
LNR= Last Known Residence

75

*MOODY, William Woolridge; b 1796, d Aug 1870 **RU:** Private, 83rd VMR, Captain L Barfoot's Company, Dinwiddie Co **CEM:** W. W. Moody Family; Dinwiddie; Sutherland **GS:** Yes **SP:** Mecca Pride Goode (1800-1881) **VI:** Son of Sterling Moody and Mary Nunnally **P:** Yes, spouse, WC-25227 **BLW:** No **PH:** No **SS:** BD pg 1307 **BS:** 245

*MOON, John Diggs; b 13 Sep 1794, d 26 Nov 1862 **RU:** Sergeant 10th or 94th VMR, Capt Abraham Buford's Co, Bedford County, attached to 8th VMR (Wall) **CEM:** Mount Air (AKA Ary, Ayr); Albemarle; 6100 Blenheim Rd, Keene **GS:** Unk **SP:** Mar (1819) Mary Elizabeth Barclay (Hanover Co, 2 Apr 1803-11 Jul 1874). Recd two pen **VI:** Son of William Moon and Charlotte Diggs; recd BLW 80 acres 1855 **P:** Widow **BLW:** Yes **PH:** No **SS:** B pg 239; M pg 244; O- pen files, Fold3 **BS:** 245

*MOORE, Jesse; b 4 Sep 1766, VA d 26 Sep 1853 **RU:** Private, 56th VMR, Captain Thomas Gregg's Company, Loudoun Co attached to 4th VMR (Beatty's) **CEM:** Moore-Hunter Family; Fairfax; GPS: 38.88537, -77.27105 **GS:** No **SP:** Catherine Brent (18 Feb 1772-19 Nov 1804), daug of Charles Brent and Anna Gunnell **VI:** Son of Jeremiah Moore (1746-1815) and Lydia French Reno (1747-1835) **P:** No **BLW:** No **PH:** No **SS:** A rec 4470, B pg 120 **BS:** 245

*MOORE, John; b Oct 1779, d 20 Apr 1838 **RU:** Private, 8th VMR, Captain Alexander's Company of Rockbridge Co, attached to 2nd Corps D'Elite (Green) **CEM:** McKee; Rockbridge; Rt 631 **GS:** Yes **SP:** Mar (17 Dec 1801) Betsy Cunningham (1780-11 May 1856), daug of James Cunningham and Mattie (__) **VI:** No further data **P:** No **BLW:** No **PH:** No **SS:** B pg 179, K pg 209 **BS:** 245

*MOORE, John; b 1792, d 1848 **RU:** Private, 20th VMR, Captain Richard H L Lawson's Company, Princess Anne Co. **CEM:** Cedar Grove; Norfolk; GPS: 36.8586, -76.2831; 238 E Princess Anne Rd **GS:** Unk **SP:** No spouse info **VI:** No further data **P:** No **BLW:** No **PH:** No **SS:** L pg 540 **BS:** 245

MOORE, Reuben; b 1769; d 20 Nov 1835 **RU: Private, 116th VMR, Capt William McMahon's Co, Rockingham County, attached to Maj Woodford's Squadron of Cavalry (Dragoons) **CEM:** Woodbine; Harrisonburg; loc cnr Ott and W. Market St **GS:** Yes Note on 27 Oct 2003 contents and Crypt were moved from Methodist church to Woodbine **SP:** Mary (__),(__-1842) **VI:** No further data **P:** No **BLW:** No **PH:** No **SS:** B, pg 62; K pg 187 **BS:** 245

MOORE, Reuben; b 25 Feb 1784; d 14 Jul 1844 **RU: Corporal, 5th VMR **CEM:** St Luke's United Church of Christ; Rockingham; 107 Short Ln, Timberville **GS:** Yes **SP:** Sarah Kingree (3 Dec 1790-28 Dec 1852), daug of Solomon Kingree and Elizabeth Jones **VI:** Son of Joseph Moore and Margaret (__), reinterred fr family cem **P:** No **BLW:** No **PH:** No **SS:** A rec 4976 **BS:** 115 pg 11; 245

*MOORE, Richard; b c1790, d c1834 **RU:** Private, 9th VMR, Capt Richard Corbin's Company of Artillery, King & Queen Co, attached to 6th VMR (Reade) **CEM:** Porporone Baptist Church; King & Queen Co; Shackelford **GS:** Yes **SP:** Frances Ware (1800-1832) **VI:** Memorialized in cemetery by descendants, actual burial place unk; Captain rank on gravestone **P:** No **BLW:** No **PH:** No **SS:** A rec 4994; B pg 112 **BS:** 245

*MOORE, William; b 7 Oct 1771, Pittsylvania Co, d 13 May 1819 **RU:** Major, Staff Officer, 18th VMR, Patrick Co **CEM:** Moore Family; Patrick; Ararat **GS:** Yes **SP:** Jane Dalton Hanby (Henry Co, 28 Feb 1783-Ararat, Patrick Co, 2 Oct 1817) **VI:** Son of Rodeham Moore (1744- 1811) and Elizabeth Gallahue (1755-1825), commissioned 20 Apr 1810 **P:** No **BLW:** No **PH:** No **SS:** B pg 157 **BS:** 245

*MORELAND, George T; b unk- d 184- **RU:** Private, 114th, Hampshire Co (Poston's) **CEM:** Goose Creek Burying Ground; Loudoun; Rt 722 Lincoln **GS:** Yes **SP:** No spouse info **VI:** No further data **P:** No **BLW:** No **PH:** No **SS:** A rec 6414 **BS:** 245

*MORGAN, John; b 25 Aug 1770, d 18 Jan 1852 **RU:** Private, 13th or 97th VMR, Shenandoah Co attached to 6th VMR (Coleman's) **CEM:** Union Church; Shenandoah; Mount Jackson **GS:** Yes **SP:** No spouse info **VI:** No further data **P:** Unk **BLW:** No **PH:** No: **SS:** B pgs 184-185; M pg 245 **BS:** 245

Key: *Additional veteran entry **Corrected veteran entry ***Deleted veteran entry
RU=Rank/Unit   CEM=Cemetery   GS=Gravestone   SP=Spousal Information   VI=Other Veteran Info   P=Pension
BLW=Bounty/Land Warrant   PH=Photo   SS=Service Source   BS=Burial Source   VMR= VA Military Regiment
LNR= Last Known Residence

76

*MORGAN, Samuel B; b 1785, d 9 Mar 1845 **RU:** Private in company attached to 1st Regt Cavalry (Holcombe's), 73rd VMR assigned to Charles Batt's Troop of Cav attached to 1st VMR (Holcombe's) **CEM:** Blandford; Petersburg; 3195 Grater St **GS:** Unk **SP:** No spouse info **VI:** No further data **P:** No **BLW:** No **PH:** No **SS:** A rec 7080; B pgs 125; 238 **BS:** 245

*MORAN, Nelson; b 1779, Patrick Co, d Dec 1862 **RU:** Private, 18th VMR, Lt Samuel Hanby's Co, Patrick County, attached to 5th VMR (Mason-Preston) **CEM:** Charity Primitive Baptist Church; Patrick; 5804 Charity Hwy, Charity **GS:** Yes **SP:** Mar 1) (20 Dec 1802) Mary Owen (1780-1838) 2) (9 Sep 1839) Nancy Brammer (1800-1840), 3) (1843) Widow, Sarah (Sally) Cannaday (Franklin Co, 1804-1882), who drew two pensions **VI:** Listed in census as Nelson Mooring, perhaps son of William Mooring/Moran **P:** Widow **BLW:** No **PH:** No **SS:** B pg 157; O- pen files fold3 **BS:** 245

*MORRIS, Charles; b 1778, d 1842 **RU:** Surgeon, VMR (not determined) Served in Colonel William Trachel's(Truchel's) Regiment (not in Butler's Guide; perhaps in Regular Army) **CEM:** The Meadows Estate Gardens; Hanover; Doswell **GS:** No (Burials determined by original owner) **SP:** Mar (1812) Emily Harris Taylor of Cherrydale farm (14 Aug 1789-11 Jan 1873), She recd pen; daug of Edmund Taylor (16 Aug 1741-28 Jan 1822) and Ann Day (1753-1835) **VI:** Medical Doctor **P:** Widow **BLW:** No **PH:** No **SS:** BD pg 1318 **BS:** 245

*MORRIS, John; b 1775, d 1845 **RU:** Private, 64th VMR, Capt John Dillard's Co, Henry County, attached to Battalion of Artillery **CEM:** Morris Family; Henry; Bassett **GS:** No **SP:** No spouse info **VI:** No further data **P:** No **BLW:** No **PH:** No **SS:** A rec 8106; B pg 101 **BS:** 245

*MORRIS, John: b 17 Jan 1791, Mathews Co, d 11 Aug 1841 **RU:** Private, 61st VMR Capt Henry Digges's Co, Mathews County **CEM:** Cedar Grove; Norfolk City; jct E Princess Anne Rd, & Salter St **GS:** Yes **SP:** No spousal info **VI:** Was Reverend, buried in plot CG, 3AW-L49-S11 **P:** No **BLW:** No **PH:** No **SS:** B pg 128; K pg 288; O- Serv Rec Index Cards **BS:** 245

*MORRIS, Samuel Coleman: b 1790, d 18 Mar 1884, Metcalfe Co, KY **RU:** Private, 64th VMR, Capt James Shelton's Co, Henry County, attached to 5th VMR (Mason-Preston) **CEM:** Morris; Henry; vic jct Wingfield Orchard Rd & Skyview Trail, Bassett **GS:** Name listed on family monument, thus memorialized in cem **SP:** Susannah Davis (1811-1888) **VI:** Son of Samuel Coleman Morris (1740-1826), and Susannah Wade. Family monument indicates family members buried in Old Morris Cem off Rt 688 **P:** No **BLW:** No **PH:** No **SS:** A rec 8399; B pg 101 **BS:** 245

*MORRIS, William; b 24 Nov 1782, d 25 Oct 1831 **RU:** Private, 47th or 88th VMR, Captain Robert McCulloch's or John Rockwell's Co, Albemarle County, attached to 7th VMR (Gray) **CEM:** Oakwood; Charlottesville; 1st St, S **GS:** Yes **SP:** No spouse info **VI:** No further data **P:** No **BLW:** No **PH:** No **SS:** A rec 8517; B pg 36 **BS:** 245

*MORRIS, William C; b 24 May 1794, d 27 Nov 1857, **RU:** Private, 40th VMR, Capt Reuben Chewning's Co, Louisa County, attached to 7th VMR (Gray) **CEM:** Morris-Payne; Louisa; Columbia Rd vic jct with Rts 615 & 617; In woods 600 yds; Zion **GS:** Unk **SP:** Elizabeth W. Allen, (2 Jun 1798-22 Sep 1885), recd pen **VI:** No further data **P:** Widow **BLW:** No **PH:** No **SS:** A rec 8517; B pg 123; BD pg 1320 **BS:** 245

*MORRISON, James; b 1776, d 1850 **RU:** Private, 18th VMR, Capt Abraham Staple's Co, Patrick County, attached to 6th VMR (Coleman) **CEM:** Old Morrison; Patrick; So side Rt 57, I mi E of jct with Rt 8 **GS:** Unk **SP:** 1) Nancy Porter (1770-1820) 2) Rebecca Neville (1785-1853) **VI:** Son of Thomas Morrison and Rebecca (__) **P:** No **BLW:** No **PH:** No **SS:** A rec 8690; B pg 157 **BS:** 245

*MULLINS, John (Holly Creek); b 1784, NC, d 19 Sep 1859, Dickerson Co **RU:** Private, 70th VMR, Captain William Smith's Artillery Company, Washington Co, attached to Battalion of Artillery **CEM:** John Mullins Family; Dickenson; Holly Creek **GS:** Yes **SP:** Ollie Cox (1792-1877) **VI:** Son of John Wesley Mullins (1752-1849), a soldier in the Revolution **P:** No **BLW:** No **PH:** No **SS:** A rec 12071; B pg 199 **BS:** 245

Key: *Additional veteran entry   **Corrected veteran entry   ***Deleted veteran entry
RU=Rank/Unit   CEM=Cemetery   GS=Gravestone   SP=Spousal Information   VI=Other Veteran Info   P=Pension
BLW=Bounty/Land Warrant   PH=Photo   SS=Service Source   BS=Burial Source   VMR= VA Military Regiment
LNR= Last Known Residence

*MURRAY, Reuben, Jr; b prob 1780's, d 22 Jun 1885 **RU:** Private, 17th VMR, Capt Allen Wilson's Co, Cumberland County, attached to 1st VMR (Trueheart) **CEM:** Freedmans; Alexandria City; cnr Church and So Washington Sts **GS:** Unk **SP:** Judith (__), recd pen **VI:** Son of Reuben Murray a Rev War soldier, recd BLW **P:** Widow **BLW:** Yes **PH:** No **SS:** B pg 64; BD pg 1335 **BS:** 245

*MYERS, Aaron; b 1798, d 25 Mar 1845 **RU:** Private, 115th VMR served in a company from Elizabeth City, Warwick or York Co, attached to Major Crutchfield at Hampton **CEM:** Hebrew; Richmond City; 4th St, Shockoe Hill **GS:** Yes **SP:** Catherine (__) recd pen **VI:** No further data **P:** Widow **BLW:** No **PH:** No **SS:** B pgs 67; 68; M pg 248; BD pg 1336 **BS:** 245

*MYERS, Michael; b 20 May 1791, York Co, PA, d 1852, New Hope **RU:** Private, Cobean's Battalion, PA Volunteers **CEM:** Knightly Mill; Augusta; on bluff on property of Edward Sites, Knightly **GS:** Unk **SP:** Margaret (__), recd pen **VI:** Son of Christian Meyers an Barbara Burkholder; applied and recd BLW **P:** Widow **BLW:** Yes **PH:** No **SS:** A rec 15209; M pg 248 **BS:** 245

*MYERS, John T; b 20 Dec 1795, d 13 Jun 1871 **RU:** Private 116th VMR, Captain Henry Welch's Company, Rockingham Co, attached to 5th VMR (Mason & Preston) **CEM:** Byerly Family; Rockingham; Pleasant Valley Rd **GS:** Yes **SP:** Lydia (26 May 1795- 9 May 1874) **VI:** No further data **P:** No **BLW:** No **PH:** No **SS:** A rec 15132, B pg 182 **BS:** 262

*NASH, William Daniel; b 12 Jul 1787, Charlotte Co, d 21 Jan 1866 **RU:** Private 24th or 100th VMR Capt Walter Fontaine's Co, Buckingham County, attached to 8th VMR (Wall). Also served in Capt Thomas Redd's Co, from Prince Edward County, attached to 8th VMR (Wall) **CEM:** William D Nash Home place; Tazewell; N side Rt 460 betw Tazewell and Springville **GS:** No **SP:** Mar (Prince Edward Co, 20 Feb 1812) Mary Frances Gaulding (Prince Edward Co, 29 Sep 1792-26 Jan 1886), daug of Joseph Gaulding and Martha (Patsy) Barnett **VI:** Source 245 indicates BLW were recd (probably 1850 and 1855) and soldier was promoted to Corporal, however military rec do not show promotion **P:** No **BLW:** Yes **PH:** No **SS:** A rec 16194; B pgs 35; 50; 239; K pg 94; L pgs 307; 665 **BS:** 245

*NEBLETT, Sterling; b 22 Sep 1792, d 16 Nov 1871 **RU:** Private/Surgeons Assistant 73rd VMR, Capt James Neblett's Co, Lunenburg County, marched to Norfolk under Lt Col Grief Green's 98th VMR, then attached to 6th VMR (Sharp); Later served as a surgeons assistant in Col Preston's Regt **CEM:** Neblett Family; Lunenburg; Rt 138, S of jct with Rt 619 **GS:** Yes **SP:** Mar (13 Aug 1821) Ann Smith MacFarland (1802-1881), recd pen **VI:** Son of Sterling (Surry Co, 23 Oct 1753-1832) and Mary Chappell. Legislator, VA House of Delegates, postmaster Brickland area, and Justice Lunenburg Co, 1840-1 **P:** Widow **BLW:** No **PH:** No **SS:** B pgs 184-5; O-Serv Index card, Fold 3 **BS:** 245

*NEFF, Abraham; b 2 Aug 1798, d 3 Sep 1880 **RU:** Private, 14th VMR, a Company of Hardy Co. **CEM:** Neff Family; Wythe; Rural Retreat **GS:** Yes **SP:** Elizabeth Repass (1798-1830) **VI:** Son of Michael Neff (1756- 1825) and Christina Kapp (1746-1830) **P:** No **BLW:** No **PH:** No **SS:** A rec 17196 **BS:** 245

*NEFF, John; b 27 Sep 1766, d 6 Sep 1828 **RU:** Private, 14th VMR, Captain Michael Yozkum's Company of Hardy Co., attached to 1st VMR (Connell) **CEM:** Neff-Kagey; Shenandoah; New Market **GS:** Yes **SP:** Barbara Kauffman (1774-1840) **VI:** Son of Jacob Neff (1742-1820) and Barbara Grabill (1742-1804) **P:** No **BLW:** No **PH:** No **SS:** A rec 17232 **BS:** 245

*NEFF, David; b 16 Jan 1785 (Bible) (Tombstone- 7 Feb 1785), d Sep 1828 **RU:** Private, 13th or 97th VMR in one of several companies of these regiments, Shenandoah Co, attached to 6th VMR (Coleman) **CEM:** Neff-Kagey; Shenandoah; New Market **GS:** Yes **SP:** 1) Barbara Kagey (1790-1816), daug of Jacob Neff and Christina Brennerman, 2) (2 Mar 1819) Mary Strickler (1785-1862), daug of Samuel Strickler **VI:** Son of Jacob Neff (1742-1820) and Barbara Grabill (1742-1804) **P:** No **BLW:** No **PH:** No **SS:** B pgs 184-5; O- fold3 Serv Index Cards **BS:** 245

*NEIGHBORS, James; b 1788, d 19 Oct 1869 **RU:** Private, 53rd or 117th VMR, Capt James Haden's Grenadiers Co, Campbell County, attached to 7th VMR (Dickenson) with MD service **CEM:** Saint Pauls Lutheran Church; Wythe; 330 St Pauls Church Rd; Rural Retreat **GS:** Yes **SP:** Mar (10 Oct 1812) Nancy Ann Mann (1778-10 Feb 1854), 2) Catherine (__) (__-aft 1869) Recd pen **VI:** No further data **P:** Widow **BLW:** No **PH:** No **SS:** B pgs 54; 240; BD pg 1343 **BS:** 245

Key: *Additional veteran entry   **Corrected veteran entry   ***Deleted veteran entry
RU=Rank/Unit   CEM=Cemetery   GS=Gravestone   SP=Spousal Information   VI=Other Veteran Info   P=Pension
BLW=Bounty/Land Warrant   PH=Photo   SS=Service Source   BS=Burial Source   VMR= VA Military Regiment
LNR= Last Known Residence

NELSON, James; b Sep 1793; d 11 Mar 1854 **RU: Private, 5th VMR **CEM:** Old Stone Presbyterian; Augusta; Rt 11, Fort Defiance **GS:** Unk **SP:** No spouse info **VI:** Ruling elder in the church **P:** No **BLW:** No **PH:** No **SS:** A rec 17741 **BS:** 2 pg 16

*NELSON, William; b 1785, d 18 Apr 1828 **RU:** Private, 58th VMR, Captain Adam Harnsberger's Co, Rockingham County, attached to 1st VMR (Trueheart) **CEM:** Tinkling Spring Presbyterian Church; Augusta; 30 Tinkling Spring Drive; Fishersville **GS:** Yes **SP:** No spousal info **VI:** No further data **P:** No **BLW:** No **PH:** No **SS:** A rec 18004; B pg 181 **P:** Fold 3, serv index card **BS:** 245

*NICHOLS, Samuel; b Oct 1774, d 27 May 1824 **RU:** Private, Regular Army, 26th Infantry **CEM:** Goose Creek Burying Ground; Loudoun; Rt 722 Lincoln **GS:** Yes **SP:** Mary Janey (9 Apr 1777-7 Mar 1856) **VI:** Son of William Nichols & Sarah Spencer; Enlisted (1 Jan 1815); Discharged (Philadelphia, PA) **P:** No **BLW:** No **PH:** No **SS:** C pg 135 **BS:** 245

***NORMAN, Charles: Deleted as service not good for Stafford County

***NORMAN, Edward: Deleted as served in Revolution, not War of 1812

NORMAN, James S; b 1777; d 17 Dec 1844 **RU: Private, 45th VMR, Capt Henry Williams's Co, Stafford Co **CEM:** Norman Family and Towson Family #1; Stafford; Rt 692 nr 98 Quarry Rd **GS:** Yes **SP:** Mar (Fauquier Co, 25 Jun 1798) Peggy Curtis. Data from *https//en.wikimedia.org* **VI:** Son of Edward Norman (1752-1814) and Jane Stewart (c1756-1814), both who died in epidemic of 1814-1815 with sons James S. and Matthew Norman **P:** No **BLW:** No **PH:** No **SS:** K pg 129 **BS:** 26 pg 295

NORMAN, Matthew; b 1779; d 24 Dec 1814 **RU: Private, 4th VMR (Peyton), Stafford Co **CEM:** Norman Family and Towson Family #1; Stafford; Rt 692 nr 98 Quarry Rd **GS:** Yes **SP:** No spouse info **VI:** Son of Edward Norman (1752-1814) and Jane Stewart (c1756-1814), both who died in epidemic of 1814-1815 with sons James S. and Matthew, Data from WPA Survey and *Magazine of Virginia Genealogy*, Vol 18 No 2 **P:** No **BLW:** No **PH:** No **SS:** A rec 20630 **BS:** 26 pg 295

NORMAN, Thomas; b 1790; d 13 Dec 1846 **RU: Corporal, 45th VMR, Capt John C Edrington's Co, Stafford Co **CEM:** Norman Family and Towson Family #1; Stafford; Rt 692 nr 98 Quarry Rd **GS:** Yes **SP:** 1) Paulina Ficklin (c1800-1830), 2) Mildred Ficklin Hill (10 Aug 1804- Jan 1886) **VI:** Son of Edward (__-1814, age 63) & Jane (__-1814 age 58) both who died in epidemic of (1814-1815) Tombstone styles him "Esquire." **P:** Spouse **BLW:** No **PH:** No **SS:** L pg 300; BD pg 1356 **BS:** 26 pg 295; 31

O'BRYHIM, Alexander; b 1788 in Ireland; d 1 Aug 1854 **RU: Corporal, 45th VMR Capt. Barton Stone's Co, Stafford County **CEM:** O'Bryhim Family; Stafford; Rt 628, 600 feet south of Ramoth Church **GS:** Yes, lists him as father of Clara **SP:** Mary Timmons (1798-1860), daug of Thomas Timmons (c1775-1842) **VI:** Son of Joseph & Ann Obryhim. Data from Stafford Co Death Register, pg 4, line 73, reported by wife Mary **P:** No **BLW:** No **PH:** No **SS:** A rec 23459 **BS:** 26 pg 88; 299

O'CONNER, John; b 1793, d 1 Mar 1832, at Millwood **RU: Sergeant, 122nd VMR, Capt Province McCormick's Co, Frederick County, attached to 5th VMR **CEM:** Old Chapel; Clarke; Millwood **GS:** Yes **SP:** Mar (Frederick Co, 3 Dec 1823) Elizabeth Wood **VI:** No further data **P:** Spouse **BLW:** Yes **PH:** No **SS:** A rec 23522; BD pg; M pg 252 **BS:** 85 pg 33

OMOHUNDRO, Richard; b 15 Apr 1777, Bremo Bluff, Fluvanna Co; d 13 Jan 1860 **RU: Captain, 111th VMR, Company Commander, Westmoreland Co **CEM:** Omohundro Family; Fluvanna; Dixie; also called Gale Hill **GS:** Yes **SP:** Mar (13 Dec 1802) Edith Seay (6 Jan 1779-14 Aug 1856), daug of John and Rebecca (__) Seay **VI:** No further data **P:** No **BLW:** No **PH:** No **SS:** L pg 616 **BS:** 95 pg 65; 234, 245

*OREBAUGH, John; b 1789, d 1872 **RU:** Private, 32nd VMR, Capt Alexander Given's Co, Augusta County, attached to 5th VMR (McDowell) **CEM:** Orebaugh; Augusta; Rt 760 **GS:** Yes **SP:** Barbara Ailor **VI:** No further data **P:** Yes **BLW:** No **PH:** No **SS:** B pg 39; O **BS:** 25; Madison's of Montpelier Chapter

*ORGAN, John; b 4 Dec 1770, d 13 Feb 1826 **RU:** Major, 53rd VMR, Staff Officer, Campbell Co **CEM:** Mt. Calvary Baptist Church; Campbell; Gladys, **GS:** Yes **SP:** Mar (20 Jan 1795) Elizabeth "Betsey" Johnson **VI:** Commissioned to Major's rank, 20 Jan 1813; Son of Samuel Organ and Sarah Simpson; County Commissioner of Revenue **P:** No **BLW:** No **PH:** No **SS:** B pg 53; O **BS:** 245

Key: *Additional veteran entry   **Corrected veteran entry   ***Deleted veteran entry

RU=Rank/Unit	CEM=Cemetery	GS=Gravestone	SP=Spousal Information	VI=Other Veteran Info P=Pension
BLW=Bounty/Land Warrant	PH=Photo	SS=Service Source	BS=Burial Source	VMR= VA Military Regiment
LNR= Last Known Residence				

*OTEY, Isaac; b 18 Oct 1765, d 18 Oct 1839 **RU**: Sergeant, 52d VMR, Capt Robert Perkin's Co, New Kent County **CEM**: Otey Street; Bedford City; GPS: 37.37170,-79.52170; Jct W Franklin St and Otey St **GS**: Yes **SP**: Elizabeth Mathews (22 Feb 1767- 4 Mar 1855) **VI**: Son of John Armistead Otey (1735-1817) and Mary Hopkins (1739-1815). Obtained the rank of Major after the war period **P**: No **BLW**: No **PH**: No **SS**: A rec 27717; B pg 144; K pg 119 **BS**: 245

*OTEY, James; b 4 May 1774, d 1850 **RU**: Sergeant; 52th VMR, Ensign Edmund Grave's Co, New Kent County, attached to 2d VMR (Sharp) **CEM**: Otey Street; Bedford City; GPS: 37.37170,-79.52170; jct W Franklin St and Otey St **GS**: Unk **SP**: No spousal info **VI**: Son of John Armistead Otey (1735-1817) and Mary Hopkins (1739-1815) **P**: No **BLW**: No **PH**: No **SS**: A rec 27719; B pg 144; L pg 374 **BS**: 245

*OTT, George; b 1770, Germany, d 18 Mar 1831 **RU**: Captain, 54th VMR, Commander of Artillery in Norfolk Borough, attached to 6th VMR (Reade) **CEM**: Elmwood; Norfolk; 238 E Princess Anne Rd **GS**: Yes **SP**: No spouse info **VI**: Native of Wittenberg, Germany **P**: No **BLW**: No **PH**: No **SS**: pg 146 **BS**: 245

*PANKEY, Edward; b 5 Sep 1787, Buckingham Co, d 13 Jun 1865 **RU**: Sergeant, 24th and 100th VMR, Capt John Morgan's Artillery Co, Buckingham County, attached to 2nd VMR (Lt Col Thomas Ballowe) **CEM**: Pankey Family; Henry; Irisburg **GS**: Yes **SP**: Nancy Branch Pankey (1787 Cumberland County-19 Nov 1865), daug of Thomas Pankey and Martha Cannon **VI**: Son of John Pankey and Keziah Chambers **P**: No **BLW** No **PH**: No **SS**: A rec 31513; B pg 242 **BS**: 245

*PARRISH, William b c1765, d c1857 **RU**: Sergeant, 73rd VMR, Lunenburg company attached to 6th VMR (Sharps) **CEM**: Parrish Family; Lunenburg; Kenbridge **GS**: Unk **SP**: Margaret Sionaker (1797-15 Nov 1872), daug of Christopher Sionaker **VI**: No further data **P**: No **BLW**: No **PH**: No **SS**: A rec 2523; B pg 125 **BS**: 245

*PATES, Chandler; b 1780, d 1860 **RU**: Private 16th VMR in one of several companies, Spotsylvania Co, attached to Lt Col Robert Crutchfield's 1st VMR **CEM**: Old Burying Grd at Cedar Grove; Spotsylvania; Turner Lane, Paytes **GS**: No **SP**: Lucy M Sullivan (Stafford Co-27 Jul 1848), daug of Francis Sullivan (1764-1830) and Frances Newton **VI**: Gr st and perhaps 1812 gr marker thought to be removed by landowner **P**: No **BLW**: No **PH**: No **SS**: B pgs188,189; O- Serv Index Cards **BS**: 245

PAYNE, James Rousseau; b 4 Jan 1788; d 1863 **RU: Private, 16th VMR Capt Gullielmus Smith's Co, Spotsylvania County **CEM**: Stark/Payne; Stafford; vic Quantico Marine Base about 0.5 miles opposite Ruby Fire Station; Rt 611 **GS**: Unk **SP**: Nancy Maria Stark **VI**: Son of Benjamin Payne & Susanna Rousseau. **P**: No **BLW**: No **PH**: No **SS**: A rec 5606 **BS**: 26 pg 365; 49

***PAYNE, Wesley: Deleted as service or burial not adequate

*PAYNE, John; b 13 Feb 1780, Orange Co, d 1833, Lancaster Co **RU**: Private, 3rd VMR, Captain William Smith's Troop of Cavalry, Orange Co, attached to 1st VMR (Crutchfield) **CEM**: Payne Graveyard; Northumberland; Hard Bargain, Hardings **GS**: Unk **SP**: Mar (Culpeper Co,1806), 1) Lucy Taliaferro, 2) (Lancaster Co, 29 Apr 1818), Harriet Eustace **VI**: Son of John Payne and Ellen Bailey **P**: No **BLW**: No **PH**: No **SS**: A rec 5624; B pg 156 **BS**: 245

*PAYNE, John; b 29 Aug 1788, d 17 Aug 1825, King George Co **RU**: Sergeant, 25th VMR, Lieutenant John Arnold's Company, King George Co **CEM**: Payne Burying Ground (AKA Red House, Cedar Hill); Westmoreland; Rt 640, Ingleside Plantation property, Horners **GS**: Yes **SP**: Mar Frances Morris (3 Mar 1795-5 Apr 1841) **VI**: Son of John Payne (1753-1824) **P**: No **BLW**: No **PH**: No **SS**: L pg 108 **BS**: 245

*PAYNE, William Francis; b 1791, d 14 Apr 1866 **RU**: Private, 85th VMR, Captain Thomas Jenning's Company, Fauquier Co **CEM**: Payne Family; Fauquier; Orlean **GS**: Yes **SP**: Frances (__) (1792-1850) **VI**: No further data **P**: No **BLW**: No **PH**: No **SS**: A rec 5785 **BS**: 245

*PAYNE, William; b Mar 1797, d 29 Apr 1884 **RU**: Private; 10th and 91st VMR, served in one of four companies of Bedford County, attached to 5th VMR **CEM**: Ayers-Payne-Brooks; Bedford; on Rt 626, nr Rt 608, nr Smith Mt Lake in open field, Huddleston **GS**: Yes **SP**: Perhaps Nancy (__) (Apr 1806-5 May 1871) with GS in cem **VI**: No further data **P**: No **BLW**: No **PH**: No **SS**: B pgs 42; 43 **BS**: 245

Key: *Additional veteran entry   **Corrected veteran entry   ***Deleted veteran entry
RU=Rank/Unit   CEM=Cemetery   GS=Gravestone   SP=Spousal Information   VI=Other Veteran Info   P=Pension
BLW=Bounty/Land Warrant   PH=Photo   SS=Service Source   BS=Burial Source   VMR= VA Military Regiment
LNR= Last Known Residence

*PEALE, Jonathan; b 9 Jun 1797, d 26 Jun 1874 **RU:** Private, 58th or 116th VMR company in Rockingham Co, attached to 9th VMR (Sharp's) at Norfolk **CEM:** Cross Keys; Rockingham; Rt 679 on left **GS:** Yes **SP:** Mar 1) Eleanor (__) (1832-6 Jan 1839), 2) Margaret Laird (9 Mar 1819-30 Jan 1891) **VI:** Builder of the Peale house at Peale Cross Roads according to historical society inscription **P:** Widow Margaret applied for pension **BLW:** No **PH:** No **SS:** B pgs 181-182; M pg 257 **BS:** 245

*PENCE, John; b 23 Feb 1791, d 9 Dec 1822 **RU:** Private, 58th VMR, Captain Robert Hooke's Co attached to Flying Camp (McDowell) **CEM:** Cross Keys; Rockingham; Rt 679 on left **GS:** Yes **SP:** No spouse info **VI:** No further data **P:** No **BLW:** No **PH:** No **SS:** B pg 181 **BS:** 245

*PENCE, Phillip; b unk, d 1851 **RU:** Private, 13th VMR, Captain William Newell's Company, Shenandoah Co **CEM:** Radar Lutheran Church; Rockingham; Timberville **GS:** Unk **SP:** No spouse info **VI:** No further data **P:** No **BLW:** No **PH:** No **SS:** B pg 185; K pg 79 **BS:** 31

*PENNYBACKER, Joel; b 9 Aug 1793, d 5 Apr 1862 **RU:** Private, 97th or 13th VMR, Captain John Bayliss's Co of Artillery, Shenandoah Co, attached to Battalion of Artillery **CEM:** Union Church; Shenandoah; Mount Jackson **GS:** Yes **SP:** No spouse info **VI:** No further data **P:** No **BLW:** No **PH:** No **SS:** A rec 9043; B pg 184 **BS:** 245

*PERRY, Nickolas; b 26 Jan 1797, d 11 May 1878 **RU:** Private, 122nd VMR, Capt James Anderson's Company, Frederick Co, attached to 1st VMR (Taylor) **CEM:** Mt Hebron; Frederick; 305 E Boscawen St, Winchester **GS:** Yes **SP:** Mar 1) (Mar 1818) Abigail Hodgson, 2) (Feb 1856) Elizabeth Campbell (__-1883) **VI:** No further data **P:** Spouse **BLW:** Yes **PH:** No **SS:** A 10925; B pg 78; BD pg 1413; M pg 258 **BS:** 81: 245

*PERRY, Thomas; b 25 Feb 1794, d 1 Jul 1860 **RU:** Private, 72nd VMR Capt John Hamon's Co, Russell County attached to 6th VMR (Coleman's) **CEM:** Wynne-Perry; Tazewell; Rt 61 near Rt 678 **GS:** Yes **SP:** No spouse info **VI:** No further data **P:** No **BLW:** No **PH:** No **SS:** A rec 11063; B pg 183 **BS:** 245

*PESCUD, Edward; b 11 Dec 1778, d 15 Jul 1840 **RU:** Captain, 39th VMR, commanded company, Petersburg City, attached to 6th VMR (Dickinson, Scott, Coleman) **CEM:** Blandford; Petersburg City; 111 Rochelle Ln **GS:** Yes **SP:** Susan Brooke Francisco (Louisa, 1797-1869), daug of Peter Francisco **VI:** Cited rank of "Colonel" obtained after war period **P:** No **BLW:** No **PH:** No **SS:** B pg 160 **BS:** 245

*PETERS, Frederick D; b c1785, Germany, d 10 Dec 1827 **RU:** Private, 19th VMR, Capt Samuel Jones's Co, Richmond City, attached to 2d VMR, (Ballowe) **CEM:** Blandford; Petersburg City; 319 S Crater st **GS:** Yes, Ward A, Sec 26 **SP:** No spousal info **VI:** No further data **P:** No **BLW:** No **PH:** Yes **SS:** B pg 175; AA pg 87 **BS:** 31

*PETERSON, Thomas; b Oct 1767, d 21 Sep 1816 **RU:** Lt Colonel, 1st VMR, Commander, Amelia Co **CEM:** Wards Chapel; Nottoway; Zozomie Rd; 300 yds fr Rt 360 **GS:** Yes **SP:** No spouse info **VI:** Gravestone was moved from his plantation located one mile East of Jetersville in Amelia County, however his remains were not removed. **P:** No **BLW:** No **PH:** No **SS:** B pg 37 **BS:** 245

*PHILLIPS, John; b 1765 Amherst Co, d 22 Nov 1822 **RU:** Private, 53rd VMR, a company of Campbell County, attached to 5th VMR (Mason & Preston) **CEM:** Old Phillips; Campbell; Evington **GS:** No **SP:** Mar (Campbell Co, 12 Dec 1808) Margaret Weber (Campbell Co, 1780-Dec 1833), daug of John (1707-1805) & Elizabeth Margaret (__) (1748-1794) Weber **VI:** No further data **P:** No **BLW:** No **PH:** No **SS:** A rec 14036-7; B pg 53 **BS:** 345

PHILLIPS, Martin; b bef 1799, d unk **RU: Private, 2nd VMR **CEM:** Hollywood; Richmond City; 412 S Cherry St **GS:** Unk **SP:** Mar (Henrico Co, 16 Sep 1822), Ann Jackson **VI:** No further data **P:** No **BLW:** No **PH:** No **SS:** A rec 13375 **BS:** 260

*PHILLIPS. Randolph B; b 3 Oct 1784, d 1 Sep 1782 **RU:** Private, 17th VMR, Capt Nicholas Faulkner's Company, Cumberland Co, attached to 4th VMR (McDowell-Knootz-Chilton) **CEM:** Mt Zion Baptist Church; Buckingham; 6277 Cartersville Rd (Rt 610) **GS:** Yes **SP:** No spouse info **VI:** No further data **P:** Applied **BLW:** No **P:** Yes **SS:** B pg 64; M pg 259; P **BS:** 31

Key: *Additional veteran entry   **Corrected veteran entry   ***Deleted veteran entry
RU=Rank/Unit   CEM=Cemetery   GS=Gravestone   SP=Spousal Information   VI=Other Veteran Info   P=Pension
BLW=Bounty/Land Warrant   PH=Photo   SS=Service Source   BS=Burial Source   VMR= VA Military Regiment
LNR= Last Known Residence

*PINKARD, John; b 1765, d 4 Apr 1844 **RU:** Captain, 110th VMR, company commander, Franklin Co. **CEM:** Jamison Family; Franklin; Snow Creek **GS:** Yes **SP:** Mary Waller **VI:** Son of John Pinkard, Rev War Patriot (Lancaster Co, 1740-28 Feb 1782), and Jane Jett (Fauquier Co 1757- __) **P:** No **BLW:** No **PH:** No **SS:** B pg 77 **BS:** 245

*PIPER, John Henry; b 7 Jul 1776, d 26 Sep 1831 **RU:** Private, in one of several companies from Shenandoah Co attached to the 6th VMR (Coleman). It should be noted that this service could apply to another person of this name born c1793, and died in Frederick Co 1869 **CEM:** Trinity Lutheran; Frederick; 810 Fairfax Pike; Stephens City **GS:** Yes **SP:** Mar (Frederick Co, 10 Mar 1807) Margaret Snapp **VI:** No further data **P:** No **BLW:** No **PH:** No **SS:** A rec 16965 **BS:** 56 pg 301; 245

*PITMAN, Lawrence; b 17 Aug 1773, d 19 Oct 1850 **RU:** Private, 13th VMR, Capt Samuel Hawkin's Company, Shenandoah Co, attached to 5th VMR (Mason-Preston) & 4th VMR (Boyd) **CEM:** Mt Jackson; Shenandoah; Rt 11, S of jct Rt 263 **GS:** Yes **SP:** Mary Catherine Will (1774-1832) **VI:** Son of Nicholas Pitman **P:** Yes **BLW:** No **PH:** No **SS:** B pg 184; BD pg 1430 **BS:** 245

*PITMAN, Phillip; b 22 Jan 1797, d 26 Mar 1876 **RU:** Private, 13th VMR, Captain Samuel Hawkin's Company, Shenandoah Co, attached to 5th VMR (Mason-Preston) & 4th VMR (Boyd) **CEM:** Snapp; Shenandoah; Edinburg **GS:** Yes **SP:** Mary Susan Houston (1800-1863) **VI:** Son of Lawrence and Mary Catherine (Will) Pitman; represented County in House of Delegates, on a list of prisoners in the war **P:** Yes **BLW:** No **PH:** No **SS:** B pg 184; BD pg 1429 **BS:** 245

*POLLARD, Benjamin; b 16 Aug 1788, d 13 Sep 1860 **RU:** Captain, 7th VMR, Company commander from Hanover Co, attached to 9th VMR (Sharp) **CEM:** Cedar Grove; Norfolk; 238 E Princess Anne Rd **GS:** Yes **SP:** Mar (1830) Eliza Nelson Page (15 Oct 1795-30 Jun 1839), daug of Mann Page and Elizabeth Nelson **VI:** No further data **P:** No **BLW:** No **PH:** No **SS:** B pg 95 **BS:** 245

*POLLARD, John; b 5 Aug 1785, d 22 Dec 1864 **RU:** Ensign, 39th VMR, Capt Edward Pescud's Co, Petersburg's County, attached to 6th VMR **CEM:** Blandford; Petersburg; 111 Rochelle Lane **GS:** Unk **SP:** No spouse info **VI:** Mayor of Petersburg; Right Eminent Grand Commander of VA, Knights Templar; Grand Master, Blandford Lodge #3; Lieutenant of Militia after war period **P:** No **BLW:** No **PH:** No **SS:** B pg 160; L pg 635 **BS:** 245

POLLOCK, Allan; b 20 Jan 1786, in Glasgow, Scotland, d 29 Jul 1816, Chelsea nr Richmond **RU: Private, 19th VMR (Ambler), Richmond City **CEM:** St John's Church; Richmond City; 24th & Broad, Church Hill **GS:** Unk **SP:** No spouse info **VI:** Merchant **P:** No **BLW:** No **PH:** No **SS:** A rec 20482 **BS:** 245

*POPE, William; b 23 Oct 1762, Louisa Co, d 19 Jul 1852 **RU:** Private, 40th VMR, Captain Davis Watson's Troop of Cavalry, Louisa Co, attached to Cocke's Detachment and Major William Armistead **CEM:** Dabney; Powhatan; Montpelier **GS:** Yes **SP:** Ann Woodson (14 Jun 1774-28 Oct 1823), daug of Charles Woodson and Nancy Trotter **VI:** Served in Rev War also was Commonwealth Attorney of Powhatan Co. After war made Captain **P:** No **BLW:** No **PH:** No **SS:** A rec 20139; B pg 124 **BS:** 245

*PRESTON, William; b 5 Sep 1770, Botetourt Co, d 24 Jan 1821 **RU:** Captain, US Army, commissioned 5 Mar 1792, perhaps recd Major rank before moving in 1814 to Louisville, KY **CEM:** Preston Family; Montgomery; GPS: 37.21577,-80.42781; off Southgate Drive, Blacksburg **GS:** Unk **SP:** Caroline Hancock (26 Mar 1785-KY, 20 Dec 1847) **VI:** Son of William Preston (25 Dec 1729-28 Jun 1823) and Susanna Smith (23 Jan 1740-19 Jun 1823). Was surveyor Montgomery Co, 1791 and Director Louisville Branch, Bank of KY **P:** No **BLW:** No **PH:** No **SS:** Fold3 Heitman's Register **BS:** 245

*PRICE, Francis, (Note this could be the same individual as Francis Pride); b 14 Feb 1793 Eng, d 3 Jul 1836 **RU:** 33rd VMR, Capt Abraham Cowley's Co, Henrico County **CEM:** Blandford; Petersburg; 111 Rochelle Ln **GS:** Yes **SP:** No spouse info **VI:** No further data **P:** No **BLW:** No **PH:** Yes **SS:** B pg 99; K pg 250 **BS:** 245

Key: *Additional veteran entry   **Corrected veteran entry   ***Deleted veteran entry
RU=Rank/Unit   CEM=Cemetery   GS=Gravestone   SP=Spousal Information   VI=Other Veteran Info   P=Pension
BLW=Bounty/Land Warrant   PH=Photo   SS=Service Source   BS=Burial Source   VMR= VA Military Regiment
LNR= Last Known Residence

82

*PRICE, John; b 1785, d Oct 1840 **RU:** Private, 31st VMR, Capt Thomas Robert's Co, Frederick County, attached to 4th VMR (Beatty) **CEM:** Mt Hebron; Frederick; 305 E Boscawen St; Winchester **GS:** Yes **SP:** Mar (21 Aug 1806) Elizabeth Cowdery **VI:** No further data **P:** No **BLW:** No **PH:** No **SS:** A rec 24674; B pg 79; S pg 106 **BS:** 245

*PRICE, Joseph F; b 29 May 1779, Hanover, Co, d 10 Nov 1862 **RU:** Captain, 74th VMR, Company commander from Hanover Co. **CEM:** Hollywood Cemetery; Richmond City; 412 S Cherry St **GS:** Yes **SP:** Elizabeth F Winston (1784-Jan 1853) **VI:** Son of Thomas Price (29 Aug 1754-21 Dec 1836) and Barbara Winston (11 Jan 1758-21 May 1831) **P:** No **BLW:** No **PH:** No **SS:** B pg 95 **BS:** 245

*PRIDE, Thomas; b c1794, d 1842 **RU:** Private; 23rd VMR. Capt Benjamin Grave's Co, Chesterfield County, attached to 4th VMR (McDowell, Koontz, Chilton) **CEM:** Pride Family; Amelia; Pridesville Rd nr Flat Creek and Deep Creek **GS:** Unk **SP:** No spousal info **VI:** Owned 395 acres Amelia Co, 1820 **P:** No **BLW:** No **PH:** No **SS:** B pg 60; L pg 373 **BS:** 245

*PURCELL, James H; b unk, d 12 Aug 1869 **RU:** Private, Capt P N Burring, Jr's Co, S.F. Regt, US Army **CEM:** Ketoctin Baptist; Loudoun; Alder School Rd (Rt 711); Eubanks **GS:** Yes **SP:** No spouse info **VI:** Son of Thomas Purcell, Jr and Lydia Vernon. He served at Sandy Point according to Muster Roll (10 Dec 1814) **P:** No **BLW:** No **PH:** No **SS:** AF **BS:** 73 pg 252

PURKINS, Thomas; b 25 Apr 1791; d 16 Jun 1855 **RU: Lieutenant, 45th VMR, probably Capt John C. Edrington's Co, Stafford County **CEM:** Hollywood; Stafford; Hollywood Farm, Rt 601 about 600 feet from house **GS:** Yes **SP:** Frances (Fannie) P Greenlaw (14 Aug 1804-11 May 1857) **VI:** Son of William & Sarah Perkins; titled "Colonel" on gravestone and in Stafford Co records, Postmaster, Monteithville, Stafford Co **P:** No **BLW:** No **PH:** No **SS:** A rec 27540 **BS:** 26 pg 248

QUARLES, John D; b 1799; d 12 Oct 1844 **RU: Private, 19th VMR, Captain Anderson Miller's Company, Richmond City **CEM:** St John's Church; Richmond City; 24th & Broad; Church Hill **GS:** Unk **SP:** Ann Mullen (1792-Feb 1854) **VI:** No further data **P:** No **BLW:** No **PH:** No **SS:** B pg 175; L pg 590-592 **BS:** 63 pg 180; 252 pg 64; 245

*RAMEY, John; b c1786, d 4 Apr 1872 **RU:** Private 13th or 97th VMR, Capt Peter Hay's Co, attached to 1st VMR (Yancey) **CEM:** Ramey Family; Warren; GPS: 38.5305,-78,1424, Dungadin Rd abt .5 mi fr jct with Eldridge Rd **GS:** Yes **SP:** Mar 24 Feb 1812, Shenandoah Co, Mary Henry **VI:** Son of Thomas Ramey and Sarah Jones **P:** No **BLW:** No **PH:** No **SS:** A rec 2018; B pg 242 **BS:** 31

RANDOLPH, William B; b 1794; d aft 1850 **RU: Private, 19th VMR (Ambler), Capt Robert Gamble's Co, Troop of Cavalry, Richmond City **CEM:** Hollywood; Richmond City; 412 S Cherry St **GS:** Unk, Sec 12 lots 1-7 **SP:** Mar (Frederick Co, 3 Jun 1805) Lydia Lupton, Mrs. William B Randolph also buried in this lot, interment #51 **VI:** Burial record #7612, interment #30 in this section. Age 56 years on 1850 census of Henrico Co **P:** Yes **BLW:** No **PH:** No **SS:** A rec 5784; BD pg 1465 **BS:** 237

RANDOLPH, William Fitzhugh; b 1795, "Chilhowie", Cumberland Co, d Millwood 16 Jul 1859 **RU: Private, Capt Miller's Co (VMR not determined) **CEM:** Strother / Jones; Frederick; Stephens City **GS:** Unk **SP:** Jane Cary Harrison of "Chilton", Cumberland Co **VI:** Death date from *Alexandria Gazette,* son of William Randolph of "Tuckahoe" and Lucy Bolling; a lawyer **P:** Spouse **BLW:** No **PH:** No **SS:** M pg 265; BD pg 1465 **BS:** 85 pg 41

*RANKIN, James S; b 4 May 1794, d 29 Jul 1849 **RU:** Private, in company of Augusta Co attached to 5th VMR (McDowell's) **CEM:** Northern Methodist; Rockingham; Mount Crawford **GS:** Yes **SP:** Kitty (__) (3 May 1790-23 Nov 1847) **VI:** No further data **P:** No **BLW:** No **PH:** No **SS:** A rec 3300; M pgs 39-40 **BS:** 245

RICE, John, Jr; b 1769, Culpeper Co; d 11 Mar 1836, Rockingham Co **RU: Sergeant, 5th VMR **CEM:** Fishback Family; Rockingham; Rt 42, Dayton **GS:** No **SP:** Mar (6 Jan 1791) Lucinda Ann Jones (1767-1843) **VI:** Died age 67 years. No stone remains. Surveyed by J. Robert Swank in 1967. Son of John Rice and Mary Finney, daug of James Finney and Elizabeth Turner; Gravestone from Rice Family Cem **P:** No **BLW:** No **PH:** No **SS:** A rec 11468 **BS:** 245

Key: *Additional veteran entry   **Corrected veteran entry   ***Deleted veteran entry
RU=Rank/Unit   CEM=Cemetery   GS=Gravestone   SP=Spousal Information   VI=Other Veteran Info   P=Pension
BLW=Bounty/Land Warrant   PH=Photo   SS=Service Source   BS=Burial Source   VMR= VA Military Regiment
LNR= Last Known Residence

83

*RICE, John Holt; b 28 Nov 1777, Bedford Co, d 3 Sep 1831 **RU:** Private, 63rd VMR, Capt Samuel Allen's, Troop of Calvary, Prince Edward Co, attached to 1st VMR (Holcombe) **CEM:** Union Theological Seminary; Prince Edward; .3 mi W of Hampton Sydney **GS:** Yes **SP:** Anne Smith Morton (1785- 1867) **VI:** Son of Benjamin Rice (1735- 1827) and Catherine Holt (Hanover Co, 1744); had degree as DD and was first professor of Christian Theology **P:** No **BLW:** No **PH:** No **SS:** A rec 11447; B pg 167; L pg 88 **BS:** 245

*RICE, Thomas; b 19 Aug 1794, d 3 Dec 1849 **RU:** Corporal, 58th VMR, Captain Robert Hooke's Company of Rockingham Co, attached to McDowell's Flying Camp **CEM:** Rice Family; Rockingham; W of Dayton **GS:** Yes **SP:** Mar (1 Sep 1828) Martha Stephens (Wythe Co, 14 Jan 1805-Wythe Co, 5 Dec 1881), daug of Lawrence Stephens and Joanne Herbert **VI:** Son of John Rice (Culpeper Co, 1767- Rockingham Co, 1836), and Lucinda Ann Jones (Madison Co, 5 1767-5 Oct 1843). Methodist Minister **P:** Yes, widow **BLW:** No **PH:** No **SS:** B pg 181; K pg 17; M pg 267 **BS:** 245

*REYNOLDS, John Devin; b 22 Feb 1783, d 18 Mar 1831 **RU:** Private, 117th VMR, Captain John Fariss's Co, Campbell County **CEM:** Worlds; Pittsylvania; Calands **GS:** Yes **SP:** Mar (Henry Co, 19 Jul 1808) Sarah Ann Philpott (1781-1852) **VI:** Son of Joseph D. Reynolds (1759-1840); also served as Private in Halifax Co **P:** No **BLW:** No **PH:** No **SS:** B pg 53; L pg 313 **BS:** 245

*RICHARDSON, George Washington; b 10 Apr 1773, d 17 Sep 1849 **RU:** Lt Colonel 12th VMR, Regimental Commander, Fluvanna Co **CEM:** Bowles Family; Goochland; Tabscott **GS:** Yes **SP:** Mar (Fluvanna Co, 21 Feb 1799) Elizabeth Barrett "Betsy" Payne **SP:** Son of Samuel Richardson & Ann Thompson; Was first Postmaster in Wilmington in County **P:** No **BLW:** No **PH:** No **SS:** B pg 75 **BS:** 245

*RICHARDSON, John D; b 31 Jan 1779, d 31 Jun 1855 **RU:** Captain, 26th VMR, Company Commander, Charlotte County, attached to 6th Artillery (Read) **CEM:** Pettus-Richardson Family; Charlotte; Keysville **GS:** Unk **SP:** Mar Elizabeth Spencer (1782-__) **VI:** No further data **P:** No **BLW:** No **PH:** No **SS:** A rec 12749; B pg 57; **BS:** 245

RICHARDSON, Robert P; b 12 May 1798, James City Co; d 30 Oct 1841 **RU: Private, 22nd VMR, Mecklenburg Co **CEM:** Shockoe Hill; Richmond City; 100 Hospital St **GS:** Yes **SP:** Mar (1838) Ann Elizabeth Carter Lyons, daug of John of Studly, VA **VI:** Doctor. Son of Allen & Elizabeth Richardson **P:** No **BLW:** No **PH:** No **SS:** A rec 12936 **BS:** 38 pg 24; 199

*RICHARDSON, William Jr; b 23 Mar 1783, Washington Co, d 21 Jan 1835 **RU:** Private, 70th or 125th VMR, Captain Fulkerson's Company, Washington Co, attached to 7th VMR (Saunders) **CEM:** Richardson Family; Smyth; Rt 610 **GS:** Yes **SP:** Rhoda Hicks (1789-9 Apr 1847) **VI:** Son of William Richardson (1748-1835) and Rebecca Hays (1750-1810), obtained rank of Captain after the war **P:** No **BLW:** No **PH:** No **SS:** A rec 13134; B pg 198 **BS:** 245

*RICHARDSON, William H; b Unk, d Jul 1867 **RU:** Captain, 19th VMR commanded company, Richmond City, attached to 1st Corps de Elite (Randolph's) **CEM:** Hollywood; Richmond City; 412 S Cherry St **GS:** Yes sec R lot 165 **SP:** No spousal info **VI:** No further data **P:** No **BLW:** No **PH:** No **SS:** B pg 175 **BS:** 245

*RILEY, William; b 1782, d 17 Dec 1825 **RU:** Private, 13th or 97th VMR, Captain Samuel Hawkin's Company, Shenandoah Co, attached to 5th VMR (Mason and Preston) **CEM:** Saint Pauls Lutheran Church; Shenandoah; Strasburg **GS:** Yes **SP:** Mary Miller (__-16 Oct 1846) VI: No further data **P:** No **BLW:** No **PH:** No **SS:** B pgs184-185 **BS:** 245

*RITCHIE, John T; b 1788, d 26 Jun 1831, **RU:** Lieutenant, US Navy **CEM:** Arlington National Cemetery; Arlington; Jefferson Davis Hwy Rt 110 **GS:** Yes **SP:** No spouse info **VI:** Commissioned, US Navy (17 Dec 1810), Commissioned Lieutenant (27 Apr 1816), originally buried at Old Presbyterian Cem, Georgetown, DC **P:** No **BLW:** No **PH:** No **SS:** AB **BS:** 245

ROBERSON, Thomas; b 1795; d 11 May 1877 **RU: Private, 45th VMR (Peyton), Stafford Co **CEM:** Roberson Family; Stafford County; Dove Road, about 200 yards beyond house and barn **GS:** Yes **SP:** Elizabeth (___) **VI:** Son of George and Fenton Roberson. ("*Virginia Deaths and Burials 1853-1912,*" *familysearch.org*) **P:** No **BLW:** No **PH:** No **SS:** A rec18068 **BS:** 26 pg 16sp

---

Key: *Additional veteran entry   **Corrected veteran entry   ***Deleted veteran entry
RU=Rank/Unit   CEM=Cemetery   GS=Gravestone   SP=Spousal Information   VI=Other Veteran Info   P=Pension
BLW=Bounty/Land Warrant   PH=Photo   SS=Service Source   BS=Burial Source   VMR= VA Military Regiment
LNR= Last Known Residence

*ROBERTS, William; b 16 Feb 1770, Mt Airy, Surry Co, NC, d 1838 **RU:** Corporal, 78th VMR one of the companies, Grayson County **CEM:** Comers Rock; Grayson Co; Town of Comers Rock **GS:** Unk **SP:** Mar 1) Mary Elizabeth Davis 2) Lydia Lewis (WV, 1785-1856 **VI:** Son of John Roberts and Elizabeth Russell **P:** No **BLW:** No **PH:** No **SS:** A box 177, mfilm M602; B pg 86 **BS:** 245

*ROBERTS, William; b unk, d 1843 **RU:** Ensign, 56th VMR Capt Thomas Gregg's Co, Loudoun County, attached to 4th VMR (Beatty's) **CEM:** Goose Creek Burying Ground; Loudoun; Rt 722 Lincoln **GS:** Yes **SP:** Rebecca (__) **VI:** No further data **P:** No **BLW:** No **PH:** No **SS:** A rec 18986; White pg 1504 **BS:** 245

*ROBERTSON, Edward, Jr.; b 1790, d 1834 **RU:** Private, Captain James Lanier's Company of Cavalry, Pittsylvania Co, attached to Capt Sale's Battalion of Cavalry **CEM:** Robertson Family; Pittsylvania; Dry Fork **GS:** Yes **SP:** Mar (29 Sep 1812) Nancy R Fuqua Shepherd **VI:** Son of Edward Robertson and Mary Pulliam Thompson **P:** Yes- Spouse, # SC-7106 **BLW:** No **PH:** No **SS:** BD pg 1505; M pg 270 **BS:** 245

*ROBERTSON, John E; b unk, d 14 Mar 1867 **RU:** Private 57th VMR a company in Loudoun Co Militia **CEM:** Pohick Episcopal Church; Fairfax; jct Rts 1 & 611, Lorton **GS:** Yes **SP:** No spousal info **VI:** No further data **P:** No **BLW:** No **PH:** No **SS:** A rec 19469 **BS** 245

ROBERTSON, William W; b 1774; d 1 Nov 1842 **RU: Private, 45th VMR (Peyton), Capt Thomas Hill's Company, Stafford Co **CEM:** Aquia Episcopal Church; Stafford; jct Rt 1 and 610; Aquia **GS:** Yes **SP:** Eleanor (__) (__)-4 Apr 1855, age 77 years) **VI:** Died age 68 years **P:** No **BLW:** No **PH:** No **SS:** L pg 428 **BS:** 26 pg 127

*ROBINSON, John; b unk, d 1832 **RU:** Sergeant, 53rd VMR, Capt James Dunningham's Artillery Company, Campbell Co, attached to Cocke's Detachment **CEM:** Presbyterian; Lynchburg City; Grace & Bailey St **GS:** No **SP:** No spouse info **VI:** No further data **P:** No **BLW:** No **PH:** No **SS:** B pg 53; K pg 168 **BS:** 245

RODGERS, Levi; b 1767, d 21 Apr 1819 **RU: Private, MD Militia 2nd Regiment, (Schucht) **CEM:** Rodgers-Boggs; Accomack; loc one mi N of Evans Wharf on E side Rt 638; **GS:** Yes **SP:** Euphony (__) (19 Mar 1778-10 Mar 1826) **VI:** No further data **P:** No **BLW:** No **PH:** No **SS:** A rec 22479 **BS:** 129

*ROGERS, Casper; b c1792, d abt 1855-1860 **RU:** Private, Frederick Co; company attached to 4th VMR (Beatty) **CEM:** Dalby-Rogers Quaker Graveyard; Frederick; nr Gainesboro **GS:** Yes **SP:** Mar 1) (24 Apr 1815) Susannah Ellis (__)-12 Apr 1831, age 31, 2) (16 Aug 1832) Elizabeth Ann Brown, daug of Adam and Christine (Zuber) Brown **VI:** Son of John and Mary (Rinker) Rogers **P:** No **BLW:** No **PH:** No **SS:** A rec 23008 **BS:** 56 pg 322

*ROGERS, Evan (Even); b 26 Dec 1793, d 12 Dec 1876 **RU:** Private, company, Frederick County attached to 4th VMR (Boyd) **CEM:** Dalby-Rogers Quaker Graveyard; Frederick; nr Gainesboro **GS:** Unk **SP:** Hannah Dalby (1800-1845) **VI:** Son of John (1759-1823) Rogers and Maria Magdalene (Rinker) (1759-1823) Rogers **P:** No **BLW:** No **PH:** No **SS:** A rec 23 **BS:** 56 pg 323

*ROGERS, Thomas III; b 1772, d 1844 **RU:** Cornet, 83rd VMR, Captain William Ross's Troop of Cavalry, Dinwiddie Co **CEM:** Butterwood; Dinwiddie; Darvills **GS:** Yes **SP:** No spouse info **VI:** No further data **P:** No **BLW:** No **PH:** No **SS:** B pg 66; L pg 678 **BS:** 245

*ROPP, Nicholas; b 14 Sep 1784, d 16 May 1859 **RU:** Private, 56th VMR of Loudoun Co **CEM:** Saint Pauls Lutheran Church; Loudoun; Neersville **GS:** Yes **SP:** Elizabeth Waltman (28 Mar 1789-11 May 1845) **VI:** No further data **P:** No **BLW:** No **PH:** No **SS:** A rec 648 **BS:** 245

*ROSE, John, Jr; b 1770, Rowan Co, NC, d 1850 **RU:** Private, 1st Regt (McDonald), NC Militia **CEM:** Vance; Dickenson; GPS: 37.07940,-82.44124; vic jcts Rt 649 and 708 **GS:** Yes **SP:** Mary Stafford or Sarah Bradford **VI:** Son of John Rose, Sr and Mary Street **P:** No **BLW:** No **PH:** No **SS:** A rec 956 **BS:** 245

Key: *Additional veteran entry   **Corrected veteran entry   ***Deleted veteran entry
RU=Rank/Unit   CEM=Cemetery   GS=Gravestone   SP=Spousal Information   VI=Other Veteran Info   P=Pension
BLW=Bounty/Land Warrant   PH=Photo   SS=Service Source   BS=Burial Source   VMR= VA Military Regiment
LNR= Last Known Residence

85

*ROSS, William; b unk, d 15 Jun 1851, Patrick Co **RU:** Private, 18th VMR, Capt Abraham Penn's Company, Patrick Co, attached to 7th VMR (Saunders) **CEM:** Daniel Ross, Jr Family; Patrick; Elamsville **GS:** Unk **SP:** No spouse info **VI:** Son of Daniel Ross (1746-1822) **P:** No **BLW:** No **PH:** No **SS:** A rec 1884; B pg 157 **BS:** 245

ROSS, William, Sr; b unk; d 3 Feb 1836 **RU: Private, 7th VMR (Saunders) **CEM:** Tinkling Spring; Augusta; 11 mi NE of Staunton **GS:** Yes **SP:** No spouse info **VI:** Died age 45 (tombstone) **P:** No **BLW:** No **PH:** No **SS:** A rec 1884 **BS:** 183

*ROWE, Jesse, b 9 Aug 1795, d unk **RU:** Private, 45th VMR, Thomas Alexander's Company, Stafford Co **CEM:** Rowe Family; Stafford; Rt 218, last driveway on north side before White Oak Run **GS:** Unk **SP:** Alsey Humphries (abt 1810-__), daug of L Raleigh Humphrey/Humphries and Polly Ferguson **VI:** Son of John Rowe (1762-1799) and Sarah Peyton (1767-1856), Justice of the Peace **P:** No **BLW:** No **PH:** No **SS:** Yes, L pg 84 **BS:** 26 pg 127

ROWE, John Gaskin; b 1 Nov 1788; d 1862 **RU: Sergeant, 45th VMR, Capt Thomas Alexander's Co, Stafford County **CEM:** Rowe Family; Stafford; Rt 218, last driveway on N side before White Oak Run **GS:** Unk **SP:** Nancy McGuire (1800-1858), daug of Thomas McGuire (1773-1821) **VI:** Son of John Rowe(1762-1799) and Sarah Peyton (1767-1856), Justice of Peace (1856) **P:** No **BLW:** No **PH:** No **SS:** A rec 2885 **BS:** 26 A

*ROWLAND, William; b 23 Mar 1783, d 7 Nov 1857 **RU:** Lieutenant, 69th or 84th VMR, Company commander from Halifax Co, attached to Pryor's Detachment **CEM:** Elmwood; Norfolk; 238 E Princess Anne Rd **GS:** Yes **SP:** No spouse info **VI:** No further data **P:** No **BLW:** No **PH:** No **SS:** B pg 90 **BS:** 245

RUFFNER, John; b 3 Dec 1792; d 23 Nov 1863 **RU: Sergeant, 6th VMR (Coleman) **CEM:** Ruffner/Bauserman; Page; Rt 615, Luray **GS:** Yes **SP:** Elizabeth Long **VI:** Son of Peter Ruffner and Elizabeth Burner **P:** No **BLW:** No **PH:** No **SS:** A rec 4219 **BS:** 115 pg 184; 245

*RUSMISEL, Adam; b Sep 1785, Berks Co, PA, d 14 Aug 1863 **RU:** Private, 121st VMR, Capt Joseph Hannah's Co, Botetourt County, attached to Flying Camp(McDowell) **CEM:** Union Presbyterian Church; Augusta; Churchville **GS:** Yes **SP:** Mar (Augusta Co, 30 Aug 1808) Salone (Sally) Sherman **VI:** Son of Adam Rusmisel and Rachel Shoemaker **P:** Spouse **BLW:** No **PH:** No **SS:** A rec 5115; B pg 45; BD pg 1531; M pg 273 **BS:** 245

*RUSMISEL, Christian; b 14 Feb 1787 d 7 Sep 1860 **RU:** Private, 32nd VMR, Capt Hannah Kirk's Co, Augusta County, attached to 5th VMR(McDowell) **CEM:** Union Presbyterian Church; Augusta; Churchville **GS:** Yes **SP:** 1) Anny Shelby 2) Martha A, McMullen (11 Feb 1805-29 Jun 1887), daug of James and Mary (Henderson) McMullen **VI:** Son of Adam Rusmisel and Rachel Shoemaker **P:** Widow Martha **BLW:** No **PH:** No **SS:** B pg 40; BD pg 1531; M pg 273 **BS:** 245

*RUSSELL, Benjamin C; b 18 Mar 1782, d 6 Oct 1836 **RU:** Private, 56th VMR, a company of Loudoun Co, attached to 5th VMR **CEM:** Old Stone Church; Loudoun; Leesburg **GS:** Yes **SP:** No spouse info **VI:** No further data **P:** No **BLW:** No **PH:** No **SS:** A rec 5432; B pgs 119-122 **BS:** 245

*RUST, George; b 4 Jan 1788, d 18 Sep 1857 **RU:** Private, 56th VMR, Loudoun Co, company **CEM:** Union; Loudoun; 323 N. King St, Leesburg **GS:** Yes **SP:** No spouse info **VI:** Son of George Rust and Elizabeth (__) **P:** No **BLW:** No **PH:** No **SS:** A rec 6081; B pg 119 **BS:** 245

RUST, John; b 8 Feb 1769; d 17 Apr 1851 **RU: Private, 116th VMR, Capt Daniel Matthew's Company, Rockingham Co, attached to McDowell's Flying Camp & 4th VMR **CEM:** Prospect Hill; Warren; Front Royal **GS:** Unk **SP:** Elizabeth Marshall (1779-1857) **VI:** No further data **P:** Both **BLW:** No **PH:** No **SS:** B pg 182; BD pg 1533 **BS:** 150

Key: *Additional veteran entry   **Corrected veteran entry   ***Deleted veteran entry
RU=Rank/Unit   CEM=Cemetery   GS=Gravestone   SP=Spousal Information   VI=Other Veteran Info   P=Pension
BLW=Bounty/Land Warrant   PH=Photo   SS=Service Source   BS=Burial Source   VMR= VA Military Regiment
LNR= Last Known Residence

RUST, Youel S; b 1789, bur 9 Jul 1863 **RU: Private,19th VMR, Capt William Murphy's Co, Richmond Light Infantry Blues, Richmond City, attached to 1st Corps d'Elite (Randolph) **CEM:** Shockoe Hill; Richmond City; 100 Hospital St **GS:** Unk **SP:** 1) Agnes Davidson (1800-1833) 2) (__-Powhatan Co, 14 Dec 1836) Elivira Jane Watkins, daug of Jabez Watkins. (Virginia Marriages (1785-1940), *"Familysearch.org"*) **VI:** Memorialized with two wives **P:** No **BLW:** No **PH:** No **SS:** L pg 607 **BS:** 63 pg 498

*RUTHERFORD, Archibald,"Sibald" b 16 Jan 1774, Shenandoah Co, d 10 Sep 1830, Harrisonburg **RU:** Major, 116th VMR staff officer, Rockingham Co **CEM:** Woodbine; Harrisonburg City; 381 East Market St, Rt 33 **GS:** Yes **SP:** Mar (24 Jun 1795) Jane Burgess (3 Aug 1768-2 Aug 1833) **VI:** Son of Robert Rutherford (1749-1811) and Mary Sevier (1753-1803). Member of VA House of Delegates (1807-1816) and (1824-1825) **P:** No **BLW:** No **PH:** No **SS:** B pg 181 **BS:** 245

*SAMPSON, Richard; b 23 May 1774, d 18 Mar 1864 **RU:** Private 33rd VMR, Captain Samuel Brown's Company, Henrico Co **CEM:** Boscobel; Goochland; ¼ mile NW at end of Pine Tree Hollow Rd **GS:** Yes **SP:** Mary Rogers (1791-1856) **VI:** No further data **P:** No **BLW:** No **PH:** No **SS:** L pg 177 **BS:** 245

SANFORD, Lawrence; b 10 Aug 1778, Westmoreland Co; d 18 Aug 1858 **RU: Private, 45th VMR, Capt John Edrington's Co, Stafford County **CEM:** Sanford Family #1; Stafford; Greenbank Rd; Rt 656 then 0.1 mile E on private rd on Celebrate Virginia site **GS:** Yes **SP:** Aphia Farmer (1774-1864), daug of David Farmer **VI:** Son of Joseph Sanford and Sarah Bunbury **P:** No **BLW:** No **PH:** No **SS:** A rec 9812 **BS:** 26 pg 352

*SAUNDERS, William L; b 1 Aug 1798, d 1 Nov 1882 **RU:** Private, 10th VMR, Capt Joshua Early & LT Nicodemus Leftwich's Companies, Bedford Co, attached to 3rd VMR (Dickinson) **CEM:** Presbyterian; Lynchburg; Grace & Bailey St **GS:** Yes **SP:** 1) Sarah L (__), 2) Mary Shields Camm (16 Sep 1806-12 Feb 1886) **VI:** No further data **P:** William applied for pen, Sarah recd **BLW:** No **PH:** No **SS:** A rec 11089; B pg 42; BD pg 1546; M pg 276 **BS:** 245

*SAYERS, Robert: b 7 Aug 1796, d 26 Nov 1876 **RU:** Private, 35th VMR, Wythe Co **CEM:** Oglesby; Wythe; Fort Chiswell **GS:** Yes **SP:** Senath Ross (12 Oct 1804-1 Oct 1886) **VI:** Son of John Thompson Sayers (1758-1816) and Susanna Crockett (1764-1828) served as a Major after the war **P:** No **BLW:** No **PH:** No **SS:** A rec 11919; B pg 204 **BS:** 245

*SCOTT, Thomas T; b 6 Dec 1765, d 1818 **RU:** Major; 63rd VMR, staff officer Prince Edward Co **CEM:** Scott Family; Prince Edward; Briery **GS:** Unk **SP:** No spousal data **VI:** Son of Thomas Scott (Gloucester Point, 21 Mar 1727-29 Nov 1804) and Catherine Tomkies (1733-1766), Commissioned Major 29 May 1811 **P:** No **BLW:** No **PH:** No **SS:** B pg 167 **BS:** 245

*SCOTT, Thompson; b unk, d 1825, Amelia Co **RU:** Captain 98th VMR, Company commander, Mecklenburg Co (raised in Amelia Co) attached to Colonel Green's Regt at Norfolk **CEM:** Scott Family; Amelia **GS:** Unk **SP:** Nancy Chaffin **VI:** Son of George Scott (__-1814) and Elizabeth Thompson **P:** No **BLW:** No **PH:** No **SS:** B pg 131 **BS:** 245

*SCOTT, William; b 29 Jan 1778, Washington Co, d 25 May 1862 **RU:** Private, 70th or 105th VMR, company of Washington Co, attached to Bradley's Regt **CEM:** Scott Family; Smyth; vic I-81 NE of Atkins **GS:** No **SP:** Elizabeth Porter (18 Jan 1794-18 Sep 1848) **VI:** Son of James Scott (1736-1817) and Rachel Holmes (1753-1833) **P:** No **BLW:** No **PH:** No **SS:** A rec 14232; B pgs 198; 194 **BS:** 245

*SCRIVNER, Benjamin; b 1796, d 12 Sep 1878 **RU:** Private, 31st VMR, Capt Michael Coyle's Company, Frederick Co, attached to 1st VMR (Taylor) **CEM:** Trone Method Church; Frederick; Timberdridge **GS:** Unk **SP:** Mar (Frederick Co, 2 Jul 1820) Barbara Brunner **VI:** Son of Vincent Scrivener & Mary Russell **P:** No **BLW:** No **PH:** No **SS:** A rec 14678; B pg 79; S pg 106 **BS:** 56 pg 328; 245

*SELDON, William Boswell; b 31 Aug 1772, d 18 Jul 1849 **RU:** Third Lieutenant, 19th VMR, served in one of three companies of Richmond City attached to 2nd VMR (Ballowe) **CEM:** Cedar Grove; Norfolk; E Princess Anne St **GS:** Yes **SP:** Charlotte Colgate (14 Jan 1781, Kent Co, Eng-16 Nov 1870), daug of Robert Colgate (1758-1826) and Sarah Ellen Bowles (__-1856) **VI:** Son of William Selden and Mary Anne Hancock **P:** No **BLW:** No **PH:** No **SS:** A rec 16746; B pg 242 **BS:** 245

Key: *Additional veteran entry   **Corrected veteran entry   ***Deleted veteran entry
RU=Rank/Unit   CEM=Cemetery   GS=Gravestone   SP=Spousal Information   VI=Other Veteran Info   P=Pension
BLW=Bounty/Land Warrant   PH=Photo   SS=Service Source   BS=Burial Source   VMR= VA Military Regiment
LNR= Last Known Residence

87

*SELLERS, John; b 10 Mar 1797. Rockingham Co, d 19 Jan 1872 Swoope, Augusta Co **RU:** Private, 58th or 116th VMR, Capt Adam Harnsberger or Capt Abraham B. Hamilton's Cos, Rockingham County, attached to 1st VMR (Trueheart) or 6th VMR (Coleman) **CEM:** Hebron Presbyterian Church; Augusta; Rt 703, 4.5 miles fr Staunton **GS:** Yes **SP:** Mar 1) (Rockingham Co, 16 Oct 1822) Catherine Brown (7 Mar 1802-17 May 1842), daug of John Brown and Elizabeth Falls, 2) (17 May 1842) Frances Sterrett Bell **VI:** Son of Henry Sellers and Magdalene Hedrick **P:** No **BLW:** No **PH:** No **SS:** A rec 16940 & 16943; B pg 181 **BS:** 245

*SETTLE, William; b 1775, d 12 Nov 1833, **RU:** Private, 41st VMR, Capt Vincent Shackleford's, Artillery Company, Richmond Co **CEM:** Settle Family; Westmoreland; Hague **GS:** Unk **SP:** Mary Polly Greenlaw (1777-1845) **VI:** Son of Bailey Settle (1731-1813) and Anne Alexander **P:** Spouse **BLW:** No **PH:** No **SS:** B pg 178; L pg 706 **BS:** 245

SEXTON, Aaron; b 3 Apr 1793; d 16 Oct 1851 **RU: Private, 3rd Regiment, East TN Militia (Johnson) **CEM:** Old Bethel Church; Smyth; Rt 627 **GS:** Yes **SP:** Mar (1816) Margaret Ann Feeley (1795-__) **VI:** Son of Thomas Campbell Sexton (1764-1849) & Charity Current (1769-1824) **P:** No **BLW:** No **PH:** No **SS:** A rec 18174 **BS:** 131 v2 pg 147

*SHOEMAKER, Abraham; b 1 Aug 1772, d 26 Nov 1842 **RU:** Lieutenant, 1st Regt Artillery (Provost's) PA Volunteers **CEM:** Goose Creek Burying Ground; Loudoun; Rt 722 Lincoln **GS:** Yes **SP:** Martha Webster (Horsham, Montgomery Co, PA 18 Jun 1778-Loudoun Co, 31 Mar 1836) **VI:** No further data **P:** No **BLW:** No **PH:** No **SS:** Fold3 General Index card **BS:** 245

*SIMMS, Richard Durrett; b 25 Nov 1788, d 15 May 1862 **RU:** 2nd Lieutenant, 88th VMR, Albemarle Co **CEM:** Watts family; Albemarle; Stoney Point **GS:** Yes **SP:** Elizabeth B. Clarkson (1798-1882) **VI:** No further data **P:** No **BLW:** No **PH:** No **SS:** A rec 802 **BS:** 245

*SHACKELFORD, George; b 4 Jul 1780, King & Queen Co, d 15 Jun 1847, Amherst Co **RU:** Captain, Virginia Inf **CEM:** Old City; Lynchburg; 401 Taylor St **GS:** Yes **SP:** No spouse info **VI:** Obtained rank of Colonel in KY after the war period and this rank inscribed on gravestone **P:** No **BLW:** No **PH:** No **SS:** D, vol 2 **BS:** 245

*SHACKLEFORD, Vincent; b 1785, d aft 13 Oct 1820 (U.S, Census) **RU:** Captain, 41st VMR Commanded an Artillery Co, Richmond County **CEM:** Shackleford; Richmond Co; Court St **GS:** Unk **SP:** Lucy Roane Brockenbrough **VI:** Son of Richard Shackleford (__-1794) and Margaret (Peggy) Landman (__)-bef 1794). Sheriff, Richmond Co 1817 and represented Co in VA Legislature. Carried a musket ball in body until his death. Plaque about him at North Farnham Episcopal Church **P:** No **BLW:** No **PH:** No **SS:** B pg 178; O -Service Index card **BS:** 245

*SHADE, Richard T; b unk,(assuming bef 1800 as person this name in county had 1812 service), d 1 Sep 1853 **RU:** Private, 31st, 51st or 122nd VMR in a company of Frederick Co, attached to 4th VMR (Boyd) or 5th VMR (Mason-Preston) **CEM:** Wesley Chapel; Frederick; 620 Chapel Hill Rd; vic jct US 522 and Rt 692, Cross Junction **GS:** Unk **SP:** No spousal info **VI:** No further data **PH:** No **SS:** A rec 1960; B pgs 78-80 **BS:** 245

*SHEETS, Henry; b 10 Feb 1785, d 25 Jun 1871 **RU:** Private, 13th VMR, Captain George Shum's Co, Shenandoah County, attached to Command of Major Nathaniel Perkin's Garrison **CEM:** Keezletown; Rockingham Co; Keezletown **GS:** Yes **SP:** Elizabeth (__) **VI:** No further data **P:** Yes & widow applied **BLW:** No **PH:** No **SS:** B pg 185; K pg 227; M pg 279 **BS:** 245

*SHEETS, John; b 8 Jan 1797, d 11 Sep 1855 **RU:** Ensign, Captain William Harrison's Co, Rockingham County, attached to 1st VMR (Trueheart) **CEM:** Mount Crawford; Rockingham Co; Mount Crawford **GS:** Yes **SP:** Sarah (__) (1810- 1852) **VI:** No further data **P:** Spouse **BLW:** No **PH:** No **SS:** B pg 181; BD pg 1573; K pg 51; M pg 279 **BS:** 245

*SHELTON, James; b unk, d 2 1841 **RU:** Private, 115th VMR, Elizabeth City Co or Warwick Co. **CEM:** Shelton; Hampton City; Rt 169; GPS: 37.03015 -76.30715 **GS:** Unk **SP:** Mary G (__) **VI:** No further data **P:** No **BLW:** No **PH:** No **SS:** A rec 22199 **BS:** 245

Key: *Additional veteran entry   **Corrected veteran entry   ***Deleted veteran entry
RU=Rank/Unit   CEM=Cemetery   GS=Gravestone   SP=Spousal Information   VI=Other Veteran Info   P=Pension
BLW=Bounty/Land Warrant   PH=Photo   SS=Service Source   BS=Burial Source   VMR= VA Military Regiment
LNR= Last Known Residence

*SHEPHERD, Royal F; b 12 Jan 1789, d 9 Jan 1850 **RU**: Private, 19th VMR, Capt John R Jones's Co, Richmond City **CEM**: Shockoe Hill; Richmond City; 100 Hospital St **GS**: Yes, in William C Smith lot, Govt gr st **SP**: No spousal info **VI**: Fr Henrico Co **P**: No **BLW**: No **PH**: Yes **SS**: B pg 175; L pg 497 **BS**: 31; 245

*SHIELDS, Mathew W; b 1785, d 7 Nov 1823 **RU**: Private, 52nd VMR, Captain William Folke's Company, Charles City Co **CEM**: Shockoe Hill; Richmond City; 100 Hospital St **GS**: Yes **SP**: Mary (__) (1799-21 Dec 1831) **VI**: No further data **P**: No **BLW**: No **PH**: Yes **SS**: B pg 143; L pg 321 **BS**: 31

*SHRIVER, Jacob, b 4 Mar 1784, d 5 Jul 1876 **RU**: Private 56th or 57th VMR a company in Loudoun Co. attached to 5th VMR (Mason and Preston) **CEM**: Saint Pauls Lutheran Church; Loudoun; Neersville **GS**: Yes **SP**: No spouse info **VI**: No further data **P**: Yes **BLW**: No **PH**: No **SS**: A rec 26194 **BS**: 245

*SHRUM, George; b 17 Sep 1774, Keezletown, Rockingham Co, d 8 Jun 1852 **RU**: Captain Company Commander, Shenandoah Co, attached to 1st VMR (Trueheart) **CEM**: Schrum Family; Shenandoah; Saint Luke **GS**: No **SP**: Sarah Alter (10 Jul 1781-17 Jan 1841) **VI**: No further data **P**: No **BLW**: No **PH**: No **SS**: A rec 26263; B pg 185 **BS**: 245

*SKINNER: Peter; b 5 May 1773, d 1 Dec 1859 **RU**: Private, 57th VMR, Loudoun County **CEM**: Skinner Family; Loudoun; Adie **GS**: Yes **SP**: No spouse info **VI**: No further data **P**: No **BLW**: No **PH**: No **SS**: A rec 3334 **BS**: 245

*SLAGLE, John; b 1777, d 1820 **RU**: Private, 77th VMR, Capt Samuel Cockerill, Hampshire Co (now WVA), attached to 7th VMR (Saunders) **CEM**: Mt Hebron; Winchester; 305 E Boscawen St **GS**: Yes **SP**: No spouse info **VI**: No further data **P**: No **BLW**: No **PH**: No **SS**: A rec 3690; B pg 92 **BS**: 86 pg 46

*SMART, William R; b 10 Jul 1784, Eng, d 10 Feb 1840 **RU**: Private, 99th VMR Capt John P Drummond, Artillery Co, 4th Arty, Accomack Co **CEM**: Ware Episcopal Church; Gloucester; 7825 John Clayton Memorial Rd **GS**: Yes **SP**: Louisa Gibbons (1794-7 Jul 1828), daug of Robert Gibbons **VI**: Emigrated early in life apparently to Accomack Co where he served; memorialized only with gravestone as buried Smart cemetery Gloucester Co **P**: No **BLW**: No **PH**: No **SS**: L pg 288 **BS**: Ware Church website burial list

*SMITH, Arthur; b 9 Oct 1779, d 6 Oct 1849 **RU**: Captain, 59th VMR, Company commander from Nansemond Co. **CEM**: Cedar Hill; Suffolk; jct Main St & Constance Rd **GS**: Yes **SP**: Susan Richardson (27 May 1781-28 Jan 1815) **VI**: No further data **P**: No **BLW**: No **PH**: No **SS**: B pg 141 **BS**: 245

*SMITH, George; b 2 Nov 1769, d 1 Oct 1844 **RU**: 51st VMR Capt James Sower's Company, Frederick Co, attached to McDowell's Flying Camp **CEM**: Smith Family; Frederick; Gore **GS**: Yes **SP**: 1) Frances Curlett, 2) Margaret McDonald, 3) (26 Nov 1819) Ann Ellis (1788-1871) **VI**: Son of Capt Jeremiah and Elizabeth Ann (Snyder) Smith **P**: No **BLW**: No **PH**: No **SS**: A rec 693; K pg 34 **BS**: 56 pg 340; 245

SMITH, George; b 19 Aug 1769; d 30 Aug 1822 **RU: Quartermaster, 36th VMR (Reno), Prince William Co **CEM**: Dumfries; Prince William; off Cameron St, SW of Dumfries Elementary School **GS**: Unk **SP**: No spouse info **VI**: No further data **P**: No **BLW**: No **PH**: No **SS**: A rec 7032 **BS**: 11 pg 21

SMITH, James; b 1769, Cahery Township, County Derry, Ireland; d 8 Feb 1832 **RU: Paymaster, 37th VMR (Downing), Northumberland Co **CEM**: Smith Family; Northumberland; Mantua Farm, Coan River **GS**: Yes **SP**: Mar (Northumberland Co, 25 Dec 1798) (Bond) to Ann Muse, daug of Daniel Muse (16 Oct 1799) in her 21st year **VI**: Died in his 64th year **P**: No **BLW**: No **PH**: No **SS**: K pg 11 **BS**: 85; 27(1)9 pg 25, 839; 49

*SMITH, Mitchell; b 1781, d 1816 **RU**: Captain, 5th VMR, Company commander in Norfolk Co **CEM**: Smith Family; Chesapeake; Mount Pleasant Farm **GS**: Yes **SP**: Olive Woodard (3 May 1785-1856) **VI**: Son of John Smith (__-1802) **P**: No **BLW**: No **PH**: No **SS**: B pg 148 **BS**: 245

*SMITH, Peter; b 27 Jul 1781, Alleghany Co, d 1858 **RU**: Captain, Company commander, 81st VMR, Bath Co, attached to McDowell's Flying Camp **CEM**: Mount Pleasant United Methodist; Alleghany; Rt 618 **GS**: Yes **SP**: Susanna Fleet **VI**: Son of William Smith (1743-1836) and Mary Wright (1759-1858) **P**: No **BLW**: No **PH**: No **SS**: B pg 41 **BS**: 245

Key: *Additional veteran entry   **Corrected veteran entry   ***Deleted veteran entry
RU=Rank/Unit   CEM=Cemetery   GS=Gravestone   SP=Spousal Information   VI=Other Veteran Info   P=Pension
BLW=Bounty/Land Warrant   PH=Photo   SS=Service Source   BS=Burial Source   VMR= VA Military Regiment
LNR= Last Known Residence

*SMITH, Philip; b 15 May 1790, d 31 Jul 1863 **RU:** Sergeant, Frederick Co VMR, in company attached to 4th VMR **CEM:** Mt Hebron; Winchester; 305 E Boscawen St **GS:** Yes **SP:** No spouse info **VI:** No further data **P:** No **BLW:** No **PH:** Yes **SS:** A rec 10019 **BS:** 31

SMITH, Phillip A; b 1788; d 1 Oct 1813 **RU: Private, Captain Robert Saunder's Troop of Cavalry **CEM:** Ware Episcopal Church; Gloucester; 7825 John Clayton Memorial Rd **GS:** Yes **SP:** Martha Tabb **VI:** Died during the war. Stone moved from Toddsbury in 1924 **P:** No **BLW:** No **PH:** Yes **SS:** L pg 131 **BS:** 31

*SMITH, Ralph; b 13 May 1792, b 22 Sep 1848 **RU:** Private. 53rd VMR, Capt John Hay's Co, Campbell County, attached to 5th VMR (Mason/Preston) **CEM:** Mount Hermon United Methodist Church; Campbell; Lynch Station **GS:** Yes **SP:** Viola (__) **VI:** No further data **P:** Widow (WC-9324) **BLW:** No **PH:** No **SS:** A rec100076; B pgs 53-54; BD pg 1617; M pg **BS:** 285

SMITH, Thomas; b 5 Mar 1785; d 13 Apr 1841 **RU: Ensign, 21st VMR, Capt Thomas Cary and Captain Catesby Jones Companies, Gloucester Co **CEM:** Ware Episcopal Church; Gloucester; 7825 John Clayton Memorial Rd **GS:** Yes **SP:** No spouse info **VI:** Eldest son of Rev Armistead & Martha Smith of Mathews Co. Stone moved from Toddsbury 1924 **P:** No **BLW:** No **PH:** Yes **SS:** K pg 266 **BS:** 82 pg 79

*SMITH, William; b 1762, d 1829 **RU:** Private, 36th VMR, Capt Richard Weedon's Co, Prince William County **CEM:** Dumfries; Prince William; off Cameron St; SW of Dumfries Elementary School **GS:** Yes **SP:** No spouse info **VI:** He was paid for 100 hours of patrolling as a citizen **P:** No **BLW:** No **PH:** No **SS:** A rec 11633; B pg 172 **BS:** 31

SMITH, William; b 6 Sep 1797, d 18 May 1887 **RU: Private, 23rd VMR, Capt John Gregory's Co, Chesterfield County **CEM:** Hollywood; Richmond City; 412 South Cherry St **GS:** Yes **SP:** Elizabeth H. Bell (1800-1879) **VI:** Major General CSA, served in Confederate Congress and was wartime Governor of VA **P:** No **BLW:** No **PH:** No **SS:** B pg 60; L pg 382 **BS:** 245

SNAPP, John, b 14 Jan 1782, d 27 Apr 1869 **RU: Captain, 58th or 116th VMR, Commanded a Troop of Cavalry, Rockingham Co **CEM:** Wisecarver; Frederick; loc vic Shenandoah Precision Gun Works and vic jct Wisecarver Ln and Rt 622 **GS:** Yes **SP:** No spousal info **VI:** Age 66 in 1850 Census, Frederick Co **P:** No **BLW:** No **PH:** No **SS:** BD pg 1623; B pg 182 **BS:** 79 pg 309; 245

SNEAD, Israel; b 30 Apr 1780; d 6 Oct 1844 **RU: Private, Capt Mask's Co, NC Militia **CEM:** Old City Cemetery; Lynchburg; 401 Taylor St **GS:** Yes **SP:** No spouse info **VI:** Inspector of tobacco, Lynchburg 1815-1839 **P:** No **BLW:** No **PH:** No **SS:** A rec 12315 **BS:** 87 pg 64; 207

*SNIDER, Daniel; b 26 Jul 1798, d 10 Feb 1872 **RU:** Private, Capt Archibald Lyle's Troop of Cavalry or Capt Isaiah McBride's Co, Rockbridge County attached to 5th VMR (McDowell's) **CEM:** Bethesda Presbyterian Church; Rockbridge; Baths **GS:** No **SP:** Elizabeth Horn (Rockbridge Co, 3 Oct 1807- Rockbridge Co, 1 Jan 1892) **VI:** Son of Johannes Schneider (1769-1838) & Anna Barbara Strickler (1774-1858) **P:** No **BLW:** No **PH:** No **SS:** A rec 12685; B pg 179 **BS:** 245

SNODGRASS, David; b unk; d 1814 (Inv) **RU: Unk, 70th VMR, Capt James Meek's Company, Washington Co, attached to 5th VMR **CEM:** Rock Spring; Washington; vic jct Rts 803 & 91 **GS:** Unk **SP:** Elizabeth Reed **VI:** No further data **P:** Applied **BLW:** No **PH:** No **SS:** M pg 286; BD pg 1624; B pg 199 **BS:** 116 pg 211; 245

*SNODGRASS, William; b 1767, d Jul 1830 **RU:** Private, 105th VMR, Lt John Gray's Co attached to Abraham Bradley's Regt **CEM:** Glade Spring Presbyterian; Washington; 33234 Lee St. Glade Springs **GS:** Unk **SP:** Mar (22 Aug 1786) Sarah Long (Washington Co,1764-Washington Co 1848) **VI:** Son of John Snodgrass (1746-1796) and Mary Miller Snodgrass (1747-1849) **P:** No **BLW:** No **PH:** No **SS:** A rec 13007; B pgs 198-199 **BS:** 245

Key: *Additional veteran entry   **Corrected veteran entry   ***Deleted veteran entry

RU=Rank/Unit   CEM=Cemetery   GS=Gravestone   SP=Spousal Information   VI=Other Veteran Info   P=Pension
BLW=Bounty/Land Warrant   PH=Photo   SS=Service Source   BS=Burial Source   VMR= VA Military Regiment
LNR= Last Known Residence

90

*SNYDER, William S; b c1795, d aft 1848 (wife's death) **RU:** Private, 20th VMR Capt John Simmons, or Capt Edward James's or Capt William Randolph's Company, Princess Anne Co, attached to 1st VMR (Clarke) **CEM:** Cedar Grove; Norfolk; cnr Salter St & E Princess Anne Rd **GS:** Yes, **SP:** Mar (Norfolk, 28 Oct 1833) Catherine Farber (1795-3 Jun 1848) **VI:** No further data **P:** No **BLW: PH:** No **SS:** A rec 13691; B pg 223 **BS:** 245

SOUTHARD, James; b by 1789; d after 5 Jun 1869 **RU: Private, 45th VMR, Capt William Fitzhugh's Company, Stafford Co **CEM:** Union Church; Stafford; Carter St, Falmouth **GS:** Unk **SP:** Mar (25 Jul 1826) Elizabeth Perry **VI:** No further data **P:** No **BLW:** No **PH:** No **SS:** L pg 325 **BS:** 26 pg 384

*SOWERS, George Jacob; b 2 Jul 1794, Montgomery Co, d 13 Apr 1880 **RU:** Private, 75th VMR, Captain William Currin's Company, Montgomery Co **CEM:** Zion Lutheran Church; Floyd; town of Floyd **GS:** Unk **SP:** Polly Epperly (1799-1874) **VI:** Son of George (1750-1834) and Elizabeth Spangler (1764-1857) **P:** No **BLW:** No **PH:** No **SS:** A rec 15026; B pg 138 **BS:** 245

*SPEAK, Nicholas; b 3 Mar 1782, Charles County, MD, d 1852 **RU:** Private 70th or 105th VMR, Capt Abraham Fulkerson's Company, Washington Co, attached first to 5th VMR (Mason, Preston) and next to 7th VMR (Saunders) both at Norfolk 1814 and 1815 **CEM:** Speaks Chapel; Lee; Nr Rose Hill **GS:** Yes **SP:** Mar (Washington Co, Aug 1803 or 1804) Sarah Faires (Washington Co, 8 Mar 1786-Rose Hill, Lee Co, 1865), daug of Gideon Faires (1748-1818) and Sarah McSpadden (1745-1820). Recd BLW (1850 or 1855); number acres not determined **VI:** Son of Charles Spakes/Speaks, GS indicates was Reverend **P:** Unk **BLW:** Widow **PH:** No **SS:** B pg 198 **BS:** 31; 245

*SPENGLER, Joseph Stover; b 13 Nov 1790, Strasburg, d 15 Dec 1876 **RU:** Sergeant, 116th VMR, Captain William McMahon's Cavalry, Company, Rockingham Co, attached to Woodford's Squadron **CEM:** Old Strasburg; Shenandoah; Strasburg **GS:** Yes **SP:** No spouse info **VI:** No further data **P:** No **BLW:** No **PH:** No **SS:** B pg 182; K pg 187 **BS:** 245

*SPEER, Robert; b 21 Jul 1787, Surry Co. NC, d 24 Nov 1890 **RU:** Ensign, 94th VMR, Captain Jeremiah Skelton's Company Lee Co, attached to Major Bradley's Regt **CEM:** Stewarts Family; Scott; GPS 36.6617,-82.7194 **GS:** Yes **SP:** Rebecca Johnson Wade (1830-1910) **VI:** No further data **P:** Yes, Self and spouse **BLW:** No **PH:** No **SS:** A rec 16160; B pg 118; M pg 287 **BS:** 245

*SPRINKEL (SPRENKEL),William; b 2 Aug 1771, York Co, PA, d 28 Mar 1825 **RU:** Private, 2d Co, 3rd Bn, York Co, Militia **CEM:** Woodbine; Harrisonburg; 381 E Market St **GS:** Yes **SP:** Mar (Rockingham Co, 5 Oct 1809) Sarah Ireland (1783-1862) **VI:** Son of William Sprenkel and Catherina (__) (1737-__) **P:** No **BLW:** No **PH:** No **SS:** See *"findagrave"* for PA source **BS:** 245

*STAFFORD, John, Sr; b 1768,Co Tyrone, Ireland, d 23 Jul 1852 **RU:** Private Enlisted for 5 yrs 1 Jan 1813, USA in Capt John Smith's Co, sick at Plattsburgh, NY 18 Apr 1815 and discharged **CEM:** Wesley Chapel; Giles; GPS: 37.24420,-80.68140; off Rt 22 (Sheffey Memorial Rd), Trigg **GS:** Yes **SP:** Elizabeth Munsey(Muncey) **VI:** Son of James Stafford and Nancy Eaton **P:** No **BLW:** No **PH:** No **SS:** O Fold3 Serv Index Card **BS:** 245

*STAFFORD, Ralph, b 1795, d 24 Jan 1879, Mechanicsburg, Bland Co **RU:** Private 86th VMR, Lt Ralph Lucas's Co, Giles Co, attached to 4th VMR (McDowell, Koontz, Chilton) **CEM:** Old Bogle; Bland; Walker's Creek, SW of Mt Zion Church **GS:** Yes **SP:** Margaret Orr (1794-1880) **VI:** Son of John Stafford and Elizabeth Brown **P:** No **BLW:** No **PH:** No **SS:** B pg 81; O- Fold3 serv index card **BS:** 245

STARK(E), John; b 1792, d 3 Mar 1817 **RU: Private, 45th VMR, Capt John Edrington's Co, Stafford County **CEM:** Union Church; Stafford; Carter St, Falmouth **GS:** Yes **SP:** No spousal info **VI:** 1810 US Census in Falmouth, Stafford Co **P:** No **BLW:** No **PH:** No **SS:** L pg 835 **BS:** 26 pg 384; 245

*STEELE, Samuel; b unk, d 10 May 1824 **RU:** Private, 8th VMR, Capt Daniel Hoffman's Co of Mounted Riflemen, Rockbridge County, attached to 4th VMR, (Mc Dowell's, Koontz's Chilton's) McDowell's Flying Camp **CEM:** Falling Spring Presbyterian Church; Rockbridge; Glasgow **GS:** Unk **SP:** Nancy McCluer (23 Dec 1802-29 Oct 1825) **VI:** No further data **P:** Yes **BLW:** Yes **PH:** No **SS:** K pgs 15, 16; B pg 179; M pg 289 **BS:** 245

Key: *Additional veteran entry   **Corrected veteran entry   ***Deleted veteran entry
RU=Rank/Unit   CEM=Cemetery   GS=Gravestone   SP=Spousal Information   VI=Other Veteran Info   P=Pension
BLW=Bounty/Land Warrant   PH=Photo   SS=Service Source   BS=Burial Source   VMR= VA Military Regiment
LNR= Last Known Residence

91

STEPHENSON, William; b 1783 Burnside, Donegal Co, IRE; d 1837 **RU**: Private, 31st VMR, Capt Eben Taylor's, Troop of Cavalry, Frederick Co, attached to 1st VMR (Taylor) **CEM**: Mt Hebron; Frederick; 305 E Boscawen St, Winchester **GS**: Yes **SP**: Mar (Frederick Co, 26 Feb 1816) (bond) Lucy Catlett (1796-1861) **VI**: Son of James W. Stephenson. He came to America, age 11 in 1795 to Charlestown, WVA **P**: No **BLW**: No **PH**: Yes **SS**: K pg 160 **BS**: 93 Frederick

*STEPTOE, James Calloway; b 10 Dec 1781, d 24 Oct 1827 **RU**: Lieutenant, 109th VMR, a company from Middlesex Co **CEM**: Calloway-Steptoe; Bedford; GPS 37.3056,-79.2947 **GS**: Yes **SP**: No spouse info **VI**: Served as Clerk of the Court and Superior Court, Bedford Co. VA historical rd sign regards cem **P**: No **BLW**: No **PH**: No **SS**: A rec 23538 **BS**: 245

*STEVENS, John; b 20 Feb 1772, d 21 Mar 1869 **RU**: 2nd Lieutenant 53rd VMR, a company in Prince Edward Co, attached to 8th VMR (Wall's) **CEM**: College Church; Prince Edward; Hampton Sydney **GS**: Yes **SP**: No spouse info **VI**: No further data **P**: No **BLW**: No **PH**: No **SS**: A rec 24283; B pgs 167-168 **BS**: 245

*STEVENS, William; b 1783, d 30 Jun 1856 **RU**: Private, 8th VMR, a company of Rockbridge Co, attached to 5th VMR (McDowell's) **CEM**: Stonewall Jackson Memorial Cemetery; Lexington City; GPS: 37.78126, -79.44585 **GS**: No **SP**: Catherine (__) (1786-6 Jun 1850) **VI**: No further data **P**: Yes **BLW**: No **PH**: No **SS**: A rec 24579; B pgs 179-180 **BS**: 245

*STEWART, James; b 29 Feb 1790, d 12 Sep 1837 **RU**: Lieutenant, 94th VMR, Capt Jeremiah Neill's Co, Lee County or Capt Abram Fulkerson's Co, Washington Co, attached to 7th VMR(Saunders) **CEM**: Joseph Carter; Scott; Rye Cove **GS**: Yes **SP**: Polly Carter (10 Jan 1793-18 Jan 1882) **VI**: Son of William Stewart and Jemina Carter **P**: No **BLW**: No **PH**: No **SS**: A rec 25459; B pgs 118; 198 **BS**: 245

*STEWART, William; b 1767, d 12 Jul 1824 **RU**: Sergeant 1st Regt DC Militia **CEM**: Presbyterian Church; Alexandria; Wilkes St **GS**: Yes **SP**: No spouse info **VI**: No further data **P**: No **BLW**: No **PH**: No **SS**: A rec 25934 **BS**: 245

*STIFF, Henry; b 23 Apr 1792, d 3 Dec 1875 **RU**: Corporal, 10th and 91st VMR, Captain John P Gray's Company, Bedford Co, attached to 2nd VMR (Ambler- Brown) **CEM**: Stiff Family; Bedford; Leftwich Lane, Thaxton **GS**: Yes **SP**: Eny Chaney Huddleston (8 Jun 1792-23 Jun 1878) **VI**: Son of James Stiff (1758-1837) and Mary Molly Lewis (1760-1841) **P**: Yes, spouse also **BLW**: No **PH**: No **SS**: B pg 42; M pg 290 **BS**: 245

*STOCKDELL, John; b unk, d 13 Jul 1833 **RU**: Private, 82nd VMR, Capt Joseph Hume's Company, Madison Co, attached to 1st VMR, Stapleton Crutchfield's Detachment **CEM**: Harrison; John Madison; Rt 712 turn left off Rt. 231, 2 miles from Rochelle **GS**: Yes **SP**: No spouse info **VI**: No further data **P**: No **BLW**: No **PH**: No **SS**: A rec 27426; B pg 126 **BS**: 30; Family Cem

STONE, Alexander S H; b 1777; d 11 Oct 1823 **RU**: Lieutenant, 45th VMR Capt Henry William's Co, Stafford County **CEM**: Edrington Family; Stafford; end of Rt 692, right 1 mi **GS**: Yes **SP**: No spouse info **VI**: Probably the son of William B Stone (8 Sep 1757-15 Oct 1793) that is buried beside him **P**: No **BLW**: No **PH**: No **SS**: A rec 28183 **BS**: 26 pg 196

*STONE, John; b 1771, d 11 May 1851 **RU**: Private, 18th VMR, Lieutenant John A Corn's Company, Patrick Co **CEM**: Stone & Johnson; Henry; Rt 798 across rd fr home of Homer Philpott **GS**: No **SP**: Susan (__) **VI**: No further data **P**: Widow drew pension (WC- 30109) **BLW**: No **PH**: No **SS**: BD pg 1664; B pg 157 **BS**: 245

*STONE, William; b 13 Mar 1766, d 1843 **RU**: Captain, 78th VMR, commanded a company in Grayson County, attached to 4th VMR (McDowell, Koontz, Chilton) at Norfolk **CEM**: Jeremiah Stone; Grayson; Comers Rock Rd; Rt 658, 1.6 mi fr jct Rt 21, behind house and barn on knoll **GS**: Yes on marker listing parents **SP**: Elizabeth (__) (__-1850) **VI**: Son of Rev War soldier, Jeremiah Stone & Susanna Hurt **P**: No **BLW**: No **PH**: No **SS**: B pg 86 **BS**: 245

***STONE, William S: Deleted as he is the same person as the one buried in Fredericksburg

Key: *Additional veteran entry   **Corrected veteran entry   ***Deleted veteran entry
RU=Rank/Unit   CEM=Cemetery   GS=Gravestone   SP=Spousal Information   VI=Other Veteran Info   P=Pension
BLW=Bounty/Land Warrant   PH=Photo   SS=Service Source   BS=Burial Source   VMR= VA Military Regiment
LNR= Last Known Residence

92

*STRIBLING, Erasmus; b 1 Jun 1784, d 2 Jul 1858 **RU:** Captain, Co of Artillery, Virginia Militia **CEM:** Trinity Episcopal Churchyard; Staunton City; 214 S Beverly St **GS:** Yes **SP:** Matilda Kinney (18 Nov 1789-17 Apr 1829) **VI:** Merchant in Staunton; Lawyer by profession, owned and operated Stribling Springs; Clerk of Old District Court of Sweet Springs; Elected Clerk of County Court of Augusta Co; Appointed clerk of the Western District; Justice of Peace for Augusta Co; Clerk of the Corporation Court; Recorder of town of Staunton, Vestryman **P:** No **BLW:** No **PH:** No **SS:** Z **BS:** 245

*STRICKLER, Daniel, Jr; b 16 Dec 1772, Hempfield, Lancaster Co, PA, d 9 Jun 1841 **RU:** Captain, Commanded a company in the 13th or 97th VMR, Shenandoah Co, attached to 6th VMR, (Coleman) **CEM:** Wilson Springs; Rockbridge; loc off Maury River Rd (Rt 39) and off Farmhouse Rd (Rt 623) on Tucker Farm in cow pasture on top of hill **GS:** Unk **SP:** Elizabeth Funkhouser (8 Jul 1775-28 May 1838) **VI:** Son of Daniel Strickler, Sr and Barbara Lehman. Burial cited in County Genealogical Society book **P:** No **BLW:** No **PH:** No **SS:** B pg 185; Fold 3, Serv Index Card **BS:** 245

*STRICKLER, Jacob M; b 2 Feb 1786, Shenandoah Co, d 7 Aug 1867 **RU:** Private, 24th US Army Regt, Captain Silas Stebbins's Co **CEM:** Strickler; Floyd; Town of Floyd **GS:** Yes **SP:** Mary Kagey (2 Feb 1792-6 May 1880) **VI:** Enlisted (Shenandoah Co, 12 Dec 1812); discharged (Mobile Bay, AL, 7 Apr 1815) **P:** No **BLW:** No **PH:** No **SS:** C pg 174 **BS:** 245

STROTHER, William Porter; b 14 Feb 1798, Stafford Co, d 23 Apr 1874 **RU: Private, 5th VMR, Also could be in a Company attached to 6th VMR (Coleman) **CEM:** Hollywood; Richmond City; 412 S Cherry St; Sec E lot 8 **GS:** No **SP:** Mar (Manchester, Chesterfield Co, 26 Aug 1823) Elizabeth Kendall Hewitt (__-19 Oct 1867, age 62) **VI:** Son of John Strother (1771-1805) and Catherine Fox Price (1772-1846). Moved to Manchester, and was employed at the Exchange Bank of VA **P:** No **BLW:** No **PH:** No **SS:** A rec 2539 **BS:** 237; 31; 63 pg 234; 245

*STUART, Hugh; b Jan 1744, d 9 Dec 1824 **RU:** Captain, 8th VMR, commanded a detachment of mounted riflemen, Rockbridge County, attached to McDowell's Flying Camp **CEM:** Walkerland; Rockbridge; on Rt 602; Walkers Creek Rd, top of hill, behind Maxwelton Cabins; bef jct with Rt 724 **GS:** Unk **SP:** Elizabeth Walker (11 Nov 1787-30 Sep 1838), daug of John Walker and Margaret Kelso (__-1818) **VI:** No further data **P:** No **BLW:** No **PH:** No **SS:** B pg 180; K pg 35 **BS:** 245

*SUBLETT, Samuel; b 15 Apr 1789, d 30 Apr 1856 **RU:** Private 19th VMR, Capt Anthony Turner's Co, detached under Ensign G.M. Carrington, Richmond City **CEM:** Shockoe Hill; Richmond City; 100 Hospital St; **GS:** Yes **SP:** Mar 1) (1817) Harriet Duval (25 Aug 1791-2 Jun 1827), daug of Benjamin Duval and Elizabeth Warrock 2) (1832) Antoinette C Carlton (21 Jun 1806-1852 MD) **VI:** Son of William Sublet and Betsey Hughes. Vestryman, St Johns Ch 1829-1843 **P:** No **BLW:** No **PH:** Yes **SS:** B pg 174; L pg 198 **BS:** 31; 245

SULLIVAN, Benjamin, Jr; b 1790, d 27 Apr 1830 **CEM: Sullivan Family; Stafford; at the end of Jacks Road in White Oak **GS:** Yes **SP:** Mar (Stafford Co, 27 Aug 1832) his cousin Lucy Fines (1792-1873), daug of James Fines & Rachel Curtis **VI:** Son of Benjamin Sullivan Sr (1803) and Susannah Kitchen, daug of James Kitchen and Mary Porch **P:** Spouse applied **BLW:** No **PH:** No **SS:** L pg 85; BD pg 1675 **BS:** 26 pg 34s

*SULLIVAN, Newton; b 1781, Stafford Co, d 1865 **RU:** Fifer, 16th VMR, Capt Thomas R Magee's Co, Spotsylvania County, attached to Major Robert Crutchfield's Battalion **CEM:** Old Cedar Grove; Spotsylvania; Turners Lane, Paytes **GS:** No **SP:** Ann McDorman (1790-1869), daug of David McDorman and Anne Tiller **VI:** Gr st and perhaps 1812 gr marker thought to be removed by landowner **P:** No **BLW:** No **PH:** No **SS:** B pg 189; K pg 383; O-Serv Index Cards **BS:** 245

***SULLIVAN(T), Jonas: Deleted as burial not verified

SUMMERS, David; b 25 Jul 1782; d 23 Feb 1857 **RU: Private, 116th VMR, Capt James Mallory's company , Rockingham Co **CEM:** Mount Tabor; Augusta; 11 mi SW of Staunton **GS:** Unk **SP:** 1) Rebecca Engleman, 2) Juliann Palmer **VI:** Son of John Summers (1746-__)) and Elizabeth (__) (1742-1832) **P:** Spouse **BLW:** No **PH:** No **SS:** K pg 176; M pg 292 **BS:** 261; vol 21 # 4, pg 7

Key: *Additional veteran entry   **Corrected veteran entry   ***Deleted veteran entry

RU=Rank/Unit	CEM=Cemetery	GS=Gravestone	SP=Spousal Information	VI=Other Veteran Info P=Pension
BLW=Bounty/Land Warrant	PH=Photo	SS=Service Source	BS=Burial Source	VMR= VA Military Regiment
LNR= Last Known Residence				

93

Addendum to Burials of War of 1812 Veterans in the Commonwealth of Virginia

*SWANN, John; b 1793, d 26 Apr 1860 **RU**: Private, 17th VMR, Capt Allen Wilson's Co, Cumberland County, attached to 1st VMR (Trueheart) **CEM**: Petersville; Powhatan; Rt 60, 1 mi E of jct 684 **GS**: Yes **SP**: No spouse info **VI**: No further data **P**: No **BLW**: No **PH**: No **SS**: B pg 64; K pg 59 **BS**: 245

*SWANN, William Thomas; b 1785, MD, d 15 Oct 1820, Alexandria **RU**: Private 1st Regt MD Militia commanded by Col Hawkins, and 5th MD Regt commanded by Col Sterrett **CEM**: Pohick Episcopal Church; Fairfax; jct Rts 1 & 611, Lorton **GS**: Yes **SP**: Frances Brown Alexander (1784-1856) **VI**: No further data **P**: No **BLW**: No **PH**: No **SS**: A recs 787; 789 **BS**: 245

*SWARTZ, George, b 1790, d 28 Aug 1864 **RU**: Private, 2nd Regt, Pennsylvania Militia (Piper's) **CEM**: Gainesboro Methodist; Frederick; Gainesboro **GS**: Yes **SP**: Phebe Mercer (1797-11 Mar 1868) **VI**: No further data **P**: No **BLW**: No **PH**: No **SS**: A rec 977 **BS**: 245

*SYDNOR, John Taylor, Sr; b 1780, d 30 Apr 1847 **RU**: Captain, 83rd VMR, company commander, Dinwiddie Co **CEM**: Sydnor & Young Family; Dinwiddie; Petersburg Battlefield cemetery at Five Forks **GS**: Unk **SP**: Mary Thweatt (1776-26 Nov 1820) **VI**: Son of Joseph Sydnor (17 Oct 1740-1787) and Ann Chowning (7 Sep 1751-__); Name on memorial on farm inside Petersburg National Cemetery **P**: No **BLW**: No **PH**: No **SS**: A rec 2287; B pg 66; L pg 752 **BS**: 245

SYDNOR, William C; b 30 Aug 1788, Frederick Co; d 29 Aug 1828 **RU: Private, 6th VMR (Sharp) **CEM**: Chichester Family; Fairfax; 6720 Newington Rd **GS**: Yes **SP**: Sarah Redman (16 Dec 1810) **VI**: Son of William F Sydnor and Sarah Chinn **P**: No **BLW**: No **PH**: N **SS**: A rec 2294 **BS**: 89 v5 LR-2; 245

*TATE, John L; b 17 Mar 1798, 28 Jul 1868 **RU**: Private, 16th Inf, US Army, Captain Greenwood's Co **CEM**: Shockoe, Richmond City, Hospital St **GS**: Yes **SP**: Elizabeth (__) (1801-1853) **VI**: Enlisted 20 Dec 1814, discharged Detroit, MI, 30 Jun 1815 **P**: No **BLW**: No **PH**: No **SS**: B pg 126; BD pg 1689; C pg 178 **BS**: 245

*TATE, William; b 20 Mar 1796, d 6 Feb 1884 **RU**: Private, 56th VMR, Capt Thomas Gregg's Co, Loudoun County, attached to 4th VMR (Beatty's) **CEM**: Goose Creek Burying Ground; Loudoun; Rt 722, Lincoln **GS**: Yes **SP**: Priscilla Jackson (1803-1865) **VI**: Son of Levi & Edith (Nickols) Tate **P**: No **BLW**: No **PH**: No **SS**: B pg 120 **BS**: 245

*TAYLOR, James; b 14 Dec 1771, d 7 Jun 1826 **RU**: Lieutenant, 7th VMR, company commander in Norfolk Co **CEM**: Cedar Grove; Norfolk; 238 E. Princess Anne Rd **GS**: Yes **SP**: Sarah Newton (11 Dec 1776-17 Jun 1855), daug of Thomas Newton and Martha Tucker **VI**: No further data **P**: No **BLW**: No **PH**: No **SS**: B pg 147 **BS**: 245

*TAYLOR, Mahlon; b 1787, d 10 Oct 1845 **RU**: Sergeant Major, Andrey's Detachment, NJ Militia **CEM**: Goose Creek Burying Ground; Loudoun; Rt 722 Lincoln **GS**: Yes **SP**: No spouse info **VI**: Son of Malhon Kirkbride Taylor & Mary Stokes **P**: No **BLW**: No **PH**: No **SS**: NARA Rec Grp 94 #654501 **BS**: 245

*TAYLOR, Robert Barraud; b 24 Mar 1774, Smithfield, Isle of Wight County, d 24 Mar 1834 **RU**: Brigadier General, Commander 9th Brigade & Commander in Chief Norfolk area to include U.S. forces **CEM**: Cedar Grove; Norfolk; 238 E. Princess Anne Rd **GS**: Yes **SP**: No spouse info **VI**: Son of Robert Taylor and Sarah Barraud, daug of Daniel Barraud; resigned (4 Feb 1814) but became Major General of State Militia, Norfolk District; member of General Assembly; appointed Judge of Circuit Court & Superior Court, First District of VA; and member of the Virginia Constitutional Convention of 1829 **P**: No **BLW**: No **PH**: No **SS**: B pg 12 **BS**: 31

*TAYLOR, Timothy II; b 11 Jan 1761, Newtown, Bucks Co PA, d 08 Jun 1838 **RU**: LT Colonel, Commander 56th VMR Loudoun Co **CEM**: Goose Creek Burying Ground; Loudoun; Rt 722 Lincoln **GS**: Yes **SP**: Mar (6 Feb 1780) Achsah Taylor (5 Feb 1759-16 May 1826) **VI**: Also was private in Rev War, commissioned Lt Col 25 May 1810 **P**: No **BLW**: No **PH**: No **SS**: B pg 119 **BS**: 245

*TAYLOR, William; b 1770, d 23 Jul 1848 **RU**: Private, 58th or 116th VMR a company of Rockingham Co, attached to 6th VMR (Coleman's) **CEM**: Cross Keys; Rockingham; Cross Keys **GS**: Yes **SP**: No spouse info **VI**: No further data **P**: No **BLW**: No **PH**: No **SS**: A rec 1964; B pgs 181-182 **BS**: 245

Key: *Additional veteran entry   **Corrected veteran entry   ***Deleted veteran entry
RU=Rank/Unit   CEM=Cemetery   GS=Gravestone   SP=Spousal Information   VI=Other Veteran Info   P=Pension
BLW=Bounty/Land Warrant   PH=Photo   SS=Service Source   BS=Burial Source   VMR= VA Military Regiment
LNR= Last Known Residence

94

*TAYLOR, William; b unk, d 1823 **RU:** Private, 35th VMR, Captain Samuel Graham's Company, Wythe Co, attached to 9th VMR (McDowell's, Koontz, Chilton) **CEM:** Bethany Community; Grayson; Independence **GS:** No **SP:** Mary Vanover (Sussex Co, NJ, 1782-__), daug of Cornelius Vanover **VI:** Son of Revolutionary War veteran, John Taylor Sr. and Mary (__) **P:** Yes, Widow, # WC 3662 **BLW:** No **PH:** No **SS:** A rec 1916; B pg 204; BD pg 1695; M pg 295 **BS:** 245

*TEBBS, Foushee G; b 28 Jan 1787, d 12 Dec 1835 **RU:** Captain, 41st VMR commanded a company, Richmond County **CEM:** Tebbsdale; Prince William; fr Possum Point Rd 2.5 mi to cem on Old Carborough Rd, Dumfries **GS:** Yes **SP:** Mar 1) Nancy Reeder Chapin (Alexandria, 2 Feb 1804-Fauquier Co, 21 Apr 1821), 2) Margaret G Tyler **VI:** Son of Willoughby Tebbs (1759-22 Oct 1803) and Elizabeth Carr (6 Oct 1771-18 Mar 1852) **P:** No **BLW:** No **PH:** No **SS:** B pg 178 **BS:** 245

*TEBBS, Thomas F; b 6 Mar 1794, Prince William Co, d 1 Apr 1828 **RU:** Sergeant/Quartermaster, 36th VMR, Staff of 36th Regt, Prince William Co **CEM:** Old Stone Church; Loudoun; Leesburg **GS:** Yes **SP:** Mar (1818) Margaret Hannah Douglas Binns **VI:** Son of Willoughby Tebbs and Betsy Carr. He was a physician **P:** No **BLW:** No **PH:** No **SS:** L pg 16 **BS:** 245

*THOM, John; b 11 Nov 1771, d 22 May 1855, Culpeper Co **RU:** Captain, 5th or 34th VMR, company commander, Culpeper Co, attached to 2d VMR (Ballowe) **CEM:** Hollywood; Richmond City; 412 S Cherry St **GS:** Yes **SP:** No spouse info **VI:** Son of Alexander Thom and Elizabeth Triplett **P:** No **BLW:** No **PH:** No **SS:** A rec 3; B pg 63 **BS:** 245

*THOMAS, George; b 17 Aug 1769, d 12 May 1839 **RU:** Private, 82nd VMR, Capt Joseph Hume's Co, Madison Co, attached to 1st VMR, Robert Crutchfield's Detachment **CEM:** Thomas # 2; Madison; 3716 Etlan Rd **GS:** Yes **SP:** No spouse info **VI:** No further data **P:** No **BLW:** No **PH:** No **SS:** A rec 607; B pg 126; **BS:** 30

*THOMAS, John; b 1779, d 20 Apr 1829 **RU:** Ensign, 65th VMR, in a company of Southampton Co **CEM:** Thomas Family; Southampton; Newson's **GS:** Unk **SP:** Elizabeth Rochelle (1794-1844) **VI:** Son of Tristan Thomas and Mary Hollingsworth **P:** No **BLW:** No **PH:** No **SS:** A rec 734; B pgs 186-187 **BS:** 245

*THOMAS, John; b 29 Sep 1798, d 29 Dec 1862 **RU:** Wagoner, 82nd VMR, Capt Joseph Hume's Co, Madison Co, attached to 1st VMR, Robert Crutchfield's Detachment **CEM:** Thomas # 2; Madison; 3716 Etlan Rd **GS:** Yes **SP:** No spouse info **VI:** No further data **P:** No **BLW:** No **PH:** No **SS:** A rec 607; B pg 126; **BS:** 30; Family Cem

*THOMAS, John; b unk, d Apr 1878 **RU:** Private, 33rd VMR, Captain Samuel Brown's Company, Henrico Co **CEM:** Hollywood; Richmond City; 4125 Cherry St **GS:** Unk **SP:** No spouse info **VI:** No further data **P:** No **BLW:** No **PH:** No **SS:** L pg 177 **BS:** 245

*THOMAS, William; b unk, d 1865 (will) **RU:** Private, 82nd VMR, Capt Joseph Hume's Co, Madison Co, attached to 1st VMR, Robert Crutchfield's Detachment **CEM:** Thomas # 2; Madison; 3716 Etlan Rd **GS:** Yes **SP:** No spouse info **VI:** No further data **P:** No **BLW:** No **PH:** No **SS:** A rec 607; B pg 126; **BS:** 30; Family Cem

*THOMPSON, Archibald W; b 16 Aug 1797, d 16 Sep 1878 **RU:** Corporal, served in Col McCobb's U.S. Volunteers **CEM:** Maplewood; Tazewell; Maplewood Ln; North Tazewell **GS:** Yes **SP:** Mary Bennett (5 Jun 1802-20 Jun 1885) **VI:** Son of John T Thompson (1764-1850) and Louisa Bowen (1768-1812) **P:** No **BLW:** No **PH:** No **SS:** Fold 3 Serv Index card **BS:** 245

THRUSTON, Robert; b 30 Mar 1782, Lansdowne, Gloucester Co; d 22 Feb 1857 **RU: Major, 21st VMR, staff officer, Gloucester Co **CEM:** Timberneck Farm; Gloucester Co; 3601 Timberneck Farm Rd; Haynes **GS:** Yes **SP:** Mar 1) (22 Dec 1804) Sarah Brown (__-8 Dec 1818), 2) (20 Dec 1820) Mary Catlett (c1791-1 Dec 1843), daug of John and Ann (Walker) Catlett **VI:** Commissioned Major (19 Mar 1814) of Lansdowne; Gloucester Co, "a plain, practical farmer one who devoted a long life solely to Agricultural pursuits" **P:** No **BLW:** No **PH:** Yes **SS:** B pg 82; L pg 776 **BS:** 82 pg 66

Key: *Additional veteran entry   **Corrected veteran entry   ***Deleted veteran entry
RU=Rank/Unit   CEM=Cemetery   GS=Gravestone   SP=Spousal Information   VI=Other Veteran Info   P=Pension
BLW=Bounty/Land Warrant   PH=Photo   SS=Service Source   BS=Burial Source   VMR= VA Military Regiment
LNR= Last Known Residence

THURMAN, John; b 1778; d 10 Dec 1855 **RU: Private, in a VMR company from Bedford or Campbell Co attached to 5th VMR (Mason-Preston) **CEM:** Old City Cemetery; Lynchburg; 401 Taylor St **GS:** Yes **SP:** Mar (Lynchburg, 1807) Elizabeth Simpson, daug of Ann Essex **VI:** Son of Richard Thurman; a saddler by trade; in 1820 was Mayor of Lynchburg, died age 77 **P:** No **BLW:** No **PH:** No **SS:** A rec 2009 **BS:** 87 pg 66; 207

*TOMPKINS, Christopher, b 24 Jan 1778, Caroline Co, d 16 Aug 1838 **RU:** Major, 61st VMR, staff officer, and regt commander, Mathews Co **CEM:** Poplar Grove Plantation; Mathews; nr Williams SW jct; Rt 14 and 613 **GS:** No **SP:** Mar 1) (3 May 1806) Elizabeth Cary Smith (1787-1814), 2) (21 Sep 1815) Maria Booth Patterson (1794-1854) **VI:** Son of Benjamin Tompkins (1732-1811) and Elizabeth Goodloe (1738-1808), served two terms in the Legislature. Daug Sally was the only female commissioned officer in the Confederate Army during the Civil War **P:** No **BLW:** No **PH:** No **SS:** B pg 128 **BS:** 245

*TILTON, Daniel; b 1767, d 10 Dec 1830 **RU:** Private, New Hampshire Militia, Waldron's Command **CEM:** Old Stone Church; Loudoun; Leesburg **GS:** Yes **SP:** No spouse info **VI:** No further data **P:** No **BLW:** No **PH:** No **SS:** A rec 958 **BS:** 245

*TINSLEY, Parke; b 1796, d 22 Dec 1824 **RU:** Private, 74th Regt, Captain William Hundley's Company, Hanover Co **CEM:** Spring Grove; Hanover; Rockville **GS:** Yes **SP:** No spouse info **VI:** Was in 74th Virginia Light Infantry (1822-1823). Son of David Tinsley **P:** No **BLW:** No **PH:** No **SS:** L pg 465 **BS:** 245

*TODD, Jared; b 16 Feb 1787, Litchfield, CT, d 16 Nov 1875 **RU:** Corporal, 3rd Artillery & Lt Infantry, NY Militia **CEM:** Chatham Burial Park; Pittsylvania; Chatham **GS:** Yes **SP:** No spouse info **VI:** No further data **P:** No **BLW:** No **PH:** No **SS:** A rec 2368 **BS:** 245

*TOLER, William F; b 1775, d 1848 **RU:** Sergeant, 40th VMR, Capt George Morris' Co, Louisa County, attached to 1st VMR (Trueheart) **CEM:** Tolersville Tavern; Louisa; 410 Old Tolarsville Rd, Mineral **GS:** Yes **SP:** Mar 1) (12 Oct 1807) Polly Walton Smith, 2) (11 Sep 1826) Elizabeth C Barrett **VI:** Son of William Toler (King William Co, c1751-1785) and Hannah Jennings Brockman (__-aft 1821) **P:** No **BLW:** No **PH:** No **SS:** A rec 173; B pg 123; K pg 54 **BS:** 245

TUCKER, Henry St George; b 1780 Virginia; d 1848 **RU: Captain, 31st VMR, company commander, Frederick Co **CEM:** Mt Hebron; Winchester; 305 E Boscawen St **GS:** Unk **SP:** Leila Skipwith, daug of Sir Peyton Skipwith (__-Williamsburg, 14 Sep 1837) (obituary) **VI:** Son of Judge St George Tucker; Nephew of Thomas Tudor Tucker; half-brother of John Randolph of Roanoke; father of John Randolph Tucker; grandfather of Henry St George Tucker (1853-1932). US Congress 1815-1819 **P:** No **BLW:** No **PH:** No **SS:** A rec 355 **BS:** VA Political Graveyards Internet 2009

VANCE, James; b 1778; d 19 Oct 1814 **RU: Corporal, 31st VMR, Capt Thomas Roberts's Co, Frederick County, attached 4th VMR (Beatty) **CEM:** Opequon Presbyterian; Frederick; 217 Opequon Church Ln, Kernstown **GS:** Yes **SP:** Mar prob Sally Vance (1782-28 Mar 1815) as her gr st next to his **VI:** Died in his 36th year **P:** No **BLW:** No **PH:** No **SS:** B pg 79; S pg 106 **BS:** 79 pg 333; 151; 245

VAN LEW, John; b 4 Mar 1790; d 13 Sep 1843 **RU: Sergeant, Swift's Det NY Militia **CEM:** Shockoe Hill; Richmond City; 100 Hospital St **GS:** Unk **SP:** Mar (Richmond City, 10 Jan 1818) Elisha Louisa Baker **VI:** No further data **P:** No **BLW:** No **PH:** Yes **SS:** A rec 374 **BS:** 199; 63 pg 238

*VANVACTOR, Solomon; b unk, d 23 Oct 1851 **RU:** Private, 57th VMR, Captain Michael Everhart's Company, Loudoun Co, attached to 1st VMR (Taylor) **CEM:** Arnold Grove; Loudoun; 37216 Charlestown Pike, Rt 9, Hillsboro **GS:** Yes **SP:** No spouse info **VI:** No further data **P:** No **BLW:** No **SS:** A rec 1354; B pg 120 **BS:** 31

*VEALE, James; b c1784, d 23 Mar 1832 **RU:** Private, 7th VMR, Capt Robert Tart's Company, Norfolk Co **CEM:** Scott-Shea family; Portsmouth; at old Shea farm destroyed 1960 for Midtown Tunnel **GS:** No **SP:** No spouse info **VI:** Son of William Crawford Veale **P:** No **BLW:** No **PH:** No **SS:** A rec 56; 58; K pg 490 **BS:** 49

Key: *Additional veteran entry   **Corrected veteran entry   ***Deleted veteran entry
RU=Rank/Unit   CEM=Cemetery   GS=Gravestone   SP=Spousal Information   VI=Other Veteran Info   P=Pension
BLW=Bounty/Land Warrant   PH=Photo   SS=Service Source   BS=Burial Source   VMR= VA Military Regiment
LNR= Last Known Residence

96

*VIRTS/WERTS, Conrad; b 26 May 1791, d 2 Dec 1881 **RU:** Private, 56th or 57th VMR, in a company of Loudoun Co, attached to 5th VMR **CEM:** Old Ebenezer Methodist Episcopal; Loudoun; Neersville **GS:** Yes **SP:** Mar Elizabeth Derry (16 Feb 1802-19 Dec 1886) **VI:** Son of Peter Wirtz (1766-1852) and Christina (__) (1771-1823) **P:** Yes, widow **BLW:** No **PH:** No **SS:** A rec 1602; M pg 304 **BS:** 245

*WALKE (WALK), Anthony; b 1778, d 13 Sep 1820 **RU:** Private, 20th VMR, Captain John Reade's Co, Princess Anne County, attached to 3rd VMR (Boykin) **CEM:** Old Dominion Episcopal Church; Virginia Beach; 4449 N Witchduck Rd **GS:** Yes **SP:** No spouse info **VI:** No further data **P:** No **BLW:** No **PH:** No **SS:** B pg 165; L pg 663 **BS:** 245

*WALKER, James; b 1786, d 12 Jun 1852 **RU:** Private, 8th VMR, Captain Alexander Campbell's Company, Rockbridge Co, attached to 2d Corps D'Elite (Ballowe) **CEM:** Walkerland; Rockbridge; Rt 602; Walkers Creek Rd, top of hill, behind Maxwelton Cabins bef jct with Rt 724 **GS:** Unk **SP:** Mary (__) **VI:** No further data **P:** No **BLW:** No **PH:** No **SS:** B pg 179; K pgs 209; 210 **BS:** 245

*WALKER, James W; b 29 Nov 1798, d 13 Jun 1873 **RU:** Private, 5th VMR **CEM:** Walker United Methodist Church; Madison; Rt 230, 7.5 miles E of US Route 29 (Orange Road) **GS:** Yes **SP:** Fannie Brownley (6 Aug 1792-8 Sep 1876) **VI:** No further data **P:** No **BLW:** No **PH:** No **SS:** A rec 2546; B pg 126 **BS:** 30; Church Cem

*WALKER, Thomas Hudson; b 31 Oct 1784, d 7 Apr 1865 **RU:** Sergeant; 8th VMR, either Captain Archibald Lyles or Captain Isiah McBrides's Company, Rockbridge Co, attached to 5th VMR (McDowell) **CEM:** Walkerland; Rockbridge; Rt 602; Walkers Creek Rd, top of hill, behind Maxwelton Cabins bef jct with Rt 724 **GS:** Yes **SP:** Betsy Culton (Jul 1792-5 Dec 1848), daug of Robert Culton **VI:** No further data **P:** No **BLW:** No **PH:** No **SS:** A rec 3144; B pg 179 **BS:** 245

WALL, Jacob; b 1799; d May 1864 **RU: Private, Hill's Regiment, PA Militia **CEM:** Mt Hebron; Winchester; 305 E Boscawen St **GS:** Yes **SP:** Harriet S. (__) (__-28 Oct 1809, age 70 years) **VI:** Died age 65, no dates. War of 1812 service engraved on stone. Son of John and Mary (__) Wall **P:** No **BLW:** No **PH:** No **SS:** A rec 55; G **BS:** 86 pg 48

*WALL, John C; b 1758, d 1879 **RU:** Sergeant, in a Frederick Co company attached to 1st VMR (Taylor) **CEM:** Mt Hebron; Winchester; 305 E Boscawen St **GS:** Yes **SP:** Mar Mary (__) **VI:** Died age 80 years with no dates. Inscription- "Volunteer in 4th VA Reg 1776. A Soldier of 1812" **P:** No **BLW:** No **PH:** No **SS:** A rec 75; B pg 78 **BS:** 86 pg 47

WALL, John F; b unk, Shawnee, Frederick Co, d aft 1813 **RU: Private, in company Frederick Co, attached to 4th VMR **CEM:** Mt Hebron; Winchester; 305 E Boscawen St **GS:** Yes **SP:** No spouse info **VI:** Died age 80 years. No dates. "Volunteer in 4th VA Reg 1776. A Soldier of 1812" (stone) **P:** No **BLW:** No **PH:** No **SS:** G **BS:** 86 pg 47

WALLACE, Thomas; b 25 Dec 1796, d 23 Dec 1882 **RU: Private, 45th VMR (Peyton), Capt Lewis Alexander's Company Stafford Co **CEM:** Liberty Hall; Stafford; Truslow Rd, 1.4 miles fr jct with Enon Rd, 0.5 miles N near Wallace Farm Ln **GS:** No **SP:** Ann Coffman (12 Jul 1820-29 Jul 1889) **VI:** Son of John Wallace (1761-1829) and Elizabeth Hooe (1765-3 Sep 1850). Was residing in Hartwood in 1880 (US Census) **P:** Spouse **BLW:** Yes **PH:** No **SS:** A rec 540; BD pg 1773; B pg 190 **BS:** 26 pg 268

WALLACE, William; b 17 Mar 1779, Retertrear, Scotland; d 26 Dec 1854 **RU: Sergeant, 74th VMR, probably Capt Benjamin Pollard's Co, Hanover County attached to 9th VMR (Sharp) **CEM:** Hollywood; Richmond City; 412 S Cherry St **GS:** Unk **SP:** Catherine Leighty (1798-1874) **VI:** No further data **P:** No **BLW:** No **PH:** No **SS:** A rec 607 **BS:** 263; v10 pg 21

WALLER, James; b 1789; d 31 Jan 1824, Aquia, Stafford Co **RU: Lieutenant, 45th VMR, Capt John C Edrington's Company, Stafford Co **CEM:** Adie / Waller; Stafford; off Widewater Rd, old portion Rt 611 **GS:** Unk **SP:** Anne Adie (1792-1870) **VI:** Died age 35, prob son of Edward Waller and Elizabeth Chadwell (1768-1818) **P:** No **BLW:** No **PH:** No **SS:** A rec 745 **BS:** 26A pg 39

Key: *Additional veteran entry **Corrected veteran entry ***Deleted veteran entry
RU=Rank/Unit   CEM=Cemetery   GS=Gravestone   SP=Spousal Information   VI=Other Veteran Info   P=Pension
BLW=Bounty/Land Warrant PH=Photo   SS=Service Source   BS=Burial Source   VMR= VA Military Regiment
LNR= Last Known Residence

97

*WARING, William Lowry; b 1783, d 19 May 1841 **RU**: Captain, 111st VMR commanded a company in Westmoreland Co **CEM**: Tuscarora Site; Essex; Dunnsville **GS**: Yes **SP**: Mary Banks (13 Sep 1788-27 Dec 1841), daug of Richard Banks and Elizabeth Smith Young **VI**: Son of Robert Payne Waring and Anne Lowry **P**: No **BLW**: No **PH**: No **SS**: B pg 202; L pg 812 **BS**: 245

*WASHINGTON, Bushrod II; b 4 Apr 1785, Oak Grove, Westmoreland Co, d 16 Apr 1831 **RU**: 2nd Lieutenant, 60th VMR, Capt George Washington Ball's Mounted Infantry Co, Fairfax County, attached to Lt Col John W Green's Mounted Infantry Battalion **CEM**: Mount Vernon Estate; Fairfax; 3200 Mt Vernon Hwy **GS**: No **SP**: Henrietta Brayne Spotswood (29 Aug 1786-10 Aug 1860), daug of Alexander Spotswood and Elizabeth Washington **VI**: Son of William Augustine Washington (25 Nov 1757-2 Oct 1810) and Jane Washington (20 Jun 1755-16 Aug 1791) **P**: No **BLW**: No **PH**: No **SS**: A rec 2795; B pg 248 **BS: 245**

*WASHINGTON, George Fayette; b 17 Jan 1790, Mount Vernon, Fairfax Co, d 16 Sep 1867, Frederick Co **RU**: First Lieutenant, 5th U.S. Army Infantry, on recruiting duty Kempsville, Princess Anne Co, (Jul 1809- Jul 1813), ordered to bring recruits to Fort Norfolk, Jul 1813 **CEM**: Mt Hebron; Winchester; 305 E Boscawen St **GS**: Yes **SP**: Maria Frame **VI**: Son of George Augustine Washington and grandson of Charles Washington, brother of President George Washington. He owned Waverly in Frederick Co **P**: No **BLW**: No **PH**: No **SS**: AF **BS**: 81; 245

*WATERS, Levi; b 19 May 1797, d 19 Feb 1872 **RU**: Private, Captain Wallace's Co, Maryland Militia **CEM**; Saint Pauls Lutheran Church; Loudoun; Neersville **GS**: Yes **SP**: No spouse info **VI**: No further data **P**: No **BLW**: No **PH**: No **SS**: A rec 412; **BS**: 245

WATKINS, John E; b 28 Feb 1789, Goochland Co, d 21 Feb 1855, "Ampthill," **RU: Private, 19th VMR, Capt Wilson Bryan's Co, Richmond City **CEM**: Watkins Family #3; Chesterfield; Spruant Plant; 5401 Jeff Davis Rd; Richmond **GS**: Yes **SP**: Mar (16 Apr 1812) Judith Eveline Watkins (her maiden name was also Watkins) (2 Apr 1794-30 Nov 1872) **VI**: Son of Benjamin Watkins and Anna Riddle. He built a mill on his property "Ampthill" and home still stands albeit in a different location on the James River **P**: No **BLW**: No **PH**: Yes **SS**: A rec 749 **BS**: 8 pg; 121; 49; 245

WATTS, Samuel; b 28 Nov 1799; d 17 May 1878 **RU: Private, 115th VMR, Capt Mile's Cary's Co, Elizabeth City Co, attached to 8th VMR (Wall) **CEM**: Cedar Grove; Portsmouth; Effington St & Fort Ln **GS**: Yes **SP**: Mar (Portsmouth, c1812) Louisa Ann Langley (c1816-__) **VI**: Son of Dempsey, Jr. and Mary (Moore) Watts; commissioned a judge of the county court and also served as Alderman in 1831; after the war in 1833 he became Captain of the Portsmouth Greys Light Infantry; represented Norfolk County in the VA Legislature; was the Whig candidate for Lieutenant Governor in 1850; enumerated on 1840 census at Old Point Comfort and on 1850 census of Portsmouth **P**: No **BLW**: No **PH**: No **SS**: A rec 1992; L pg 205 **BS**: 65 pg 119

*WEBB, John; b 20 Jul 1792, d 8 Jun 1876 **RU**: Private, 78th VMR, Captain James Anderson's Co, Grayson County & Captain Timothy Dalton's Co, Grayson County, attached to 4th VMR (McDowell-Koontz-Chilton) **CEM**: Webb; Carroll; jcts Rts 744 & 705, Hillsville **GS**: Yes, Govt **SP**: Hannah Cocke (1800-1890) **VI**: Son of Henry Webb (1762-1845) & Susanna Cocke (1763-1847) **P**: Yes **BLW**: No **PH**: No **SS**: A rec 986; 1001; B pg 78; **BS**: 245

*WEBB, Lewis N; b 15 Jul 1789, d 6 Mar 1873 **RU**: Private, 19th VMR, served in either Capt William McCabe's, Capt John McPherson's, or Capt Anthony Turner's companies, Richmond City **CEM**: Shockoe Hill; Richmond City; 100 Hospital St; **GS**: Yes **SP**: Anne New (8 Jun 1796-8 Mar 1837) **VI**: Railroad stock agent and commissioner **P**: Applied for **BLW**: No **PH**: Yes **SS**: B pg 175; L pg 573; M pg 309 **BS**: 31; 245

*WEST, John; b 28 Feb 1770, Bradford, Essex Co, MA, d 16 Apr 1818 **RU**: Private, 20th Inf, USA, Capt Robert Love's Co, Norfolk **CEM**: Eastern Shore Chapel; VA Beach; 2000 Laskin Rd **GS**: Yes **SP**: Elizabeth (__) **VI**: Inscription on GS indicates late war service; member of Companions in Arms; member of Common Council; delegate to General Assembly **P**: Spouse **BLW**: No **PH**: No **SS**: BD vol II; pg 1804 **BS**: 245

Key: *Additional veteran entry   **Corrected veteran entry   ***Deleted veteran entry
RU=Rank/Unit   CEM=Cemetery   GS=Gravestone   SP=Spousal Information   VI=Other Veteran Info   P=Pension
BLW=Bounty/Land Warrant   PH=Photo   SS=Service Source   BS=Burial Source   VMR= VA Military Regiment
LNR= Last Known Residence

*WHEELER, James; b 1791, d 12 Jul 1883 **RU:** Private, 72nd VMR, Captain Andrew Caldwell's Company, Russell Co, attached to 4th VMR (McDowell, Knootz Chilton) or Captain John Hamon's Company, Russell Co attached to 6th VMR (Dickerson, Scott, Coleman) **CEM:** Wheeler Family; Russell; Castlewood; GPS: 36.9073196, -82.2970827 **GS:** No **SP:** Mary Amburgey (1791-18 Dec 1862) **VI:** No further data **P:** No **BLW:** No **PH:** No **SS:** A rec 1120; B pg 183 **BS:** 245

*WHITE: Jacob; b 1789, d 13 Nov 1862 **RU:** Corporal, 31st, 51st or 122nd VMR, Frederick Co in company attached to 4th VMR (Boyd) **CEM:** Fairview Lutheran; Frederick; 464 Fairview Rd; Gore **GS:** Yes **SP:** Mar (Frederick Co, 28 Feb 1816) Elizabeth Duncan (1792-2 Aug 1870) **VI:** No further data **P:** No **BLW:** No **PH:** No **SS:** A rec 810 **BS:** 56 pg 371; 245

*WHITE, James Buford; b unk, d 24 Apr 1839 **RU:** Private, 10th or 91st VMR, company in Bedford Co, attached to 5th VMR **CEM:** Old Ebenezer Church; Amherst; 882 Ebenezer Rd **GS:** Yes **SP:** No spouse info **VI:** No further data **P:** No **BLW:** No **PH:** No **SS:** A rec 711 **BS:** 245

WHITE, James; b 30 Jan 1783, d 23 Sep 1825 **RU: Major, Staff officer, 105th VMR Washington Co Militia; commissioned Major 10 Jul 1810 **CEM:** Rehobeth United Methodist; Loudoun; jct Rt 691 & Bollington Rd, Rt 692, Morrisonville **GS:** Yes **SP:** Mar (Shelburne Parish, Loudoun Co, 15 May 1833) Eliza R Best. Enos Best attested to bride's age **VI:** No further data **P:** No **BLW:** No **PH:** No **SS:** B pg 198 **BS:** 73 pg 338; 245

*WHITE, John; b c1767, d 3 Mar 1851 **RU:** Sergeant, 56th VMR, Capt James Cochran's Co, Loudoun County **CEM:** Burnt Factory United Methodist Church; Frederick; 1943 Jordan Springs Rd; Stephenson **GS:** Yes **SP:** No spouse info **VI:** No further data **P:** Yes **BLW:** No **PH:** No **SS:** A rec 999; B pg 121; BD pg 1815 **BS:** 245

*WHITE, John; b 1795, d 26 Jan 1873 **RU:** Private, Capt James Sower's Co, Frederick County, attached to Flying Camp (Mc Dowell's) **CEM:** St Johns Lutheran Church; Frederick; Hayfield **GS:** Yes **SP:** Mar (8 Dec 1821) Sarah McIlwee (17 Aug 1800-26 Jan 1884), daug of John McIlwee **VI:** No further data **P:** No **BLW:** No **PH:** No **SS:** B pg 80; K pg 22 **BS:** 245

*WHITE, John; b 16 Nov 1798, d 6 Oct 1861 **RU:** Private, 57th VMR, Captain Michael Everhart's Company, Loudoun Co, attached to 1st VMR (Taylor) **CEM:** North Fork Baptist; Loudoun; 38139 N Fork Rd, Purcellville **GS:** Yes **SP:** No spouse info **VI:** No further data **P:** No **BLW:** No **PH:** No **SS:** A rec 908; b pg 120 **BS:** 245

*WHITE, John; b 16 Dec 1797, d North Garden, Albemarle Co, 8 Feb 1877 **RU:** Private, 47th VMR, a company of Albemarle Co, attached to 7th VMR (Yancey) **CEM:** Linden; Albemarle; Ivy **GS:** No **SP:** Caroline Perkins Moore (15 Aug 1801-16 Nov 1865), daug of Stephen Moore (1751-1835) and Eliza Royster (1764-1844) **VI:** No further data **P:** No **BLW:** No **PH:** No **SS:** A rec 969; B pg 36 **SS:** 245

WHITE, John A; b 1795; d 14 May 1848 **RU: Private, 61st VMR, Capt Francis Jarvis's Company, Mathews Co **CEM:** White Haven; Mathews; Pine Haven Rd **GS:** Yes **SP:** Martha W James (1800-16 Aug 1849, age 49, yrs 5 mos), daug of Thomas James and Betsy Davis **VI:** Son of Sgt. John White (1758-1834), who served in Revolution and Elizabeth Davenport (1771-1845) **P:** No **BLW:** No **PH:** No **SS:** K pg 299 **BS:** 54 pg 184; 245

*WHITE, John Baker; b 4 Aug 1794, Winchester, d 19 Oct 1862 **RU:** Ensign, 74th VMR, Capt Robert Mallory's Co, Hanover County, attached to 1st VMR (Crutchfield) **CEM:** Hollywood; Richmond City; 412 S Cherry St **GS:** Yes **SP:** Francis Streit **VI:** Service record at NARA indicates rank as Private, however GS is assumed correct. It indicates "Ensign, War of 1812". Naval records indicate he was a Sailing Master (2 Dec 1813) and he resigned 20 Apr 1814. It also indicates he was "Clerk of Hampshire County" (Note: Mineral County, WVA was formed from Virginia's Hampshire County in 1866.) He was son of Robert and Arabella (Baker) White **P:** Yes **BLW:** No **PH:** Yes **SS:** B pg 95; BD pg 1815; G; BG **BS:** 31

*WHITE, Sampson; b 1797, d 20 Apr 1867 **RU:** Corporal; 71st VMR Capt John Velvin's Co, Surry County **CEM:** Ivy Hill; Isle of Wight; 451 N Church St; Smithfield **GS:** Yes **SP:** Julia E. Hayden (11 Oct 1833-15 Dec 1865) **VI:** Had only 11 days service, thus not eligible for pen or BLW **P:** No **BLW:** No **PH:** No **SS:** B pg 193; L pg 799 **BS:** 245

Key: *Additional veteran entry   **Corrected veteran entry   ***Deleted veteran entry
RU=Rank/Unit   CEM=Cemetery   GS=Gravestone   SP=Spousal Information   VI=Other Veteran Info   P=Pension
BLW=Bounty/Land Warrant   PH=Photo   SS=Service Source   BS=Burial Source   VMR= VA Military Regiment
LNR= Last Known Residence

*WHITFIELD, Richard: b 12 Sep 1777, Whitby, North Yorkshire, Eng, d 4 Aug 1866 **RU:** Private, 32d VMR Capt Briscoe G Baldwin's Co of Mounted Infantry, Augusta County, attached to McDowell's Flying Camp, (Note: A person this name from either Albemarle, Fauquier or Orange Co was in Major Crutchfield's Detachment) **CEM:** Shockoe Hill; Richmond City; 100 Hospital St **GS:** Yes **SP:** Ann Booker (26 Sep 1792-21 Sep 1846) **VI:** More than 60 years was a resident of Richmond as a businessman **P:** No **BLW:** No **PH:** No **SS:** A rec 44 and/or 45; B pg 49 **BS:** 245

WHITWORTH, Thomas; b 4 Jan 1794, England; d 24 Jun 1874 **RU: Captain, 83rd VMR (Byrne), company commander, Dinwiddie Co **CEM:** Sweeden; Dinwiddie; town of Sweeden **GS:** Yes **SP:** Mar (10 Jan 1833) Eliza Harrison, daug of Col Peterson Harrison. She is buried in an above-ground granite vault as she had a horror of being buried underground The granite was quarried at "Mayfield" and hauled to Sweeden for this purpose **VI:** "Captain in the War of 1812" on his tombstone **P:** No **BLW:** No **PH:** No **SS:** G **BS:** 210

WHORTON, William; b 1793; d 1850 **RU: Private, 16th VMR, Capt James Fox's Co, Spotsylvania County, attached to Major Crutchfield's Battalion **CEM:** Whorton Family; Stafford; end of Norman Rd, Rt 661 **GS:** Unk **SP:** Rebecca (___) (1797-1882) **VI:** No further data **P:** No **BLW:** No **PH:** No **SS:** K pg 380 **BS:** 26A pg 37

WICKER, Francis; b 1793; d (bur) 14 Oct 1849 **RU: Captain, 33rd VMR, company commander Henrico Co **CEM:** St John's Church; Richmond City; 24th & Broad, Church Hill **GS:** Unk **SP:** 1) Elizabeth Hopkins (4 Sep 1789-15 Aug 1837), 2) (Richmond 29 Mar 1838) Lucy Ann Lipscomb **VI:** No further data **P:** No **BLW:** No **PH:** No **SS:** B pg 100; K pg 879 **BS:** 63 pg 242; 352; 252 pg 67; 245

WILKINSON, Jesse; b 1790; d 23 May 1861, Norfolk **RU: Lieutenant, Schooner *Hornet* **CEM:** Cedar Grove; Portsmouth; Effington St & Fort Ln **GS:** Yes **SP:** No spouse info **VI:** Promoted to Captain 1829; commanded the *John Adams*, (1835-1840); commanded the frigate *United States;* served on the flagship *Macedonian* (1840-1842); fr (1843 to 1847) was commandant of the Norfolk Navy Yard; made Commodore, and in 1848 to 1849 commanded the West Indies squadron, with the *Raritan* as his flagship **P:** No **BLW:** No **PH:** No **SS:** A pg 267 **BS:** 49

*WILKINSON, Thomas; b Feb 1770, d 18 Sep 1843 **RU:** Captain, 57th VMR, company commander of Artillery, Loudoun Co, attached to 6th VMR (Reade) **CEM:** New Jerusalem Lutheran Church; Loudoun; Lovettsville **GS:** Yes **SP:** No spouse info **VI:** No further data **P:** No **BLW:** No **PH:** No **SS:** B pg 122 **BS:** 245

*WILKINSON, William; b 1792 Sussex Co, d 1855 **RU:** Private, Capt Philip Pryor's Troop of Cavalry, Brunswick Co, attached to 1st VMR (Byrne) **CEM:** Wilkinson Farm: Sussex: Stony Creek **GS:** No **SP:** Susan W (___) **VI:** Son of John and Martha (Rives) Wilkinson **P:** Spouse **BLW:** No **PH:** No **SS:** A rec 2178; B pg 49; BD vol 2; pg 1830; M pg 313 **BS:** 245

*WILKINSON, William; b 10 Jul 1796, d 19 Jan 1871 **RU:** Private, 56th VMR Capt Thomas Gregg's Co, Loudoun County, attached to 4th VMR (Beatty) **CEM:** South Fork Meeting House; Loudoun; Unison **GS:** Yes **SP:** Sidney (___) (20 Oct 1796-27 Sep 1871) **VI:** No further data **P:** No **BLW:** No **PH:** No **SS:** A rec 2170 B pg 120 **BS:** 245

*WILLIAMS, John; b Stafford Co, d 29 Sep 1812, FL **RU:** Captain, commanded a detachment of Marines in Florida **CEM:** Arlington National; Arlington; off George Washington Parkway **GS:** Unk **SP:** No spouse info **VI:** Promoted to Second Lieutenant, 20 August, 1805, 1st Lieutenant, 2 March 1807 and Captain 31 January 1811. He died of wounds received in action in August 1812 in a skirmish with East Florida Indians. He is buried in section 27, site 3317 (another source WS lot 158) **P:** No **BLW:** No **PH:** No **SS:** AL **BS:** 53 pg 19

WILLIAMS, Nathaniel Pope; b by 1793, d aft 1831 **RU: Ensign, 45th VMR, Capt Thomas Fristoe's Company, Stafford Co **CEM:** William's Family #1; Stafford County; on dirt road 0.5 miles beyond Bethlehem Baptist Church, Rt 602 **GS:** Yes **SP:** Mary Combs, daug of Joseph H. Combs and Mary Rousseau (c1760-___) **VI:** Son of Person Williams. Justice of Peace 1820; 1822; 1827; 1830; 1831 **P:** No **BLW:** No **PH:** No **SS:** L pg 342 **BS:** 26 pg 397

Key: *Additional veteran entry   **Corrected veteran entry   ***Deleted veteran entry
RU=Rank/Unit   CEM=Cemetery   GS=Gravestone   SP=Spousal Information   VI=Other Veteran Info   P=Pension
BLW=Bounty/Land Warrant   PH=Photo   SS=Service Source   BS=Burial Source   VMR= VA Military Regiment
LNR= Last Known Residence

WILLIAMS, Samuel; b c1790; d 25 Apr 1855 **RU: Private, 56th VMR, Capt Thomas Gregg's Co, Loudoun County, attached to 4th VMR (Boyd) **CEM:** Rehoboth United Methodist; Loudoun; jct Rt 691 & Bollington Rd, Rt 692, Morrisonville **GS:** Yes **SP:** No spouse info **VI:** No further data **P:** No **BLW:** No **PH:** No **SS:** A rec 310; B pgs 120, 226 **BS:** 73 pg 342

*WILLIAMS, Thomas, b 13 Sep 1787, d 8 Jul 1857 **RU:** Private, 23rd VMR, Capt John Hix's Company, Chesterfield Co **CEM:** Hollywood Cemetery; Richmond City; 412 S Cherry St **GS:** Yes **SP:** No spouse info **VI:** He was a lumber merchant **P:** No **BLW:** No **PH:** No **SS:** L pg 430 **BS:** 245

*WILLIAMS, Thomas S; b 1785, d 1886 **RU:** Private, 73rd VMR, Lt James Bragg's Co, Lunenburg County, attached to 4th VMR (Boyd) **CEM:** Meherrin Baptist Church; Lunenburg; Lunenburg town **GS:** Yes **SP:** Judith G (__) (1815-1890) **VI:** "War of 1812" on gravestone **P:** Both **BLW:** No **PH:** No **SS:** A rec 125; B pg 125; BD pg 1792; M pg 314 **BS:** 245

*WILLIAMS, William; b Oct 1795, Ireland, d 24 Oct 1849 **RU:** Private, 33rd VMR, Captain Abraham Cowley's Company, Henrico Co **CEM:** Hollywood; Richmond; 412 S Cherry St **GS:** Yes **SP:** No spouse info **VI:** No further data **P:** No **BLW:** No **PH:** No **SS:** K pg 250 **BS:** 245

*WILLIS, William Champe; b 1770, d 1843 **RU:** Private, 3rd VMR, Captain Lawrence T Dade's Artillery Company, Orange Co, attached to 1st VMR (Yancey) **CEM:** Willis; Orange; off Rt 612 at jct S of jct with Shady Grove Ln **GS:** No **SP:** Mar 1) Lucy Taliaferro (1775-1812), daug of Robert Taliaferro (1716-__) and Jane Bankhead (1738-__), 2) Elzira (__) **VI:** Son of Lewis Willis (1734-1812 and Mary Champe (1735-__) **P:** Yes and widow Elizira **BLW:** No **PH:** No **SS:** A rec 1805; B pg 156; BD pg 1841; M pg 314 **BS:** 245

*WILLS, Charles; b c1775, Isle of Wight Co, d aft Jun, bef Dec 1820 **RU:** Private, 19th VMR, Capt Andrew Stephen's Artillery Co, Richmond city, attached to 2nd VMR, (Ballowe) **CEM:** St John's Church; 24th and Broad; Church Hill **GS:** No **SP:** Sally (__) (__-1826 **VI:** Was titled Captain in a newspaper marriage notice of 1828. Son of Josiah and Martha (Milner) Wills; moved to Richmond 1801 (Tax records). Conducted business (1805) on Main Street vic today's Farmers Market. In 1812, he built a house standing today on Church Hill and another business building next door. His family is listed in the Annals of Henrico Co and his daughters were married in St John's Church. Although he is not listed in the church cemetery records evidence that he is buried there exists. **P:** No **BLW:** No **PH:** No **SS:** B pg 175; L pg 740 **BS:** 49

*WILLS, John; b 21 Apr 1793, d 5 Jan 1853 **RU:** Corporal, in company Bedford or Campbell Co, attached to 5th VMR in Norfolk **CEM:** Presbyterian Cemetery; Lynchburg City; 2020 Grace St **GS:** Yes **SP:** Mary R (__) (19 Apr 1797-8 Oct 1877) **VI:** No further data **P:** No **BLW:** No **PH:** No **SS:** A rec 2094; B pgs 42- 43; 53- 54 **BS:** 245

*WILLS, Thomas; b c1795, d unk **RU:** Private, 64th VMR, Captain Jesse Canter's Company, Henry Co, attached to 4th VMR (Huston and Wooding) **CEM:** Wills Cemetery; Henry Co; the home place of Henry Wills **GS:** No **SP:** Mar Henry Co, 7 Nov 1816, Bethania King **VI:** No further data **P:** No **BLW:** No **PH:** No **SS:** A rec 2155; B pg 101 **BS:** 245

*WILLSON, Thomas; b 17 Feb 1794, d 11 Dec 1857 **RU:** Sergeant, 8th VMR a company of Rockbridge Co, attached to 5th VMR (McDowell's) **CEM:** New Providence Presbyterian Church; Rockbridge; Raphine **GS:** Yes **SP:** Elizabeth H (__) (1801-27 Aug 1854) **VI:** No further data **P:** No **BLW:** No **PH:** No **SS:** A rec 2653; B pg 179 **BS:** 245

WILSON, Matthew, Sr; b 2 Jul 1796 d 23 Feb 1849 **RU: Drummer, 68th VMR, Capt Robert P Taylor's Company, James City Co **CEM:** Bethel Church; Augusta; 11 mi SW Staunton **GS:** Unk **SP:** No spouse info **VI:** No further data **P:** No **BLW:** No **PH:** No **SS:** K pg 377 **BS:** 183; 245

*WILSON, Thomas, b 1791, d 4 Apr 1864 **RU:** Private, 8th VMR Either Capt Isaiah McBride's Co or Capt Archibald Lyle's, Troop of Cavalry Rockbridge County, attached to 5th VMR (McDowell) **CEM:** High Bridge Presbyterian Church; Rockbridge; Rt 11, 15 mi S of Lexington **GS:** Unk **SP:** No spousal info **VI:** No further data **P:** No **BLW:** No **PH:** No **SS:** O-Serv Index Card **BS:** 183; 245

Key: *Additional veteran entry   **Corrected veteran entry   ***Deleted veteran entry
RU=Rank/Unit   CEM=Cemetery   GS=Gravestone   SP=Spousal Information   VI=Other Veteran Info   P=Pension
BLW=Bounty/Land Warrant   PH=Photo   SS=Service Source   BS=Burial Source   VMR= VA Military Regiment
LNR= Last Known Residence

101

WINN, John; b 9 Sep 1782, Hanover Co; d 13 Nov 1837 **RU:** Ensign, 47th or 88th VMR, Captain Robert McCullough's Co or Captain Rothwell's Co, Albemarle Co. attached to 7th VMR **CEM:** Riverview; Albemarle; 1701 Chesapeake St, Charlottesville **GS:** Yes **SP:** Mary Johnson (3 Jan 1789-30 Sep 1857) **VI:** Was on Committee with Jefferson that would establish UVA; was postmaster in Charlottesville (1803- 1837); was a founding member of Albemarle Agricultural Society and on a committee that established Charlottesville's first library; Death notice in *Richmond Whig* (21 Nov 1837), pg 2 **P:** No **BLW:** No **PH:** No **SS:** A rec 1122; B pg 36 **BS:** 31; 245

*WISE, John; b 1786, d 1856 **RU:** Private, 116th VMR, Captain William Newell's Company, Shenandoah Co. **CEM:** Radar Lutheran Church; Rockingham; Timberville **GS:** Yes **SP:** No spouse info **VI:** No further data **P:** No **BLW:** No **PH:** No **SS:** B pg 181 **BS:** 31; 245

*WOLFE, John; b 27 Mar 1781, d 13 Feb 1861 **RU:** Lieutenant, Adjutant 105th VMR, Captain William Smith's Company of Artillery, Washington Co, attached to Battalion of Artillery **CEM:** Wolfe Family; Scott; Weber City **GS:** Yes **SP:** No spouse info **VI:** No further data **P:** No **BLW:** No **PH:** No **SS:** A rec 331; B pg 199 **BS:** 245

*WOLFE, John; b 1793, PA, d 1855 **RU:** Private, 48th VMR, Captain James Rowland's Company, Botetourt Co, attached to 4th VMR (Boyd's) **CEM:** Wolfe-Sively; Alleghany; Potts Creek **GS:** No **SP:** Mar (21 Jun 1813) Sarah Rayhill (1792-1855), daug of Mathew Rayhill and Esther Stull **VI:** Son of Jacob and Mary Wolfe **P:** No **BLW:** No **PH:** No **SS:** A rec 330; B pg 46 **BS:** 245

WOMACK, William; b 1769; d 26 Mar 1828 **RU:** Lt Colonel, Regimental commander, 23rd VMR, Chesterfield Co **CEM:** Locust Bottom; Botetourt; vic jct Rts 696 & 622, 1 mi E of Glen Wilton **GS:** Yes **SP:** Jane Kyle (24 Feb 1783-22 Jun 1858) **VI:** No further data **P:** No **BLW:** No **PH:** No **SS:** B pg 59 **BS:** 155 pg 12

*WOMBLE, John; b 21 May 1799, d 15 Feb 1854 **RU:** Private, 74th VMR, Captain Thomas Jones's Company, Hanover Co, attached to 8th VMR (Magnien's) **CEM:** Hollywood; Richmond City; 412 S Cherry St **GS:** No **SP:** Alice (___) **VI:** No further data **P:** No **BLW:** No **PH:** Yes **SS:** A rec 626; B pg 95 **BS:** 31

*WOOD, Benjamin: b c1792, d 22 Apr 1872 **RU:** Private, 47th VMR, Captain John Rothwell's & Capt Robert McCulloch's Co, Albemarle County, attached to 7th VMR (Gray) **CEM:** Locust Grove; Albemarle; Rt 678, Ivy Village **GS:** Yes **SP:** No spouse info **VI:** No further data **P:** Yes **BLW:** No **PH:** No **SS:** B pg 36; BD pg 1858; K pg 354 **BS:** 245

***WOOD, James-born 1775: Deleted as source 245 indicates death was before the war period

*WOOD, James: b 1781, d 1853 **RU:** Sergeant, in a company from Culpeper Co, attached to 6th VMR **CEM:** Masonic; Culpeper; Radio Lane & Rt 29 **GS:** Yes **SP:** No spouse info **VI:** No further data **P:** No **BLW:** No **PH:** No **SS:** A rec 1145; B pgs 62-3 **BS:** 245

*WOOD, Joshua; b 24 Aug 1792, d 9 Nov 1859 **RU:** Private: 57th VMR, Captain Thomas Wilkerson's Artillery Company, Loudoun Co, attached to 6th VMR (Reade) **CEM:** Fairfax Friends; Loudoun; Waterford **GS:** Yes **SP:** Eleanor H Stone (5 May 1801-30 Jan 1873) **VI:** No further data **P:** No **BLW:** No **PH:** No **SS:** A rec 1455; B pg 122 **BS:** 245

*WOOD, William; b 1784, d 1862 **RU:** Private, 51st VMR, Captain James H Sower's Company, Frederick Co, attached to 5th VMR (McDowell's Flying Camp) **CEM:** Gainesboro Methodist; Frederick; Gainesboro **GS:** No **SP:** Margaret Ridgeway (1794-23 Jul 1866), daug of David Ridgeway and Martha Beeson **VI:** No further data **P:** Yes, widow **BLW:** No **PH:** No **SS:** A rec 1817 **BS:** 245

*WOODHOUSE, Henry Barnes; b 9 Jun 1793, d 24 Jan 1879 **RU:** Private, 20th VMR, Captain Thomas Keeling's Company, Princess Anne Co **CEM:** Woodhouse Family; VA Beach; Southern Blvd, Lynnhaven **GS:** Yes **SP:** No spouse info **VI:** Obtained rank of General after war **P:** Applied for **BLW:** No **PH:** No **SS:** L pg 518; M pg 318 **BS:** 245

Key: *Additional veteran entry   **Corrected veteran entry   ***Deleted veteran entry
RU=Rank/Unit   CEM=Cemetery   GS=Gravestone   SP=Spousal Information   VI=Other Veteran Info   P=Pension
BLW=Bounty/Land Warrant   PH=Photo   SS=Service Source   BS=Burial Source   VMR= VA Military Regiment
LNR= Last Known Residence

*WOODSON, John; b 25 Nov 1778, d 31 Mar 1856 **RU**: Private, 23rd VMR Capt Thomas Cheatham's Co, Chesterfield County **CEM**: Shockoe Hill; Richmond City; 100 Hospital St **GS**: Yes **SP**: Elizabeth W Coleman (6 May 1807-31 Mar 1856) **VI**: Son of Archibald Pleasants of Goochland Co. Came to Richmond in 1795 **P**: No **BLW**: No **PH**: Yes **SS**: B pgs 59; 226; L pg 211 **BS**: 31; 245

*WRENN, Thomas; b 1774, d 1817 **RU**: Private, 7th or 95th VMR, Captain Tubman Law's Company of Norfolk Co. attached to 3rd VMR (Boykin) **CEM**: Elmwood; Norfolk City; 238 Princess Anne Rd **GS**: Yes **SP**: No spouse info **VI**: No further data **P**: Yes **BLW**: No **PH**: No **SS**: A rec 134; B pg 148; M pg 319 **BS**: 245

WRIGHT, George; b 11 Sep 1792, Dunnington, North Yorkshire, England, d 27 Feb 1859 **RU: Sergeant, in company, Frederick Co attached to 4th VMR **CEM**: Mt Carmel; Frederick; 3rd & High St, Middletown **GS**: Yes **SP**: Also possible spouse buried here is Catherine Senseney Wright (1792-8 Jul 1843) **VI**: Son of Rachel Butler (1850) **P**: No **BLW**: No **PH**: No **SS**: A rec 485 **BS**: 79 pg 360

*WRIGHT, John: b 22 Feb 1790, d 20 Nov 1863 **RU**: Private, 93rd VMR, Capt Archibald Stuart's Co, Augusta County, attached to McDowell's Flying Camp **CEM**: Bethel Presbyterian Church; Augusta; 563 Bethel Green Rd, Middlebrook **GS**: Yes **SP**: Christina McComb (1796-1887) **VI**: No further info **P**: Spouse **BLW**: No **PH**: No **SS**: B pg 40; BD pg 1872; M pg 319 **BS**: 245

*WRIGHT, Joseph; b 1777, d 18 Sep 1840 **RU**: Private, 59th VMR company, Nansemond Co **CEM**: Wright Family; Suffolk; Bennett's Creek in residential area nr Nansemond River **GS**: Yes **SP**: No spouse info **VI**: No further data **P**: No **BLW**: No **PH**: No **SS**: A rec 929 **BS**: 245

*WYATT, Joseph; b c1763, d 9 Apr 1843 **RU**: Lt Colonel, 26th VMR, Regimental commander, Charlotte Co **CEM**: Joseph Wyatt; Charlotte; Evergreen Rd **GS**: Yes **SP**: No spouse info **VI**: Son of Joseph, Sr (1728-1767) and Dorothy Smith Peyton. Was a Justice of Charlotte Co Court (1793) and county Sheriff fr 1813 to1815 and again in 1843 **P**: No **BLW**: No **PH**: No **SS**: B pg 57 **BS**: 31

*WYATT, Thomas E; b 13 Dec 1790, d 24 May 1824 **RU**: Private, 8th VMR, in company of Rockbridge County attached to 4th or 5th VMR commanded by McDowell, Mason, Preston or Koontz **CEM**: Turley; Rockingham; Turleytown **GS**: Yes **SP**: No spouse info **VI**: No further info **P**: No **BLW**: No **PH**: No **SS**: A rec 1763; 1764; 1765; 1766; B pgs 181-182 **BS**: 245

WYNNE, Samuel; b 31 Jul 1789, d 18 Sep 1817 **RU: Private, 112th VMR, Lieutenant Rees B Thompson's Company, Tazewell Co attached to 4th VMR (McDowell, Koontz, Chilton) **CEM**: Wynn-Litton; Lee; Dryden **GS**: No **SP**: No spouse info **VI**: Wounded in war. Son of William Wynn **P**: No **BLW**: No **PH**: No **SS**: A rec 2254; B pg 118; BD pg 1877 **BS**: 245

*YAGER, Salathiel Wayland; b 12 Dec 1792, d 4 Jul 1878 **RU**: Corporal, 82nd VMR, Capt Elliott Fink's Co, Madison County, attached to 6th VMR (Coleman) **CEM**: Cedar Hill; Madison; Clore Rd **GS**: Yes **SP**: Mar Madison Co, 6 May 1819, Anna M. Miller (14 Aug 1795-4 Jul 1880), daug of Adam and Mary (Wilhoit) Yager **VI**: He was licensed to keep an ordinary in Madison 14 May 1818 **P**: Both **BLW**: No **PH**: No **SS**: A rec 2357; B pg 126; BD pg 1877; W pg 320 **BS**: 30; Community Cem

*YEATMAN, Thomas Robinson; b 5 Jan 1789, d unk **RU**: Lieutenant. 61st VMR, Capt Thomas Teagle's Co, Mathews County **CEM**: Ware Episcopal Church; Gloucester; 7825 John Clayton Memorial Rd **GS**: Yes **SP**: Elizabeth Tabb Patterson (20 Jun 1796-19 Oct 1868) **VI**: Son of Thomas Muse Yeatman & Mary Tomkins; memorialized in cem as stone only moved from Yeatman Plantation, Gloucester Co **P**: No **BLW**: No **PH**: No **SS**: K pg 304 **BS**: Ware church website burial list

*YOUNG, David: b 23 Aug 1774, d 1 Feb 1829 **RU**: Captain, Calvary Company commander, Albemarle Co **CEM**: Sunny Bank; Albemarle; North Garden **GS**: Unk **SP**: Mar (Albemarle Co, 27 Apr 1809) Mary Ann Hart (1787-1825) **VI**: Son of Capt John Young (1737-1824) and Mary White (1744-1779) **P**: No **BLW**: No **PH**: No **SS**: A rec 1100; B pg 36; **BS**: 245

Key: *Additional veteran entry **Corrected veteran entry ***Deleted veteran entry
RU=Rank/Unit   CEM=Cemetery   GS=Gravestone   SP=Spousal Information   VI=Other Veteran Info   P=Pension
BLW=Bounty/Land Warrant   PH=Photo   SS=Service Source   BS=Burial Source   VMR= VA Military Regiment
LNR= Last Known Residence

*YOUNG, George; b c1795, d unk RU: Sergeant, Capt Thomas Wilkinson's Artillery Co, Loudoun Co, attached to 5th VMR (Reade's) CEM: Goose Creek Burying Ground; Loudoun; Rt 722 Lincoln GS: Yes SP: No spouse info VI: No further data P: No BLW: No PH: No SS: A rec 1189; B pg 122 BS: 245

**YOUNG, Henry; b 1780; d 1841 (Will) RU: Private, 99th VMR (Bagwell), in a company in Accomack Co CEM: Young Plot; Accomack; Bloxom GS: Yes SP: Mar (25 Aug 1823) Ann Wessells of Ephraim VI: Son of George and Susan (__) P: No BLW: No PH: No SS: A rec 1249 BS: 6 pg 302

**YOUNG, James; b 2 Mar 1795, Dinwiddie Co; d 17 Sep 1857 RU: Private, 39th VMR, Capt Cadwallader J Claiborne's Co, Petersburg CEM: Blandford; Petersburg; 111 Rochelle Ln GS: Yes SP: Rosina Fenn (Prince George Co, 1811-Petersburg, Dec 1853) VI: No further data P: No BLW: No PH: No SS: K pg 136 BS: 200

*YOUNG, William S; b 11 Oct 1782, d 3 Feb 1860 RU: Private 5th VMR (McDowell's) CEM: Tinkling Springs Presbyterian Church; Augusta; Fisherville GS: Yes SP: No spouse info VI: Son of Capt John Young (1737-1824) and Mary/Polly Sitlington (1759-1838) P: No BLW: No PH: No SS: A rec 1976 BS: 245

*YOUNG, Samuel Wade; b unk, d 28 Jan 1825 RU: Captain, 56th VMR, Company commander, Loudoun Co CEM: Ketoctin Baptist Church; Loudoun; Purcellville GS: Yes SP: Ruth M (__) (29 Nov 1775-12 Jan 1842) VI: No further data P: No BLW: No PH: No SS: B pg 122 BS: 245

**ZIRKLE, George; b 10 Jan 1797; d 19 Jul 1869 RU: Private, 13th or 97th VMR, in company attached to 6th VMR (Coleman) CEM: Zirkle Family; Shenandoah; New Market GS: Yes SP: Mar (22 May 1816) Elizabeth Howbert (1798-1870) VI: Son of John O Zirkle & Rosanna Roush P: No BLW: No PH: No SS: A rec 2833 BS: 217; 245

**ZIRKLE, George; b 1780, d 9 Jun 1857 RU: Private, 13th or 97th VMR, in company attached to 6th VMR (Coleman) CEM: Zirkle family; Shenandoah; New Market GS: Yes SP: Mar (29 Mar 1803) Barbara Kagey VI: Son of George Adam Zirkle & Elizabeth Ridenour P: No BLW: No PH: No SS: A rec 2834 BS: 217; 245

---

*CHANDLER, William; b unk, d 1825 RU: Private 52d VMR, Capt John Merry's Co, New Kent County CEM: Chandler; New Kent; Laurel Springs house off Cook's Mill Road near Pamunkey River GS: No SP: Frances Apperson (ca1786-27 Oct 1847) VI: No further data P: No BLW: No PH: No SS: B pg 144; L pg 584 BS: 245

**CHEWNING, Reuben; b 4 May 1773; d 9 Jul 1837 RU: Captain, 40th VMR, Commanded a company, Louisa County, attached to 7th VMR (Gray) CEM: Westland Farm; Louisa; Springfield parcel, Green Springs area off Rt 22 on farm rd owned by Ned Gumble GS: N SP: Mar 1) Louisa Co,18 Dec 1799, Anne Dickinson, 2) 24 Oct 1816, Louisa Co, (return by William Cooke) Louisa Anderson. Tarleton Henley was surety to the bond, witnessed by Nathaniel Thompson, Jr. VI: He was also styled "Captain" on his marriage bond. A trunk was presented to him by officers of 7th VMR P: No BLW: No PH: N SS: B pgs 123, 239 BS: 49- 2d Gr Grandson John Starke

*COOKE, William, Jr; b 1788, d 4 Jul 1831 RU: Private,21st VMR Capt John R Cary's Co, Gloucester County, CEM: Cooke Family; New Kent: Town of North Green SP: Rebecca Hayes (Cumberland, New Kent Co 12 Jan 1826) VI: Son of William Cooke, Sr (__-1709/1) P: No BLW: No PH: No SS: B pg 82; K pg 268 BS: 31 Barham, Nellie & Marius, *Families are Forever* from the Diary of Col R.P. and Anne Elizabeth Cooke

*COX, David Leach; b 2 Jun 1783, d 8 Jun 1866 RU: Private 11th VMR Capt George Davis's Co, Harrison County (now WVA), attached to 1st VMR (Connell) CEM: Boatright; Scott; GPS 36.7225, -82.6961 GS: Yes SP: Mar 22 Jan 1807, Rebecca Boggs (1787-1864) VI: Son of David C Cocke (1748-28 Feb 1828) and Jemima Leach (abt 1752-1834) P: No BLW: No PH: No SS: A rec 116; B pgs 98, 221 BS: 245

Key: *Additional veteran entry   **Corrected veteran entry   ***Deleted veteran entry
RU=Rank/Unit   CEM=Cemetery   GS=Gravestone   SP=Spousal Information   VI=Other Veteran Info   P=Pension
BLW=Bounty/Land Warrant   PH=Photo   SS=Service Source   BS=Burial Source   VMR= VA Military Regiment
LNR= Last Known Residence

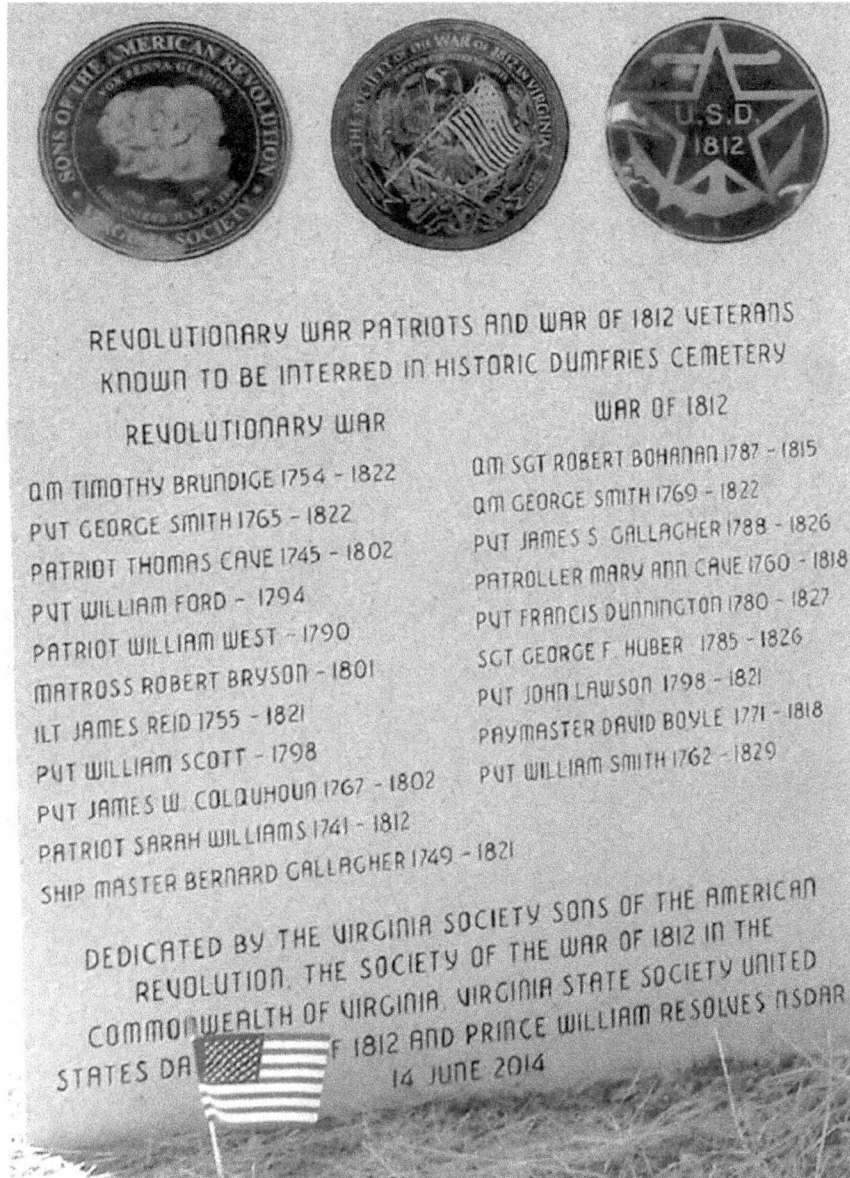

REVOLUTIONARY WAR PATRIOTS AND WAR OF 1812 VETERANS
KNOWN TO BE INTERRED IN HISTORIC DUMFRIES CEMETERY

REVOLUTIONARY WAR

QM TIMOTHY BRUNDIGE 1754 - 1822
PVT GEORGE SMITH 1765 - 1822
PATRIOT THOMAS CAVE 1745 - 1802
PVT WILLIAM FORD - 1794
PATRIOT WILLIAM WEST - 1790
MATROSS ROBERT BRYSON - 1801
ILT JAMES REID 1755 - 1821
PVT WILLIAM SCOTT - 1798
PVT JAMES W. COLQUHOUN 1767 - 1802
PATRIOT SARAH WILLIAMS 1741 - 1812
SHIP MASTER BERNARD GALLAGHER 1749 - 1821

WAR OF 1812

QM SGT ROBERT BOHANAN 1787 - 1815
QM GEORGE SMITH 1769 - 1822
PVT JAMES S. GALLAGHER 1788 - 1826
PATROLLER MARY ANN CAVE 1760 - 1818
PVT FRANCIS DUNNINGTON 1780 - 1827
SGT GEORGE F. HUBER 1785 - 1826
PVT JOHN LAWSON 1798 - 1821
PAYMASTER DAVID BOYLE 1771 - 1818
PVT WILLIAM SMITH 1762 - 1829

DEDICATED BY THE VIRGINIA SOCIETY SONS OF THE AMERICAN
REVOLUTION, THE SOCIETY OF THE WAR OF 1812 IN THE
COMMONWEALTH OF VIRGINIA, VIRGINIA STATE SOCIETY UNITED
STATES DA___ F 1812 AND PRINCE WILLIAM RESOLVES NSDAR
14 JUNE 2014

Monument mounted in the Dumfries Cemetery, Prince William County

105

WAR OF 1812 VETERANS INTERRED OR MEMORIALIZED
IN THIS HISTORIC CEMETERY

Private Thomas Bradley 1774-1826   Private Rix Jordan 1791-1857

Sergeant Major Mark L. Chevers 1795-1875   Captain Walter Kating 1779-1822

Sergeant William Face 1770-1855   Sergeant Edward King 1792-1879

Lieutenant James M. Glassell 1790-1838   Private Thomas Latimer 1776-1837

First Sergeant Timothy Green 1782-1847   Private James Powell 1778-1853

Private Richard S. Hicks 1793-1868   Major Arthur Simpkins UNK-1820

Private William P. Hope 1792-1845   Captain John Simpkins 1773-1860

*Dedicated by the War of 1812 Society in the Commonwealth of
Virginia and the Virginia State Society, National Society,
United States Daughters of 1812
JUNE 21, 2013*

Plaque mounted on the cemetery wall of the St. John's Episcopal Church, Hampton

FOUNDERS & VETERANS OF LYNNHAVEN PARISH CHURCH
INTERRED IN THIS CEMETERY AND SURROUNDING AREA

CAPT ADAM THOROWGOOD (1604-1640) FOUNDER
COL THOMAS WALKE I (1642-1694) COLONIAL WAR
COL EDWARD MOSELEY (1661-1736) COLONIAL WAR
COL ANTHONY WALKE I (1692-1768) COLONIAL WAR
COL EDWARD HACK MOSELEY (1717-1783) COLONIAL WAR
LT COL EDWARD HACK MOSELEY, JR (1740-1814) REVOLUTIONARY WAR
CPT JONATHAN SAUNDERS (1726-1765) COLONIAL WAR
COL ANTHONY WALKE II (1726-1779) COLONIAL WAR
COL ADAM THOROUGHGOOD (1750's-1780's) REVOLUTIONARY WAR
CAPT THOMAS WALKE IV (1760-1797) REVOLUTIONARY WAR
PVT JOHN HENDERSON (1769-1825) WAR OF 1812
PVT ANTHONY WALKE (1778-1820) WAR OF 1812
SGT JOHN BROWNLEY (1780-1853) WAR OF 1812

DEDICATED MAY 17, 2014 BY VA. SOCIETY ORDER OF FOUNDERS & PATRIOTS
OF AMERICA; VA. SOCIETY COLONIAL WARS; LYNNHAVEN PARISH CHAPTER, NSDAR;
NORFOLK CHAPTER, SONS OF THE AMERICAN REVOLUTION; COLONIAL DAMES
XVIIC SUFFOLK CHAPTER; FT. NORFOLK CHAPTER US DAUGHTER'S WAR OF 1812;
SOCIETY OF THE WAR OF 1812 IN THE COMMONWEALTH OF VA.

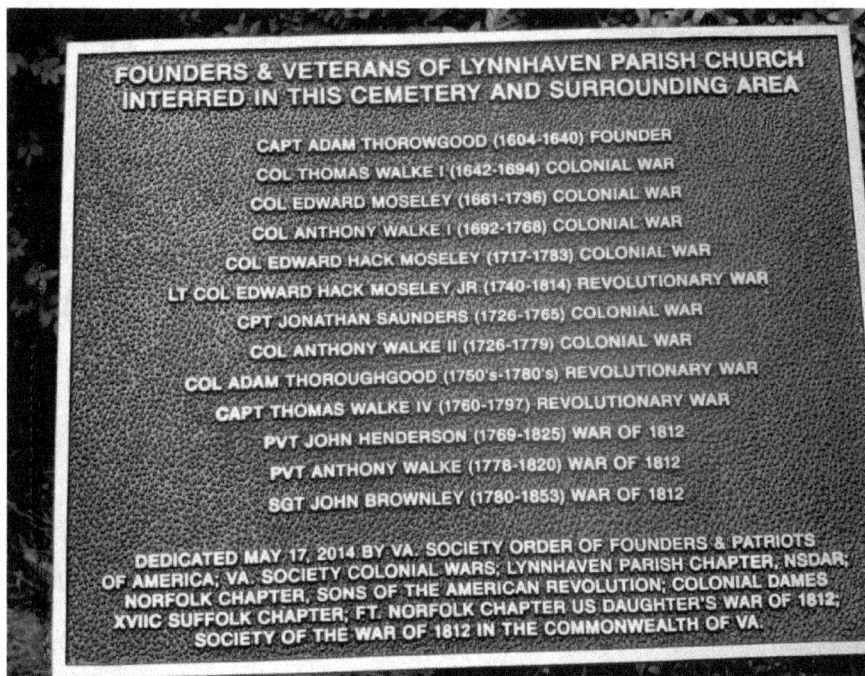

Plaque mounted in the cemetery of the Lynnhaven Parish Church, Virginia Beach

THE FOLLOWING WAR OF 1812 VETERANS ARE INTERRED
OR MEMORIALIZED IN THIS HISTORIC CEMETERY

PRIVATE BENNETT ALDRIDGE
PRIVATE JAMES M ANDERSON
ENSIGN ALLEN ARCHER
PRIVATE JOHN C ARMISTEAD
CORPORAL JOHN M BANNISTER SR
PRIVATE GEORGE D BASKERVILLE
PRIVATE JOHN P BOLLING
LIEUTENANT JOHN B BOTT
PRIVATE WILLIAM B BRITTON
PRIVATE DANIEL BROCKWELL
PRIVATE LARKIN BRUMLEY
PRIVATE WILLIAM P BURTON
PRIVATE ROBERT W BUCK
MAJOR ARMISTEAD BURWELL
PRIVATE WILLIAM A CHIEVES
PRIVATE JOHN G CLAIBORNE
PRIVATE REUBEN CLEMENTS
MIDSHIPMAN HARRISON H COCKE
PRIVATE WILLIAM COOKE
DOCTOR CHARLES CORLING
PRIVATE ATHIEL CROWDER
CORPORAL EDWARD DAVENPORT
PRIVATE WILLIAM E DAVIS
PRIVATE JAMES S DEATON

PRIVATE GRIEF DRUMMOND
PRIVATE JAMIE S ENNISS
DRUMMER JOHN S EPPES
SERGEANT HENRY GEE
PRIVATE WILLIAM B GOODWYN
LT COLONEL WILLIAM C GREENHILL
PRIVATE BENJAMIN E GRIFFITH
PRIVATE JESSE HEATH
PRIVATE ARTHUR JOHNSTON
CORPORAL WILLIAM G JONES
PRIVATE JAMES KIDD
SERGEANT EDWARD D LEE JR
PRIVATE WILLIAM LOWNES
PRIVATE MATTHEW MABEN
PRIVATE JAMES MACFARLAND
PRIVATE ROGER MALLORY
PRIVATE WILLIAM O MALLORY
PRIVATE THOMAS MARSH
PRIVATE JOSEPH MASON
PRIVATE BENJAMIN H MAY
PRIVATE JOHN F MAY
CAPTAIN RICHARD MCRAE
PRIVATE JOHN C MEADE
PRIVATE JOHN MICHAELS

Monument dedicated June 7, 2014, erected adjacent to the visitor's center parking lot in Blandford Cemetery, Petersburg. The remaining names are on the reverse side (shown on opposite page) which was updated in a ceremony on June 30, 2016

108

PRIVATE JOHN MICHAELS
PRIVATE JOHN MITCHELL
PRIVATE HUGH NELSON
PRIVATE THOMAS NELSON
CORPORAL JOSEPH C NOBLE
PRIVATE CHARLES OHARA
CORPORAL JOSEPH PANNELL
PRIVATE WILLIAM PANNELL SR
CORPORAL EDMOND PARRISH
ENSIGN JOHN POLLARD
PRIVATE FRANCIS PRICE
ADJUTANT RICHARD R RANDOLPH
SERGEANT JOHN B READ
PRIVATE JAMES RIDDLE
FIRST SERGEANT ROBERT RITCHIE
ENSIGN FRANCIS E RIVES
PRIVATE WILLIAM ROBERTSON
PRIVATE THOMAS ROBINSON
QUARTERMASTER FRANCIS RUFFIN
SERGEANT ABRAHAM B SEAY
ENSIGN WILLIAM SHANDS
PRIVATE JONATHAN SMITH
PRIVATE ALDEN B SPOONER

CORNET BAINBRIDGE SPOTSWOOD
ENSIGN GEORGE W STAINBACK
SERGEANT SAMUEL STEVENS
PRIVATE JOHN B STRACHAN
PRIVATE JAMES STUART
PRIVATE SEGUH THAYER
PRIVATE JOHN D TOWNES
SERGEANT JAMES TYRER
PRIVATE JOHN B UNDERHILL
PRIVATE NATHAN VINCENT
LIEUTENANT THOMAS WALLACE
PRIVATE WILLIAM WALLACE
PRIVATE JOHN WALTHALL
PRIVATE PEYTON WELLS
RIDING MASTER STEPHEN G WELLS
PRIVATE WILLIAM G WEST
PRIVATE THOMAS WHITE
PRIVATE HERBERT WHITMORE
PRIVATE JOHN V WILCOX
PRIVATE HENRY WILKINSON
PRIVATE JOHN WORSHAM
PRIVATE FRANCIS G YANCEY
PRIVATE JAMES YOUNG

DEDICATED JUNE 7 2014 IN MEMORY OF THESE WAR OF
1812 VETERANS SERVICE TO OUR COUNTRY BY THE
GENERAL SOCIETY WAR OF 1812 AND THE SOCIETY
OF THE WAR OF 1812 IN THE COMMONWEALTH OF VIRGINIA
AND THE VIRGINIA STATE SOCIETY UNITED STATES DAUGHTERS OF 1812

Reverse side of monument in Blandford Cemetery, Petersburg

WAR OF 1812 VETERANS
KNOWN TO BE INTERRED OR MEMORIALIZED IN THIS HISTORIC
BURIAL GROUND

PRIVATE FREDERICK A AULICK
PRIVATE FREDERICK BARLEY
PRIVATE JOHN CARTER
PRIVATE MICHAEL COPENHAVER
PRIVATE JAMES HOLLIDAY
PRIVATE ISAAC KURTZ
PRIVATE JOHN PRICE
CAPTAIN HENRY S G TUCKER
PRIVATE PHILIP YOUND
PRIVATE HENRY BAKER
LT COLONEL HENRY BEATTY
PRIVATE JOSEPH COLLINS
CAPTAIN MICHAEL COYLE
GOVERNOR DAVID HOLMES
PRIVATE JAMES LITTLE
BRIG GEN JAMES SINGLETON
PRIVATE GEORGE WALL
QM SERGEANT JACOB BAKER
2D LIEUTENANT GEORGE BRENT
PRIVATE CHARLES CONRAD
PRIVATE ROBERT BERKELEY

PRIVATE STEPHEN JENKINS
LIEUTENANT JOSIAH MASSIE
PRIVATE JOHN SINGLETON
PRIVATE JACOB WALL
PRIVATE JOHN BAKER
CORPORAL ANDREW BUSH
PRIVATE GEORGE COPENHAVER
PRIVATE SAMPSON GLAIZE
SERGEANT ISAAC KIGER
PRIVATE WILLIAM MYERS
MAJ GEN JOHN SMITH
SERGEANT JOHN C WALL
ENSIGN WILLIAM A BAKER
PRIVATE WILLIAM BUSH
PRIVATE JOHN COPENHAVER
PRIVATE JOHN HOFF
PRIVATE GEORGE KREMER, SR.
PRIVATE NICKOLAS PERRY
PRIVATE WILLIAM STEPHENSON
1ST LT GEORGE WASHINGTON
PRIVATE JOHN FLETCHER

DEDICATED BY THE SOCIETY OF THE WAR OF 1812
IN THE COMMONWEALTH OF VIRGINIA AND
THE VIRGINIA STATE SOCIETY, UNITED STATES DAUGHTERS OF 1812
AUGUST 24, 2013

Monument in the Mt. Hebron Cemetery, Winchester

Plaque mounted on a pole just inside the entrance to the historic
Masonic Cemetery, Fredericksburg

WAR OF 1812 VETERANS MEMORIALIZED IN THIS
HISTORIC CEMETERY

PRIVATE JOSEPH BEARD
SECOND LIEUTENANT JAMES BELL
MAJOR WILLIAM BELL
PRIVATE JAMES CRAIG, JR
PRIVATE GEORGE CRAWFORD
PRIVATE JAMES CRAWFORD
LIEUTENANT SAMUEL CRAWFORD, SR
CORPORAL SAMUEL CURRY
SERGEANT ROBERT GUY
PRIVATE SILAS HUNTON
PRIVATE JOSEPH KERR
PRIVATE JAMES NELSON
PRIVATE WILLIAM POAGE
PRIVATE JACOB STOVER
SERGEANT JOHN A. TATE
PRIVATE JACOB VAN LEAR
CORPORAL JOHN WALKER
CORPORAL JAMES WILSON DIED 1828
SURG. MATE JAMES WILSON DIED 1836
PRIVATE JAMES WILSON DIED 1854
PRIVATE ANDREW YOUNG

DEDICATED JULY 25 2015

BY THE SOCIETY OF THE
WAR OF 1812 IN THE
COMMONWEALTH OF
VIRGINIA

Plaque mounted on a pole in the Old Stone Presbyterian Church cemetery,
Fort Defiance, Augusta County

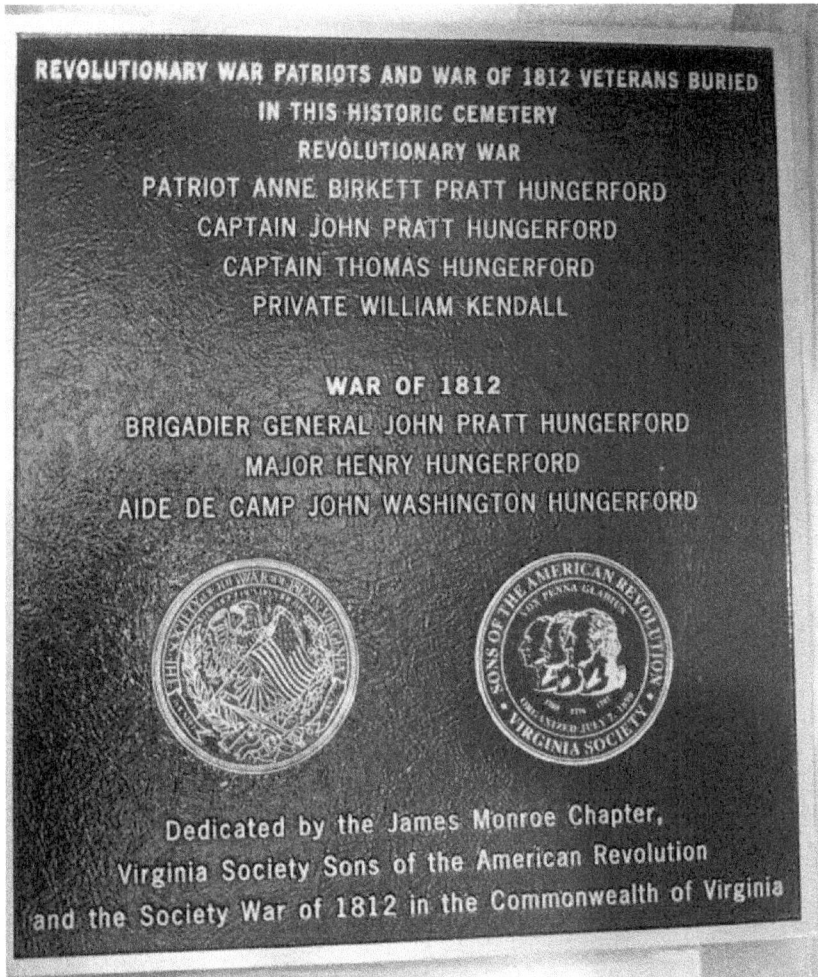

REVOLUTIONARY WAR PATRIOTS AND WAR OF 1812 VETERANS BURIED
IN THIS HISTORIC CEMETERY
REVOLUTIONARY WAR
PATRIOT ANNE BIRKETT PRATT HUNGERFORD
CAPTAIN JOHN PRATT HUNGERFORD
CAPTAIN THOMAS HUNGERFORD
PRIVATE WILLIAM KENDALL

WAR OF 1812
BRIGADIER GENERAL JOHN PRATT HUNGERFORD
MAJOR HENRY HUNGERFORD
AIDE DE CAMP JOHN WASHINGTON HUNGERFORD

Dedicated by the James Monroe Chapter,
Virginia Society Sons of the American Revolution
and the Society War of 1812 in the Commonwealth of Virginia

Plaque, dedicated May 7, 2016, mounted on the entrance gate of the
Hungerford Cemetery, Leedstown, Westmoreland County

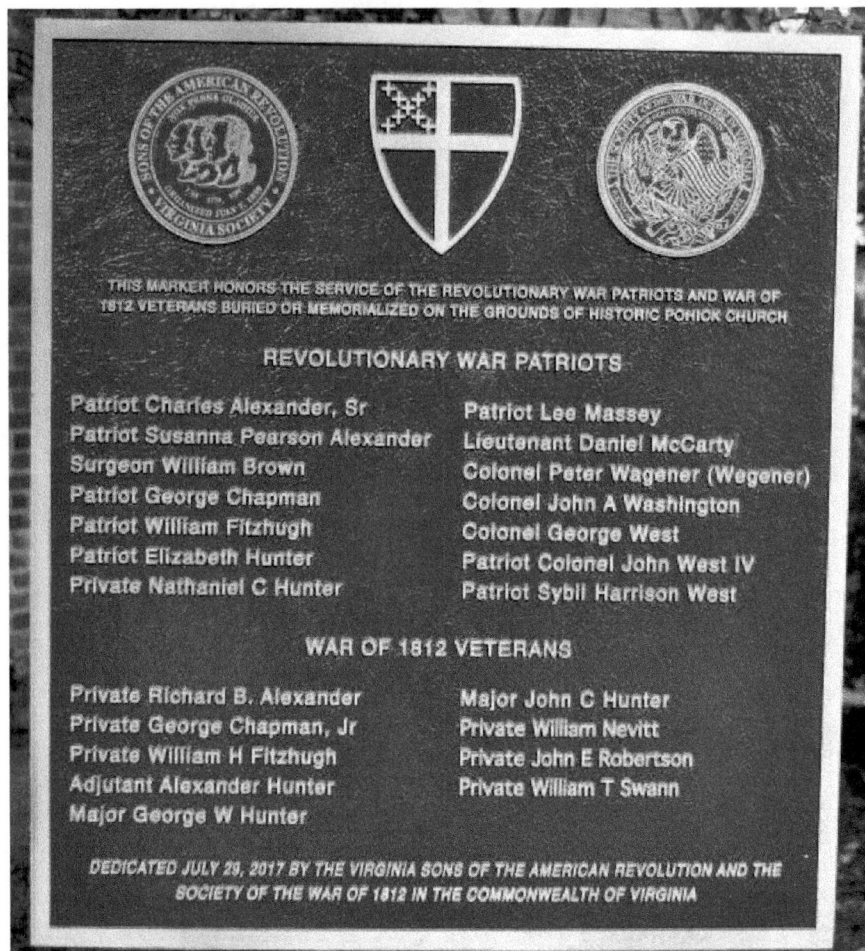

Plaque mounted on a pole along the pathway beside the historic
Pohick Church, Lorton, Fairfax County

APPENDIX A: ADDITIONS TO
VETERAN LIST BY COUNTY / INDEPENDENT CITY

ACCOMACK COUNTY

Annis, Custis
Ashby, James

Boogs, James
Fletcher, John Goldsmith

Kelly, John
Rogers, Levi

ALBEMARLE COUNTY

Amiss, John S
Gianniny, Nicholas
Gilmer, George

Moon, John Diggs
Simms, Richard Durrett
White, John

Winn, John
Wood, Benjamin
Young, David

ALEXANDRIA COUNTY

Allison, Robert, Jr
Cooper, Samuel
Dixon, John B
Isaacs, Samuel
Garner, William

Green, William
Gregory, William
Howard, Beal
Kerr, James
Mankin, Charles

Mankin, David
Mason, John
Mills, William (b 1783)
Mills, William (b 1797)
Stewart, William

ALLEGHANY COUNTY

Burk, James
Keyser, Fleming

Smith, Peter
Wolfe, John

AMELIA COUNTY

Anderson, Joseph
Archer, Richard Thompson
Bailey, Jeremiah
Chappell, John Sr

Jesup, Thomas Sidney
Jeter, John
Jeter, Tilmon (Tilman)
Lingan, James McCubbin

Pride, Thomas
Scott, Thompson
Williams, John

AMHERST COUNTY

Goodwin, Thomas Cary

Lavender, William

APPOMATTOX COUNTY

Hunter, Benjamin

Mathews, William

ARLINGTON COUNTY

Auld, Hugh, Jr
Cassin, Stephen
Custis, George Washington
   Parke

Graham, George
Hutchinson, Franklin S
Kisner, Samuel
Minor, George

Ritchie, John T

AUGUSTA COUNTY

Batis, Charles
Bell, James
Berry, Henry Jr
Berry, John
Bickle, Adam
Blackburn, Thomas Rolander
Bosserman, Frederick
Bosserman, George
Brooks, John
Brownlee. John
Caldwell, John

Clarke, Samuel W
Coiner, Michael
Cook, Jacob
Crawford, James
Cupp, Frederick
Dold, William
Frazier, James A
Guy, Robert
Hall, Robert
Hamilton, James
Hanger, Peter

Hiser, Samuel
Keller, George
Keller, Lewis
Kunkle, Jacob
Landes, Daniel
Landes/Landis, Samuel
Lawrence, William
Lohr, John
McComb, William
McDowell, John
McDowell, William

(Augusta, continued)

McClung, Henry
McCue, John
McCue, Moses
McClure, Hugh
McCune, James
Miller, Abraham
Miller, Martin

Myers, Michael
Nelson, James
Nelson, William
Orebaugh, John
Ross, William, Sr
Rusmisel, Adam
Rusmisel, Christian

Sellers, John
Stribling, Erasmus
Summers, David
Wilson, Matthew
Wright, John
Young, William S

BATH COUNTY

Bradley, Solomon
Brockenbrough, John

Cleek, John
Eakle, John

BEDFORD COUNTY

Anderson, Jesse
Board, John III
Campbell, Robert
Carter, Littleberry (Littlebury)
Claytor, Robert Mitchell
Crouch, John Thomas
Goggin, Pleasant Moorman

Harris, William
Hopkins, Price Wiley
Hurt, William
Leftwich, James
Leftwich, Thomas Lumpkin
Lynch (Linch), James
Otey, Isaac

Otey, James
Payne, William
Steptoe, James Calloway
Stiff, Henry
White, James Buford

BLAND COUNTY

Stafford, Ralph

BOTETOURT COUNTY

Lake, Timothy

BRUNSWICK COUNTY

Abernathy, Raleigh

Bishop, John

BUCKINGHAM COUNTY

Booker, George
Flood, William

Garnett, Thomas
Horsley, John Jr

Isbell, William
Phillips, Randolph B

CAMPBELL COUNTY

Lee, John
Leftwich, John "Jack"

Organ, John
Phillips, John

Smith, Ralph

CAROLINE COUNTY

Anderson, Thomas Bates
Buckner, William Aylette
Cole, John Gatewood
Collins, Charles

Downing, Rufus
Green, Theophilus
Hill, Robert Garlick
Hoomes, John

McLaughlin, Jno (John or
  Jonathan)

CARROLL COUNTY

Blair, John
Blair, Thomas
Dalton, Timothy

Dickerson, John Bethel
Duncan, Greenberry
  (Greenbury)

Jennings, Thomas
Montgomery, Robert
Webb, John

CHARLES CITY COUNTY

Christian, James H

Harrison, William Allen

CHARLOTTE COUNTY

Barksdale, Claiborne	Henry, John III	Wyatt, Joseph
Bouldin, Robert	Read, Issac	
Hancock, John	Richardson, John D	

CHARLOTTESVILLE CITY

Benson, John	Garrett, Ira	Morris, William
Davis, John B	Lindsay, Henry	

CHESAPEAKE CITY

Hodges, John	Land, Peter	Smith, Mitchell

CLARKE COUNTY

Allen, David Hume	Burwell, William Nelson	Hose, Phillip
Allen, William T	Byrd, Francis Otway	Jenkins, William K
Anderson, Joseph Edward	Castleman, Alfred	LaRue, Jabez
Burwell, Robert C	Furr (Fur), Moses	
Burwell, Nathaniel	Gray, Zebedee	

CRAIG COUNTY

Bell, Buchanan	Huffman, Jacob Trolinger
Huffman, Andrew	Kessler, John

CULPEPER COUNTY

Fishback, Martin	Wood, James

CUMBERLAND COUNTY

Anderson, James Baker

DICKENSON COUNTY

Mullins, John	Rose, John Jr

DINWIDDIE COUNTY

Bishop, James, Jr	Rogers, Thomas III	Sydnor, John Taylor
Gunn, Burwell	Moody, William Woolridge	Whitworth, Thomas

ESSEX COUNTY

Brockenbrough, Austin	Garnett, James Mercer	Waring, William Lowry
Brooks, Lewis Durham	Garnett, Robert Selden	
Clarkson, Richard	Layton, Charles Grimes	

FAIRFAX COUNTY

Alexander, Richard Brown	Gunnell, George	Moore, Jesse
Darne, Robert	Hunter, George Washington	Robertson, John E
Ellmore, John	Mason, George	Swann, William Thomas
Gary, James	Millan, John	Washington, Bushrod II

FAUQUIER COUNTY

Armistead, John Baylor	Edmonds, James	Glascock, William
Byrne, William	Embrey, Robert	Harper, Joel Zane
Edmonds. Elias	Glascock, John	Payne, William Frances

FLOYD COUNTY

Blackwell, Moses	Harter, Adam	Sowers, George Jacob
Booth, George	Howell, Daniel	Strickler, Jacob M
Boyd, James	Lester, John Jr	

FLUVANNA COUNTY

Lack, John	Omhohundro, Richard

FRANKLIN COUNTY

Allen, William Jr	Claiborne, Ferdinand Leigh	Law, Nathaniel
Arthur, John A Jr	Hancock, Benjamin Peter	Pinkard, John
Board, Henry	Jones, Abraham	
Bradley, Philip	Kesler, Henry	

FREDERICK COUNTY

Abbott, William	Fries (Fry), Michael	O'Conner, John
Aby, Jonas	Fry, Jacob	Perry, Nickolas
Atkinson, William Mayo	Glaize (Glaze), Sampson	Piper, John Henry
Aulick, Frederick A	Good, Felix	Price, John
Bageant, William	Good, Jacob	Randolph, William Fitzhugh
Baker, Henry	Griffith, John M	Rogers, Casper
Baldwin, Robert Thomas	Grove, John	Rogers, Even (Evan)
Barley, Frederick	Hillman, Simeon	Scrivener, Benjamin
Barley, Jacob	Hoff, John	Shade, Richard
Beatty, Henry	Holliday, James	Slagle, John
Bell, John Jr	Holmes, David	Smith, George
Berkeley, Robert	Hoover, John	Smith, Philip
Brent, George	Jenkins, Stephen	Snapp, John
Brown, Samuel	Johnston, Joseph	Stephenson, William
Brown, William H	Kiger, Isaac	Swartz, George
Bush, Andrew	Kremer, George Sr	Sydnor, William
Carter, John	Kurtz, Isaac	Tucker, Henry St George
Cather, James	Mason. Jonathan	Vance, James
Cooper, Henry	Massie, Josiah	Wall, Jacob
Copenhaver, George	McIntire, Charles	Wall, John C
Copenhaver, Michael	McIntire, William	Wall, John F
Coyle, Michael	McKee, Joseph	Washington, George Fayette
DeHaven, John	Mesmer, Jacob	White, Jacob
Dillon, William	Meyers, Martin Sr	White, John
Elliott, William Aquilla Sr	Miller, John	White, John
Evans, John	Miller, Peter	Wood, William

FREDERICKSBURG CITY

Beale, William Churchill	Evans, George	Hall, John Byrd
Browne, William	Fitzgerald, James Henderson	Hull, John C
Carter, William Fitzhugh	Gordan, Alexander	Minor, Hubbard
Crump, John	Gordan, Samuel	

GILES COUNTY

Brown, William	Lipscomb, James
Carr, William Crow	Stafford, John Sr

GLOUCESTER COUNTY

Cooke, Warner T
Dixon, John
Eastwood, William
Jones, Catesby

Smart, William R
Smith, Phillip
Smith, Thomas
Stubbs, John S

Thruston, Robert
Yeatman, Thomas Robinson

GOOCHLAND COUNTY

Gathright, William Jr
Lacy, Fleming

Richardson, George
    Washington

Sampson, Richard

GRAYSON COUNTY

Bourn(e), Stephen "Grey"
    Boyer
Boyer, John
Cornett, Reuben
Cox, David "River Dave"

Cox, James Robertson
Cox, John (d 1850)
Cox, John (d 1856)
Hale, Lewis II
Long, William

Lundy, Amos
Roberts, William
Stone, William
Taylor, William

HALIFAX COUNTY

Carter, Thomas Sr
Crews, Issac Seat
Edmondson, Richard

Henderson, William
Holt, William
Leigh, William

Scott, John Baytop

HAMPTON CITY

Jones, Thomas

Shelton, James

HANOVER COUNTY

Digges, Dudley

Morris, Charles

Tinsley, Parke

HENRICO COUNTY

Brown, Samuel
Chamberlayne, Lewis Webb

Green, William
Hix, Richard J

HENRY COUNTY

Bassett, Alexander Hunter
Boaz, Robert
Campbell, David
Carter, Cary
Goode, Jacob
Gravelly (Gravely), Joseph
    Jefferson

Gravely, Lewis
Hairston, George
King, John "Jack"
Marshall, James Douglas
Martin, Joseph
Martin, William
Morris, John

Morris, Samuel Coleman
Pankey, Edward
Stone, John
Wills, Thomas

HIGHLAND COUNTY

Arbogast, Michael Jr

ISLE OF WIGHT COUNTY

Jordan, Merit

White, Sampson

JAMES CITY COUNTY

Lee, Robert H

KING & QUEEN COUNTY

Atkins, John
Bagby, John
Gresham, Thomas
Moore, Richard

KING GEORGE COUNTY

Dishman, John

KING WILLIAM COUNTY

Aylett, John
Baxton, Charles Hill
Hill, William
Lipscomb, Sterling

LEE COUNTY

Ball, Moses III
Bishop, Elijah
Campbell, James Jr
Ewing, Alexander
Fitts, Cornelius
Fletcher, John Goldsmith
Fulkerson, Peter
Jones, Stephen
Speak, Nicholas
Wynne, Samuel

LEXINGTON CITY

Dold, Samuel Miller
Fuller, John W

LOUDOUN COUNTY

Adams, William
Aldridge, John
Allen, James
Ball, George Washington
Barnhouse, Richard
Brown, John
Brown, William Cox
Butcher, John Humphrey
Butts, William Esq
Carr, David
Cockrell, Samuel
Coleman, William
Compher, Peter
Cooper, George
Cooper, John
Cooper, John Sr
Cooper, John Sr
Cridler, John
Demory, Wilham
Drake, Francis T
Dulin, John
Edwards, Joseph
Elgin, Charles West

Elgin, Robert
Fleming, William
Gibson, William E
Gray, James
Gregg, James B
Griffith, Thomas
Harding, Edward
Harris, Isaac
Hawling, Isaac Wilcoxson
Hill, James
Holmes, William
Hughes, Thomas
Johnson, John
Johnson, Robert
Johnson, William
Klein, Lewis Allen
Love, Richard H
Miller, George
Miller, Jacob
Mills, William
Moreland, George T
Nichols, Samuel
Purcell, James H

Roberts, William
Ropp, Nicholas
Rust, George
Russell, Benjamin C
Shoemaker, Abraham
Shriver, Jacob
Skinner, Peter
Tate, William
Taylor, Mahlon
Taylor, Timothy II
Tebbs, Thomas F
Tilton, Daniel
Young, George
Vanvactor, Solomon
Virts/Werts, Conrad
Waters, Levi
White, James
White, John
Wilkinson, Thomas
Wilkinson, William
Williams, Samuel
Wood, Joshua
Young, Samuel Wade

LOUISA COUNTY

Bronaugh, Thomas
Burk, Edmond
Chewning, Reuben
Goodwin, William
Harris, Edward
Morris, William C
Toler, William F

LUNENBURG COUNTY

Ellis, William
Love, Henry Hicks
Neblett, Sterling
Williams, Thomas S
Parrish, William

LYNCHBURG CITY

Bailey, James	Garland, James	Lee, John R
Blackford, Thomas	Hare, Jesse	Mason, Daniel
Thornburgh	Harrison, Samuel Jordan	Robinson, John
Burch, Samuel L	Harvey, John	Saunders, William L
Daniel, William Sr	Hix, William	Shackelford, George
Davenport, Martin	Holbrook, Josiah	Snead, Israel
Davis, Henry	Holmes, Joshua Rathborne	Thurman, John
Davis, William Jr	Johnson, Gerard E	Wills, John

MADISON COUNTY

Banks, William Tunstall	Crigler, Lewis	Thomas, George
Barnes, Henry	Garnett, Robert	Thomas, John
Barnett, Lawson	Goodall, Philander	Thomas, William
Carpenter, John	Hume, Joseph	Walker, James W
Conway, John F	Stockdell, John	Yager, Salathiel Wayland

MATHEWS COUNTY

Foster, Issac	Tompkins, Christopher	White, John

MECKLENBURG COUNTY

Andrews, Anderson	Evans, William	Jones, William
Burwell, Armistead	Hayes, James	Love, Chappell

MONTGOMERY COUNTY

Barrnett, Robert	Francis, Miles	Kent, James Randal
Davis, Thomas	Gardener, John	Preston, William
Flanagan, William	Jennelle, Lewis M	

NELSON COUNTY

Carter, Thomas	Fortune, Thomas Eubank	
Coleman, William Harris	Massie, William Wright	

NEW KENT COUNTY

Chandler, Robert Bradenham	Chandler, William	Cooke, William, Jr

NORFOLK CITY

Armstrong, William Morris	Jameson, William	Morris, John
Banks, John	Keeling, Henry	Ott, George
Bell, David	King, Miles Jr	Pollard, Benjamin
Cunningham, James B	Klien, Jacob	Rowland, William
Davis, William	Lee, John Hite	Seldon, William Boswell
Dornin, Thomas Alysuis	Maurice, James	Snyder, William S
Dove, Elisha	McCandlish, Thomas C	Taylor, James
Henley, William D	Middleton, John	Taylor, Robert Barraud
Henry, John H	Montague, William Valentine	Wrenn, Thomas
Ironmonger, Charles B	Moore, John	

NORTHAMPTON COUNTY

Jarvis, William Sr

NORTHUMBERLAND COUNTY

Blackwell, Samuel B	Fousbee, Griffin Henry	Payne, John
Cralle, Darius Griffin	Harding, William	Smith, James

NOTTOWAY COUNTY

Allen, John C	Dickenson, Robert	Miller, Anderson
Bland, Edward	Jones, Wood	Perkinson, Thomas

ORANGE COUNTY

Gordan, James	Willis, William Champe

PAGE COUNTY

Bailey, John	Jones, Thomas	Miller, Peter
Burner, John Rhodes	Kauffman, Daniel	Ruffner, John
Dovel, William	Kibler, Jacob	
Fleming, Andrew	Mathias, Miller	

PATRICK COUNTY

Bennett, Bartlett	Harris (Harriss), Moses	Morrison, James
Bowman, Archelaus	Lyon, Elisha	Ross, William
Carter, Bane (Baynes/Bains)	Moore, William	
Cruise, John Mortimer	Moran, Nelson	

PETERSBURG CITY

Allison, John	Heath, Jesse	Lewis, William
Andrews, Bullard R	Hinton, John	Martin, Peter
Bass, Thomas William	Jones, Benjamin	Morgan, Samuel B
Bragg, John	Jones, Thomas	Pescud, Edward
Butts, Daniel Claiborne	Kent, Charles	Peters, Frederick D
Drinkard, Beverly	Knox, James	Pollard, John

PITTSYLVANIA COUNTY

Bennett, John	Davis, Thomas	Reynolds, John Devin
Blair, Samuel Brittain	Doss, Stephen	Robertson, Edward Jr
Cocke, John E	Jones, Richard	Todd, Jered

PORTSMOUTH CITY

Carr, Dabney B	Harrison, William	Watts, Samuel
Cocke, John	Livingston, Samuel	Wilkinson, Jesse
Cocke, William Irby	Martin, Alexander A	
Davis, Samuel	Veale, James	

POWHATAN COUNTY

Dabney, Robert Kelso	Pope, William	Swann, John

PRINCE EDWARD COUNTY

Rice, John Holt	Scott, Thomas T	Stevens, John

PRINCE GEORGE COUNTY

Best, John Henry

Eppes, Francis Alexander

PRINCE WILLIAM COUNTY

Arrington, Richard
Boyle, David

Hutchison, John
Smith, George

Smith, William
Tebbs, Foushee

PULASKI COUNTY

Cloyd, Thomas

Hall, David

RAPPAHANNOCK COUNTY

Atkins, John

Dearing, Alfred S

Jackson, Daniel Franklin

RICHMOND CITY

Allan, John
Allen, Joseph
Ambler, Edward
Anderson, Richard
Andrews, Erasmus Granger
Archer, Robert
Bagby, Bennett
Baker, John
Barret, William
Bigger, Thomas Bibb
Blair, Walter Dabney
Bland, Richard
Booth, William
Bootwright, William
Bosher, William
Brooks, Alexander
Brown, Ellis
Chevallie, Peter Joseph
Cottrell, Samuel Smith Sr
Cottrell, William
Coulling (Couling), James
  Mathias
Crouch, Thomas H
Crump, Joshua
Davidson, William M
Davis, Levi
Dibrill, Edwin
Dill, Adolph
Drewey, Henry
Dunnford, Philip Tennyson
Dunn, John

Edwards, James
England, John
Francis, Joseph E
Francisco, Peter
Graves, George
Gray, James
Greaner, William
Griffin, James
Harwood, William Jr
Hawkins, Samuel
Heath, James Ewell
Hicks, Benjamin A
Higginbotham, David
Hill, John
Hix, Samuel
Howard George
Howard, John
Jarvis, William
Keeling, Henry
Keesee, Jacob Burton
Lester, John
Lynch, James
Mayo, Philip
McBarty, John
McCarty, William Mason
McGruder, Sublett
Meredith, Reuben
Miller, John
Mitchell, William Jr
Monroe, James
Myers, Aaron

Phillips, Martin
Pollock, Allan
Price, Joseph F
Quarles, John
Randolph, William
Richardson, Robert
Richardson, William H
Rust, Youel
Shields, Mathew
Shepherd, Royal F
Smith, William
Strother, William Porter
Sublett, Samuel
Sullivan, Benjamin
Tate, John L
Thom, John
Thomas, John
Van Lew, John
Wallace, William
Watkins, John E
Webb, Lewis N
White, John Baker
Whitfield, Richard
Wicker, Francis
Williams, Thomas
Williams, William
Wills, Charles
Womble, John
Woodson, John

RICHMOND COUNTY

Shackleford, Vincent

ROANOKE CITY

Burrwell, Nathaniel

Cooper, Jacob

ROCKBRIDGE COUNTY

Albright, John
Alexander, James B
Alexander, William Reid
Armstrong, William J
Ayers, John Grigsby Sr
Bell, Joseph Jr
Brown, William
Bryan, Edward Jr
Campbell, Alexander
Carter, John Jr
Clark, Nelson H
Clark, Robert Jr
Crawford, Robert

Cunningham, William
Daniel, John
Dixon, Thomas
East, David C
Foster, James Coulter
Hight, Joel
Kerr, Thomas
Larew, Jacob
McBride, Isaiah
McClung, James
McKee, John Telford
Miller, Samuel
Mohler, John (b 1789)

Mohler, John (b 1772)
Montgomery, James
Montgomery, John
Moore, John
Snider, Daniel
Steele, Samuel
Stevens, William
Strickler, Daniel Jr
Stuart, Hugh
Walker, James
Walker, Thomas Hudson
Willson, Thomas

ROCKINGHAM COUNTY

Anderson, James
Anderson, William
Baker, John
Bare, Jacob
Beahm, Martin
Berry, John S
Bible, John
Bowman, George
Brown, Petter
Bushong, Phillip
Carr, John
Cootes, Samuel
Conrad, Jacob
Conrad, John
Davis, Robert
Dean, William
Dove, Reuben
Effinger, John Frederick

Fisher, William
Ford, Tipton
Fulton, James
Hopkins, James
Huffman, John
Keezel, George
Koogler, Jacob
Lawson, John
Lewis, William
Lincoln, David
Linebaugh, William
Long, Frederick
Maiden, William
Mauzy, Joseph
May, Andrew
May, James
Miller, Christian
Miller, Jacob (b 1769)

Miller, Jacob (b 1780)
Miller, George
Moore, Reuben
Myers, John T
Peale, Jonathan
Pence, John
Pence, Phillip
Rankin, James S
Rice, John
Rice, Thomas
Rutherford, Archibald
Sheets, Henry
Sheets, John
Sprinkel (Sprenkel), William
Taylor, William
Wise, John
Wyatt, Thomas E

RUSSELL COUNTY

Bickley, John
Campbell, Henry
Carrell, James P

Keen, William Sr
Litton, John Whitley Sr
McFarlane, Alexander

McFarlane, James
Wheeler, James

SALEM CITY

Jeter, Ira

SCOTT COUNTY

Bailie, Robert Bruce
Bledsoe, Issac
Boatright, Meador
Carter, John
Carter, John R Jr

Cox, David Leach
Freeman, David
Jones, Thomas Sr
Kilgore, Hiram
McKinney, James

Speer, Robert
Stewart, James
Wolfe, John

SHENANDOAH COUNTY

Allen, Israel Sr
Allen, Reuben
Allison, Henry
Anderson, Samuel
Baker, Isaac
Baker, Jacob
Biller, Christian
Bowers, John
Bowers, Philip
Boyer, Peter
Coffman, Daniel
Cooper, George
Cooper, Joseph
Cooper, Samuel
Cullers, Henry Samuel
Eberly (Eberley), George

Eberly, Jacob
Fadely, Henry
Foltz, George
Frye, Jacob
Grabill (Grabeel), Daniel
Hawkins, Samuel
Hickman, John T
Hoover, Jacob
Hupp, George Franklin
Keller, George
Keller, Jacob
Keller, Jacob B
Kerlin, David
Kniseley (Knicely/Knisley),
  John
Koontz, John

Lichliter, Adam
Marshall, Henry
McCann, James
Moore, Reuben
Morgan, John
Neff, David
Neff, John
Lary, John
Pennybacker, Joel
Pitman, Lawrence
Pitman, Phillip
Riley, William
Shrum, George
Spengler, Joseph Strover

SOUTHAMPTON COUNTY

Council, James

Kello, Samuel

Thomas, John

SPOTSYLVANIA COUNTY

Davenport, John
Gardener, James

Herndon, Joseph
Leavell, Burwell

Pates, Chandler
Sullivan, Newton

SMYTH COUNTY

Campbell, John II
Cox, William
Davidson, Robert

Freeman, William Nance
Richardson, William Jr
Scott, William

Sexton, Aaron

STAFFORD COUNTY

Bell, William
Berry, Richard J
Briggs, David
Bruce, Charles
Bussell, Charles
Byram, John
Carter, Sanford Jr
Conway, Thomas Barrett
Cox, Berryman
Curtis, George
Daffin, William
Embrey, Daniel
Finnel/Finnall, Jonathan
Fitzhugh, William Henry
Fritter, Barnett

Gordon, William Richards
Greenlaw, William
Holmes, Jeremiah
Kendall, Joshua
Kendall, William
Knight, Bailey Alexander
Milstead, Samuel
Moncure, Edwin Conway
Moncure, John
Moncure, John Jr
Moncure, John Sr
Norman, James S
Norman, Matthew
Norman, Thomas
O'Bryhim, Alexander

Payne, James Rousseau
Purkins, Thomas
Roberson, Thomas
Robertson, William
Rowe, Jesse
Rowe, John Gasking
Sanford, Lawrence
Southard, James
Stark(e), John
Stone, Alexander
Wallace, Thomas
Waller, James
Whorton, William
Williams, Nathaniel Pope

SUFFOLK COUNTY

Holladay, Joseph
Jenkins, John Cole

Smith, Arthur
Wright, Joseph

SURRY COUNTY

Baird, Peter

Edwards, William

SUSSEX COUNTY

Bailey, Phillip	Chappell, James	Wilkinson, William

TAZEWELL COUNTY

Bowen, Arthur	Kirk, William	Thompson, Archibald W
Bowen, Henry	Nash, William Daniel	
Doake, Robert	Perry, Thomas	

VIRGINIA BEACH CITY

Cornick, Lemuel III	Lovett (Lovitt), Reuben	Woodhouse, Henry Barnes
Keeling, William S or T	West, John	

YORK COUNTY

Hopkins, Charles	James, Cyrus Basye

WARREN COUNTY

Allen, Robert Millar	Cox, William	Ramey, John
Blakemore, Joseph	Funk, Michael	Rust, John
Boyd, John	King, Augustine	

WASHINGTON COUNTY

Allen, John	Denton, William D	Leonard, Frederick Jr
Blackwell, Joseph	Edmondson, Andrew	Lewark, Joseph
Buchanan, John	Fleenor, Issac Blackhawk	Snodgrass, David
Clark, James	Fulkerson, Adam	Snodgrass, William
Clark, James Sr	Grant, Gardener	

WAYNESBORO CITY

Brooks, William

WESTMORELAND COUNTY

Dashiell, Thomas Bennett	Payne, John	Settle, William P

WILLIAMSBURG CITY

Coleman, Thomas	Hill, Robert F

WISE COUNTY

Evans, Meredith Nathaniel

WYTHE COUNTY

Brown, Christopher Strophel	Kinder, George	Neighbors, James
Clark, John	King, John	Sayers, Robert
Jonas, Daniel	Mathews, John	
Keesling, Peter	Neff, Abraham	

APPENDIX B: ADDITIONS TO
CEMETERY LIST BY COUNTY / INDEPENDENT CITY

ACCOMACK COUNTY

Heron Hill	Locustville, jct Rt 615 & 614
Tehern	Town of Guilford
Kelly Grave	Town of Keller
Fletcher Family	Near Jenkins Bridge, Grotons
Rodgers-Boggs	Located one mi north of Evans Wharf on east side county rd 638

ALBEMARLE COUNTY

Gianniny Family	Buck Island Creek area by rd btw Monticello and Carter's bridge
Linden	Town of Ivy
Mount Air	Town of Keene
New Bethel	Town of Woodridge
Sunny Bank	Town of North Garden
Watts Family	Stoney Point

ALEXANDRIA CITY

Presbyterian Cemetery	601 Hamilton Lane
Trinity United Methodist Church	Cameron Mills Rd

ALLEGHANY COUNTY

Emory United Methodist	7401 Jackson River Rd, Covington
Wolfe-Sively Cemetery	Potts Creek area

AMELIA COUNTY

Anderson Family	Rt 620
Bailey Family	Poorhouse Rd, beyond jct rt 701 at WC Golden Farm
Jeter Family	3 mi N of Jetersville
Jeter Family	Town of Dodophil
Pride Family	Pridesville Rd Nr Flat Creek
Scott	Contact Ameilia County Historical Society for location

AMHERST COUNTY

Goodwin Family	Off Rt 60, west of Amherst at Forks Buffalo River
Lavender Family	Stone Wall Creek Rd

APPOMATTOX COUNTY

Hunter-Marshall Family	In town of Appomattox
Walker's Presbyterian Church	Hixburg

ARLINGTON COUNTY

Arlington National	Rt 110, Jefferson Davis Highway

AUGUSTA COUNTY

Emmanuel Church	Mount Solon area
Frazier Family	West of Lone Mountain Rd on Rt 250 at Jennings Gap
Keller	On Dr Knobbs Farm, Churchville
Knightly Mill	Town of Knightly
Landes Cemetery	Landes Family 1002 Fadley Rt 646
Mount Zion	Not determined
Oak Lawn	Rt 617 New Hope
Orebaugh Family	Rt 760
Sangerville Church	26 Vance Rd, Rt 755
Trinity Episcopal Churchyard	Town of Crimora
Union Presbyterian	Town of Churchville
Van Lear	Rt 613 south 1/3 mile from jct of Rt 742 on farm 300 yds from rd
Western State Hospital Cemetery	Vic jct Statler Blvd and Greenville Ave, Staunton
Zion Lutheran Church	297 Zion Church Rd

BATH COUNTY

Eakle Family	Little Valley Rd, Bolar

BEDFORD COUNTY

1131 Park Street	
Ayers-Payne-Brooks	Rt 626
Anderson Family	Town of Goode
Board Family	Rt 655, Moneta
Calloway-Steptoe	Rt 460, New London
Carter Family	Sheep Creek Rd nr Prospect Baptist Ch on farm of Eugene Boyer
Ewing-Patterson	Penicks Mill
Finch Family	Town of Moneta
Goggin Family	Bunker Hill area
Leftwich Family	Nr Bunker Hill
Lynch Family	High Point Rd
Old Ebenezer Church	GPS 37.476,-79.374
Overstreet-Crowder-Foster	Rt 24 at jct with Crowder Pt Rd, Chestnut Fork
Thomas Campbell Family	GPS 37.060,-79.557

BLAND COUNTY

Old Bogle	Walkers Creek

BRUNSWIICK COUNTY

Bethel Methodist Church	Town of Alberta
Bishop Family	Rt 644

BUCKINGHAM COUNTY

Garnett Family	Family property see tax records for location
Mount Zion Baptist Church	6277 Cartersville Rd, Rt 610
Red Oak Hill	David's Creek
Rose Hill Plantation	Town of Andersonville

CAMPBELL COUNTY

Lee Family	200 yards west of Rt. 626 in the Lynch Station Quadrant
Mount Calvary Baptist Church	Gladys area
Mount Hermon United Methodist Church	Town of Lynch Station
Old Phillips	Evington

CAROLINE COUNTY

Hill Family Cemetery	Town of Mt Airy
Ruther Glen	West Mount Vernon
Old Salem Baptist Church Yard	Near Alps
Topping Castle	Houston's Corner

CARROLL COUNTY

Blair Cemetery	Cliffview
Greenberry Duncan Family	West of Rt 613 Greasy Creek Rd, Dudspar
Quaker-Nester	Rt 624, Nester School Rd
William Dalton Family	Dugspur, route 221, 5 miles NW of New Hope Church

CHARLES CITY COUNTY

Berkley Plantation	12602 Harrison Landing Road
Manoah	Rt 615 Glebe Lane, Ruthsville

CHARLOTTE COUNTY

Henry (AKA Red Hill)	Red Hill Plantation, near Brookneal
Joseph Wyatt	Evergreen Rd
Martin Hancock	Red House
Mount Carmel United Methodist Church	Town of Brookneal
Pettus-Richardson Family	Town of Keysville
South Isle	2471 Ridgeway Rd, Brookneal

CHARLOTTESVILLE CITY

Maplewood	Lexington Ave and Maple St
Oakwood	First St
University of Virginia and Columbarium	Alderman and Cemetery Rd

CHESAPEAKE CITY

Smith Family	Mount Pleasant Farm
Walker	2417 Water Mill Way

CLARKE COUNTY

Franks	Rt 607 Ebenezer Rd
Runnymeade Farm	3838 Shepherds Mill Rd, Berryville

CRAIG COUNTY

Bell Family	Rt 632 Upper John Creek Rd
Fairview	On McClover Hill Lane
Ross	Rt 42 between jct Rts 629 &641

CUMBERLAND COUNTY

James Anderson Family	Town of Reeds

DICKENSON COUNTY

John Mullins Family	Town of Clintwood
Vance	Caney Bridge Rd behind old service station

DINWIDDIE COUNTY

Butterwood Church	Town of Darvills
Historic Bishop Family	Jct Sapony Rd & Mealy Branch
Sydnor & Young Family	On Petersburg Battlefield cemetery at Five Forks
W.W. Moody Family Graveyard	Sutherland

ESSEX COUNTY

Brooks Family	North side of Rt 689, Howerton Rd
Champlain Estate	Town of Lloyds
Clarkson Family	Town of Rexburg
Garnett Family Burial Ground	Town of Loretto
Hundley Family	Rose Hill Center Cross
Tuscarora Site	Town of Dunnsville

FAIRFAX COUNTY

Gary Family	Town of Centerville
Moore-Hunter Family	GPS 38.88537, -77.27105
Mount Veron Estate	3200 Mt Vernon Hwy
Strickler Cemetery	Disregard
Wren-Darne	Hillsman Drive & Mahala Lane

FAUQUIER COUNTY

Payne Family Cemetery	Town of Orlean

FLOYD COUNTY

Weddle Cemetery	At Shelors Mill
Lester Family	Route 738
Old Topeco	Route 221 abt .25 mi fr Topeco Church
Red Oak Grove	Rt 684 to top of hill past Dobbin's farm another 1/2 mi
Reed	South across stream from Spirit Wind Industries
Zion Lutheran Church	Town of Floyd

FRANKLIN COUNTY

Arthur Family	John Arthur Rd, 500 yds behind old home place
Board Family	Off unnamed road vic jct Rt 40 7 Rt 703
Bradley/Bird Family	Town of Scruggs
Claibrook Plantation	Town of Rocky Mount
Hancock Family	Novelty Rd, at Uniom Hall
Jamison, W.O. Family	Town of Snow Creek
Jones Family Cemetery	Town of Endicott
Law Family	Off McNeil Rd, Rt 718

FREDERICK COUNTY

Back Creek Quaker	Town of Gainesboro
Bageant Family	Town of Cross Junction
Barley Family	1392 Martz Rd 100 yds up hill
Brown Cemetery	see Winchester City
Dalby-Rogers Quaker Graveyard	Nr Gainsboro
Johnson Family	Nr Trone Methodist Church on Timber Ridge
McIntire Family	4 mi fr Hinkle nr Morgan County, WV line
Middleton	1/2 mi W of Middleton on a hill
Ridings Family	Town of Middleton
Wesley Chapel	620 Chapel Hill Rd, Cross Junction

FREDERICKSBURG CITY

Gordans of Kenmore	Washington Ave behind Washington monument
Saint George's Episcopal Burial Ground	905 Princess Anne St

GILES COUNTY

Fairview	Between Riverside Ave and Fletcher St, Narrows
Shannon-King	GPS 37.2182,-80.7415 nr Staffordsville

GLOUCESTER COUNTY

Bellamy United Methodist Church	Rt 615 Chestnut Fork Rd, Bellamy
Valley Front	Valley Front Ln, off Hickory Fork Rd, Sassaffras

GOOCHLAND COUNTY

Boscobel Cemetery	1/4 mil NW of end of Pine Tree Hollow Rd
Bowles Family	Town of Tabscott
Woodson Family Cemetery #2	Rt 609, 1.5 mi fr East Lake

GRAYSON COUNTY

Bethany Community	Town of Independence
Churchwell-Boyer Family	1.2 mi north jct Rt 611 & US 21, on pvt rd 3/4 ml, left in field
Comers Rock	Town of Comers Rock
Cox Chapel	Town of Independence
James Cox Family	Town of Bridle Creek
Jeremiah Stone	Corners Rock Rd, Rt 658
Long-Phipps Family	West of Rt US 21 & south of Peach Bottom Creek, Long Gap
Lt. David Cox cemetery	Town of Baywood
Nuckolls	Beyond end of Wild Turkey Rd, vic jct US 58
Potato Creek	Mouth of Wilson
Reuben Cornett Family	Corners Rock Rd, 1.6 mi to private road on left, 400 yds in field
Stephen G Bourne	Rt 777 in pasture field, Liberty Hill

HALIFAX COUNTY

Halifax Town Cemetery	GPS: 36.7639,-78.9269
Crews Family (AKA William R Carr Farm)	1 mi west of Ellis Creek Church
Saint John's Episcopal Church	Rt 360 E of jct Rt 501
Scott Homesite	Town of Scottsburg

HAMPTON CITY

Shelton	GPS: 37.03015,-76.30715

HARRISONBURG CITY

Woodbine	Fr Harrisonburg Court Square take Rt 33E gate on Reservoir St

HANOVER COUNTY

The Meadows Estates Gardens	Town of Doswell

HENRICO COUNTY

Brook Hill	Town of Montrose
Cauthorn Family Cemetery	Short Pump
Old Graveyard	5.6 mi SE of Richmond on Darbytown Rd, then N on dirt rd .4 mi
Smith Family	Town of Tuckahoe

HENRY COUNTY

Dillion	North Mount Herman Church on Philpott Rd
(AKA Daniel-Dobson-Shelton Families)	
Hairston Family	Beaver Creek Plantation, N of Martinsville on SR 108
Goode Home Place	Oak Level
Hodges Family	Rt 627 4.5 mi west of Fieldale
Horsepasture Christian Church	1146 Horsepasture Price Rd, Rt 692
King	Leatherwood Creek
Leatherwood	Town of Leatherwood
Martin Family	Near jct Rts 57 & 710, Leatherwood
Morris Family	Town of Bassett
Old Hickey-Martin	GPS: 36.81719,-79.97507; Oak Level
Pankey Family	Town of Irisburg
Stone & Johnson	Rt 798 across rd fr home of Homer
William Marshall Gravely	.1 mi east of jct Rt 620 & Pebble Brook Rd
Wills Cemetery	Homeplace of Benjamin Wills-check property records

HIGHLAND COUNTY

Arbogast Family	Crab Bottom Pauper Farm

ISLE OF WIGHT COUNTY

Jordan Family	Moonlight Rd, Rt 627

KING & QUEEN COUNTY

Bruington Baptist Church	4784 The Trail
Dew Family	Town of Owenton
Oak Spring Farm	Located on Pamunkey River
Porporone Baptist Church	Town of Shackelford
Shackelford's Chapel United Methodist Church	Town of Plain View

KING WILLIAM COUNTY

Mount Columbia	Off Rt 649, Town of Manquin
Springfield	2 mi south of King William Court House Rt 621
Sweet Hall	Rt 634 at Pamunkey River

LEE COUNTY

Bishop Family	Rt 665 behind Mt Moriah Church Jonesville
Campbell Family	SW of Rose Hill
Chadwell Baptist Chruch	11 mi E of Cumberland Gap, Rt 58
Ewing-McClure	Jonesville
Fitts Family	Rt 758 to Fitts Gap, bottom of hill behind barn on right, 300 yds up hill
Jonesville	Town of Jonesville
Fletcher Family	Town of Jonesville
Jones #293	Alt US 58, Town of Dryden
Powell Valley	Town of Dryden
Speaks Chapel	Town of Rose Hill
Wynn-Litton	Town of Dryden

LEXINGTON CITY

New Monmouth Presbyterian Church	2348 W Midland Trail

LOUDOUN COUNTY

Allen Family	Town of Watson
Bethel United Methodist Church	Town of Stumptown
Dulin-Kenne	Town of Sterling
Ebenezer Baptist Church	Town of Bluemont
Elgin Family	On Kingdom Farm, Evergreen Mill Rd
Fairfax Friends	Town of Waterford
New Jerusalem Lutheran Church	Town of Lovettsville
North Fork Baptist Church	38130 North Fork Rd
Old Presbyterian	South Church Street
Salem Church	Harpers Ferry Rd, Hillsboro
Skinner Family	Town of Aldie
Saint Paul's Lutheran Church	Town of Neersville

LOUISA COUNTY

Goodwin Family	Ellisville Dr, Rt 669
Morris-Payne	Columbia Rd, jct RTs 615 & ^17, Zion
Oak Grove	Rt 208 Courthouse Rd
Tolersville Tavern	410 Old Tolersville Rd, Mineral
Trinity Baptist Church	135 Mansfield Rd, Mineral

LUNENBURG COUNTY

Ellis Family	Town of Rehoboth
Love Family	Town of Victoria
Meherrin Baptist Church	Rt 719 0.4 mi fr Rt 623
Neblett Family	Rt 138 South of jct Rt 619

MADISON COUNTY

Barnett Family	4.4 miles NW of Wolftown on Rt 662
Carpenter and Lohr	Rt 616, Oak Park Rd about one mi from Rt 636
Cedar Hill	Clore Rd, Madison
Conway, John F	Rt 230 before the Conway River bridge
Crigler, Lewis	Rt 636 0.2 miles E of intersection with Rt 637
Goodall, Philander	1.1 miles up river from Graves Mill on Rt 662
Harrison, John	Rt 712 turn left off Rt. 231, 2 miles from Rochelle
Hume	Check property records of Hume Family
Piedmont Episcopal Church	214 Church St, Madison
Thomas #2	3716 Etlan Rd
Walker United Methodist Church	Rt 230, 7.5 Miles E of US Route 29 (Orange Road)

MATHEWS COUNTY

Friendship	Jct Rts 14 and 697
Poplar Grove Plantation	265Poplar Grove Rd, Williams

MECKLENBURG COUNTY

Easters United Methodist Church	Town of Boydton
Love-Young Family	Town of Baskerville
Oakwood	Town of South Hill
William Andrews Family	Whittles Mill Rd Rt 654 past Bridge Rd
Zion Methodist Church	Busy Bee Rd, South Hill

MONTGOMERY COUNTY

Francis Family	Jct Rts 666 & 1245, Christiansburg
Jennelle Family	SW jct Rt 613 and Old Yellow Sulphur Rd
Kent Family	
Piedmont – Otey	Town of Piedmont
Preston Family	Southgate Dr, Blacksburg
Sunset	Jct South Franklin St and Elliott Dr
(AKA Christianburg Municipal)	

NELSON COUNTY

Coleman Family	Town of Wintergreen
Fortune	Town of Lovingston
Level Green Estate	524 level Green Rd, Rt 679, Roseland
Wilkerson-Carter Family	Off Rt 721, Norwood Rd, vic Carter Creek

NEW KENT COUNTY

Chandler Family	Laurel Springs house off Cook's Mill Road near Pamunkey River

NEWPORT NEWS CITY

Lee, Robert H.	Curtis Drive nr Ripley St

NORFOLK CITY

Basilica of Saint Mary Churchyard	5201 Kennebeck Ave
Forest Lawn	8100 Granby Street
Elmwood Cemetery	238 East Princess Anne Rd
Fort Tar Mass Burial Site	Fort Tar Lane

NORTHUMBERLAND COUNTY

Cralle's Bank	Contact Northumberland Co Historical Society for location
Gibeon Baptist Church	48 John Deere Rd, Gibeon
Wicomico United Methodist Church	Rt 200, Wicomico Church

NOTTOWAY COUNTY

Bland Family	Town of Centerville
Fowles Family	6808 Courthouse Rd (Rt 625), Burkeville
Lakeview	Town of Blackstone
The Grove	Nr Burkeville at Miller's Hill
(AKA Locust Grove and Miller Hill)	

ORANGE COUNTY

Willis Cemetery	on Meadows farm Golf Course at Locust Grove

PAGE COUNTY

Atwood Family	Town of Rileyville
Evergreen Memorial Gardens	Massanutten Ave, Luray
Green Hill	Town of Luray
Huffman	Vic Mill Creek
John Bailey Family	Overlook Drive, Rileyville
Kibler Family	Town of Big Spring
Rileyville	Town of Rileyville
William Dovel	off Stroll Farm Rd (Rt 613)

PATRICK COUNTY

Bowman Family	On hill north side of Rt 631 at Squirrel Creek
Carter/Murphy/Tatum Families	Near Taylorsville
Charity Primitive Baptist Church	5804 Charity Hwy, Charity
Cruise Family	Town of Vesta
Moore	Town of Ararat
Old Harris	Rt 691 on ridge near Billy Martin Cemetery
Old Morrison	South side of Rt 57, 1 mile East of Jct Rt 8
Ross, Daniel Jr Family	Elamsville via North side Rt 57 vic jct Rt 635
Whitlow Family	Charity Rd NE Woolmine, Charity

PETERSBURG CITY

Andrews Family	River Rd on private property in lower district
Bass Family	Bass St at Exeter Mills

PITTSYLVANIA COUNTY

Bennett-Lewis	Off Rt 715 toward Grassy Branch, Keeling
Chatham Burial Park	Town of Chatham
Dalton-Adams-Doss	End of Georges Creek Rd, Gretna
Davis Family at Cherrystone Plantation	Nr Chatham at Cherrystone Creek
Jones (Richard)	Town of Sheva
Robertson Family	Town of Dryfork
Samuel Blair Family	Rt 790 Piney Rd
Wagner Community	GPS: 36.91862,-79.45871
Worlds, Calands	Rt 969 nr Christian Tabernacle, Calands

PORTSMOUTH CITY

Portlock	Part of Oak Grove Cemetery, jct Peninsula Ave & London Blvd
Scott-Shea family	Old Shea farm, destroyed 1960 for Midtown

POWHATAN COUNTY

Dabney Cemetery	At Montpelier
Petersville	Rt 60, 1 mi E of jct Rt 684

PRINCE EDWARD COUNTY

Scott Family	Town of Briery

PRINCE GEORGE COUNTY

Atwood Farm	5 mi back of home Disputanta

PRINCE WILLIAM COUNTY

Arrington Family	Between Mineola Ct and Arrington Farm Ct

RAPPAHANNOCK COUNTY

Atkins Family	On Rt 211 W of Sperryville, just inside Shenandoah National Park
Dearing Family	Caledonia Farm, 47 Dearing Rd, Flint Hill
Richard Jackson Family	1 mile from Washington at Rosemont

RICHMOND CITY

Forest Lawn	4000 Pilots Lane
Hebrew	300 Hospital St

RICHMOND COUNTY

Shackeford	Court St, Warsaw

ROCKBRIDGE COUNTY

Bell Family	Cameron Hall across from Iron Bridge, Goshen
Bethesda Presbyterian Church	Rockbridge Baths
Carter Family	Kerrs Creek
Clark Family	Off Blue Ridge Parkway
Collierstown Presbyterian Church	Town of Collierstown
Daniel-Turpin	Vic of Rappa Mill
Egypt-Cunningham	Past end of Rt 629, Waterloo, Kerns Creek
Haines Chapel	Town of Vesuvius
McKee Family	Rt 631
Old Ebenezer Church	Rockbridge Baths
Stonewall Jackson Memorial	Rt 11, Jct with White St, Lexington
Walkerland	Vic jct Rts 602 & 724
Wilson Springs	Rt 39 off Rt 623, Tucker Farm, site of Bethesda Church

ROCKINGHAM COUNTY

Byerly Family Cemetery	Pleasant Valley Road (Rt 679/704) Koontz Farm
Central	Main St, Port Republic
Cook Creek Presbyterian Church	Rt 726 Mt Clinton
Cross Keys	GPS: 38.35817,-78.84124
East Point	Town of Elkton
Friedens United Methodist Church	Town of Friedens
Greenmount	Rt 772
Lacey Springs	Rt 608, Lacey Springs Rd, jct Rt 805
Linville Creek Church of the Brethren	Town of Broadway
Meyers Family	Vic Old Linville Creek
Mount Olive	Rt 843 Cemetery Rd, McGaheysville
Oak Lawn	Jct Rts 704 & 257, Bridgewater
Pleasant Grove Methodist Church	1393 Hasty School Rd
Reedy	Rt 878, Wenger Mill
Rice Family	Knicely property, Town of Dayton
Saint Jacob's-Spaders Lutheran Church	Mount Crawford
Saint Peter's	Near Massanutten United Methodist Church
Thomas Q Wyant	Town of Beldor
Turley	Town of Turleytown
Wise Family	Dancing Bear Lane

RUSSELL COUNTY

Bickley	Town of Castlewood
Keen William	at Town of Belfast
Litton Hill	Town of Elk Garden, Route 1
McFarlane	End of McFarlane Lane
Wheeler Cemetery	GPS: 36.9072,-82.2972

SALEM CITY

East Hill	Jct Rt 460 and Lynchburg Turnpike

SCOTT COUNTY

Bledsoe Family	Rt 600 before Neely Dr, end of road to the right
Boatright	GPS: 36.7225,-82.6961
Carter Family	Hunters Valley Rd
Enoch P Lane Family	Off Rt 665, Manville Rd nr Smith Chapel & Williams Cem
Freeman Family	North side Rt 614, 1 mi east of jct with Rt 636
Green Family	South of Lick Creek and east of Rt 777, Duncan Mill
Joseph Carter Family	Town of Rye Cove
Jones Family	Possum Creek Rd GPS: 36.3701,-82.4109
McKinney-Carter Family	Town of Duffield
Rolllins Family	Town of Clinchport
Stewart Cemetery	In Rye Cove just E of J P Hill's house on Rt 717
Taylor Family	Town of Rye Cove
Wolfe Cemetery	Town of Dungannon, Weber City

SHENANDOAH COUNTY

Benjamin Bowman Family	Town of Narrow Passage
Isaiah Clem	Edith, Columbia Furnace
Emanuel Lutheran Church	Town of Woodstock
Frye Family	Town of Wheatfield
Henry Culler Family	Off Rt 678 to East, north of Seven Fountains
Knisley Family	Town of Detrick
Lichliter Family	Fort Valley Rd, south of Detrick
Massanutten	Town of Woodstock
Miller Marshall	Narrow Passage, 3 mi SW Woodstock
Mount Jackson	Rt 11, south of Jct Rt 263
Neff-Kagey	Town of New Market
Peter Boyer Family	Town of Oranda
Shrum Family	Town of Saint Luke
Snapp Family	Town of Edinburg
Saint Paul's Lutheran Church	Town of Strasburg
Saint Stephen's	Town of Strasburg Junction
Tomahawk Pond (AKA Barb Schoolhouse)	Bayse area
Union Church	Town of Mount Jackson
Walnut Springs Christ Church	Town of Oranda
Zion Lutheran Church	Town of Hamburg

SOUTHAMPTON COUNTY

Council Famly	Franklin City
Thomas Family	Town of Newsons

SPOTSYLVANIA COUNTY

Davenport Family	Town of Post Oak
Gordon-Herndon	Town of Post Oak
Leavell Family	6107 Sweetwater Drive
Old Burying Ground, Cedar Grove	Turner Lane, Paytes
Rose Valley Farm	Rose Hill Subdivision

SMYTH COUNTY

Cox Family	Vic jct rts 610 & 622
Richardson Family	Rt 610, east of jct of Rt.16
Rich Valley Presbyterian Church	3811 Valley Rd, Saltville
Scotts	Teas Rd abt 2 mi E of Slab Town Rd

STAFFORD COUNTY

Briggs Family (AKA Stoney Hill)	Stoney Hill road, adjacent to Curtis Memorial Park
Elijah Sullivan Family (AKA Martin Sullivan)	Located at entrance to Lucks Drive
Embrey Family (AKA Embrey Cem #3)	South side of Ramoth Church Rd, near jct I-95
Knight-English Family	Logging road on ext Rt 709 off Winding Creek Rd-Rt 628
Miller Family	Rt 692 near 98 Quarry Road
Norman and Towson Family #1	Rt 692 near 98 Quary Road on hill
Rowe Family	Rt 218, last driveway on north side before White Oak Run

SUFFOLK CITY

Wright Family	By Bennetts Creek in residental area nr Nansemond River

SURRY COUNTY

Peter Baird Family	Off Rt 40 on field with fence Spring Grove

SUSSEX COUNTY

Chappell Family	Fowlkes Bridge Road, between Burtons Bridge & Rt 644
Wilkinson Farm	Stony Creek area

TAZEWELL COUNTY

Bowen Family	Town of Cove Creek
Brushy Hill	Thompson Valley area
Henry Bowen Family Farm	Maiden Spring Cove
Maiden Spring	Maiden Spring Cove
Maplewood	Maplewood Lane, North Tazewell
Thompson- Buchanan Family	Thompson Valley area
William D Nash Homeplace	Rt 460, see property records for location

VIRGINIA BEACH CITY

Atwood-Cornick Family	Board Bay Farm, Great Neck Road
Eastern Shore Chapel	2000 Laskin Road
Lovitt Family	Across from Virginia Beach Courthouse
Woodhouse Family	Southern Blvd

WARREN COUNTY

Cox Homestead	Howellsville
Prospect Hill	Front Royal
Ramey Family	Dungadin Rd
Reliance	Rt 627 north of Reliance Lane

WASHINGTON COUNTY

Blackwell Chapel United Methodist Church	Vic jct Rts 700 & 746
Fleenor Memorial Baptist Church	Near 11547 Rich Valley Rd
Malone	GPS: 36.6325,-82.2242
Montgomery	Abingdon, 24139 Denton Valley Road

WESTMORELAND COUNTY

Payne Graveyard	Horners, GPS 38.13357,-76.97069, on grounds of Ingleside Winery off Rt 640
Settle Family	Town of Hague
Yeocomico Episcopal Church	1233 Old Yeocomico Rd

WILLIAMSBURG CITY

Cedar Grove	809 S Henry St

WINCHESTER CITY

Brown	Valley Pike Rd vic Safeway store

WISE

Hamm	Jct Rts 657 & 658, Dry Fork

WYTHE

East End	East Goodwin Lane, Wytheville
Keesling	Town of Rural Retreat
Kings Grove United Methodist Church	Town of Crockett
McMillan	Town of Clarke's Summit
Neff	Town of Rural Retreat
Oglesby	Town of Fort Chiswell
Saint John Lutheran Church	Town of Wytheville
Saint Paul's Lutheran Church	Town of Rural Retreat
Saint Peter's Lutheran	Rt 619, Cripple Creek

YORK

Ironmonger Family	611 A Shirley Rd Grafton

APPENDIX C: ADDITIONS TO
CODE TO AND BIBLIOGRAPHY OF SERVICE SOURCES

F.   Ancestry.com. *A dictionary of all officers who have been commissioned in the Army of the United States* [online database]. Provo UT: Ancestry.com Operations Inc, 2004. Original data: Gardner, Charles K. *A dictionary of all officers who have been commissioned, or have been appointed and served, in the Army of the United States, since the inauguration of their first president in 1789, to the first January, 1853: with every commission of each, including the distinguished officers of the volunteers and militia of the states, who have served in any campaign or conflict with an enemy since that date, and of the Navy and Marine Corps, who have served with the land forces: indicating the battle in which every such officer has been killed or wounded, and the special words of every brevet commission.* New York: G.P. Putnam and Co., 1853.

R.   Biographical Directory of the American Congress 1974-1949. United States Government Printing Office, 1950.

S.   Cartmell, T.K. *Shenandoah Valley Pioneers and Their Descendants: A History of Frederick County, VA.* Berryville VA: Chesapeake Book Company, 1963.

T.   Southampton Co, *Court Order Book (1814-1816)*, and *Minute Book (1830-1835)*.

U.   Rowland, Mrs. Dunbar. *Mississippi Territory in War of 1812.* Reprinted from *Publications of the Mississippi Historical Society, Centenary Series*, Vol IV (1921). Baltimore: Clearfield Publishing Company, 2005.

V.   Generations Gone By. Website www.generationsgoneby.com.

Y.   Marine, William M. *British Invasion of Maryland, 1812-1815.* Baltimore: Society of the War of 1812 in Maryland, 1913. Reproduced in *Maryland Settlers & Soldiers 1700-1800* (Family Tree Maker CD #521).

Z.   *Northern Neck of Virginia Historical Society Magazine*, Vol 3 (Dec 1953).

AA.   Find A Grave. www.findagrave.com.

AB.   Douthat, James L. *Roster of War of 1812 Southside Virginia.* Signal Mountain TN: Mountain Press, 2007. (Mountain Press, P.O. Box 400, Signal Mountain, TN 37377-0400, www.mountainpress.com).

AC.   General Navy Register.

APPENDIX D: ADDITIONS TO
CODE TO AND BIBLIOGRAPHY OF BURIAL SOURCES

30. Madison County Historical Society, Cemetery Project CD, November 2010.

53. Eshelman, Ralph E., Scott S. Sheads, and Donald R. Hickey. *The War of 1812 in the Chesapeake: A Reference Guide to Historic Sites in MD, VA & DC.* Baltimore: Johns Hopkins University Press, 2010.

55. Arlington National Cemetery Interment form, War Department, Q.M.C. Form 14.

56. Kerns, Wilmer L. *Historic Records of Old Frederick and Hampshire Counties, VA.* Bowie MD: Heritage Books, Inc., 1992.

81. Mt. Hebron Cemetery Burial Records, Winchester VA. www.mthebroncemetery.org.

85. Hughes, Charles Randolph, *"Olde Chapel" Clarke County, Virginia.* Berryville VA: Blue Ridge Press, 1906.

129. *The Virginia Genealogist*, 24:4 (Oct-Dec 1983).

159. "Cemeteries by Tax Number." Unpublished listing by Stafford County, updated 02 Nov 2001.

167. Green, Laurie Boush and Virginia Bonney West. *Old Churches, Their Cemeteries and Family Graveyards of Princess Anne County, Virginia.* Virginia Beach VA: self-published, 1985.

170. *Virginia Military Records.* Baltimore: Genealogical Publishing, 1983.

APPENDIX E: ADDITIONS TO
BIBLIOGRAPHY OF ADDITIONAL SOURCES

Butler, Stuart I. *Defending the Old Dominion, VA and its Militia in the War of 1812.* Charlottesville: University of Virginia Press, 2013, pgs 513-525.

Kegley, Mary B. *Early Adventures on the Western Waters: The New River of Virginia in Pioneer Days, 1745-1800, Vol. 2.* Orange VA: Green Publishers, 1982.

Magazine of Virginia Genealogy, Vol 18 No 2.

Richmond Compiler, Marriage Notices, 21 Oct 1819, pg 3.

Virginia Deaths and Burials 1853-1912. www.familysearch.org.

Virginia Marriages (1785-1940). www.familysearch.org.

Wikimedia Commons. https://commons.wikimedia.org.

Wikipedia: The Free Encyclopedia. https://en.wikipedia.org.

APPENDIX F: ADDITIONS TO INDEX OF NAMES
OTHER THAN VETERANS AND UNIT COMMANDERS

Adams, Emmaline Young 47
Adams, Jane 37
Adams, Mary 24
Adham, Mary 11
Adie, Anne 97
Ailor, Barbara 79
Akers, Delilah 41
Alexander, Anne 88
Alexander, Charles 1
Alexander, Frances Brown 94
Alexander, Joseph 1, 2
Alexander, Maria 16
Allen, Charles 2
Allen, Elizabeth W 77
Allen, Judith Churchill 52
Allen, Mary 48
Allen, Matthis 48
Allen, Sarah 9
Allen, Thomas 2
Allen, William 2
Allen, William 3
Allison, William 3
Alsop, Lucy 50
Alter, Sara 89
Ambler, John 3
Amburgey, Mary 99
Amick, Elizabeth 74
Amiss, Lucy 39
Amiss, William 39
Anderson, Ann 55
Anderson, Elizabeth 72
Anderson, Joseph Edward 4
Anderson, Jordan 6
Anderson, Mary Watkins 6
Anderson, Nelson 3
Anderson, Sarah 3
Anderson, Sarah 43
Anderson, Susannah 16
Anderson, Thomas 3
Andis, Matthias 39
Andis, Susannah 39
Andrews, Bullard 4
Andrews, Varney 4
Anthony, Jane 68
Arborgast, Michael 4
Archer, Jane S 3
Archer, William 4
Armistead, John 5
Armistead, Mary A 20
Arnold, Christina 73
Arnold, Tabitha 64
Ashcraft, Mary 1
Ashwell, Elizabeth 67
Ashwell, John 67
Aspinwall, Hannah 75

Atkins, Ambrose 5
Atkinson, Caroline Elizabeth 70
Atkinson, Jane 58
Atkinson, Robert 5
Atkinson, Roger 70
Aulick, Charles 6
Avery, Martha T 1
Aylett, Philipp 6
Ayres, William 6
Baer, Anna Barbara 70
Bageant, John Dick 6
Bailey, Benjamin 6
Bailey, Ellen 80
Bailie, William Law 6
Baker, Arabella 99
Baker, Elisha Louisa 96
Baker, Elizabeth 30
Baker, Phillip Peter 7
Baker, Susanna 34
Baldwin, Margaret 31
Ball, Burgess 7
Ball, Elizabeth 69
Ball, Elizabeth Lucy 22
Ball, George 7
Ball, Spencer Mottrom 22
Ball, William 69
Bane, Elizabeth 21
Bankhead, Jane 101
Banks, Mary 98
Banks, Richard 98
Barber, Frances 3
Barber, Mary 35
Barclay, Mary Elizabeth 76
Barclay, Mary Margaret 54
Barger, Anna Margaret 25
Barksdale, Sarah 73
Barnes, Ann Maria 4
Barnett, A G 8
Barnett, Martha (Patsy) 78
Barnett, Rebecca 26
Barnhouse, George 8
Barrack, Mary Ann 75
Barraud, Daniel 94
Barraud, Sarah 94
Barret, John 8
Barrett, Elizabeth C 96
Barrett, Jane A 47
Bass, William 8
Bassett, Burwell 9
Baughman, Christiana 9
Bayley, Charles 5
Bayley, Sarah 5
Baylor, Lucy 5
Beach, Phebe 24

Beam, Jacob 9
Bedsaul, Elisha 67
Bedsaul, Mary 67
Beeson, Martha 102
Belcher, Elizabeth 28
Belcher, Elizabeth Eppes 4
Belches, Margaret 60
Bell, Elizabeth H 90
Bell, Frances Sterrett 88
Bell, Jeremiah 9
Bell, Joseph 9
Bell, Nancy 18
Belsher, Sarah 41
Bennett, John 10
Bennett, Mary 95
Berkeley, Nelson 10
Bernard, Martha Ann 51
Bernard, Sarah R 15
Berry, Benjamin 10
Berry, Henry 10
Berry, Johanna 10
Berry, Oscar 10
Best, Elizabeth 99
Best, Enos 99
Best, Frances 49
Bible, Johann Adam 10
Bickley, Charles 11
Bingham, Eliza M 7
Binns, Margaret Hannah Douglas 95
Birch, Ann 74
Bishop, James Sr 11
Bishop, Mathew 11
Bixler, Mariah 63
Black, Frances 19
Black, Mary 19
Blackburn, William 11
Blackwell, Elizabeth 36
Blackwell, Moses Sr 12
Blackwell, William 12
Blackwell, William B 11
Blair, (----) 12
Blair, John 12
Blair, John Durbarrow 12
Blair, Thomas 12
Blake, Frances 16
Bland, Edward 12
Bland, Richard 12
Blankenbaker, Mildred 21
Blue, Hannah D 4
Boatright, Valentine 13
Bolton, Harriet, 10
Bolling, Lucy 83
Bonner, Margaret 50
Boogs, John 13

Clarkson, Elizabeth B 88
Clarkson, James 24
Clarkson, Mary Letitia 72
Claytor, Frances G 53
Clem, Christina 12
Clifton, Sarah Cocke 56
Clopton, Anne 72
Clopton, Mildred 51
Cloyd, Joseph 24
Cloyd, Gordon 60
Cloyd, Mary 60
Clyee, Mary Elizabeth 26
Coalter, Sarah 34
Cock, Andrew 33
Cock, Mary 33
Cocke, Chastain 21
Cocke, Hannah 98
Cocke, Martha Ann 49
Cocke, Sarah 28
Cocke, Susanna 98
Cockram, Hannah 30
Cockrell, Sally Hudnall 48
Coffman, Ann 97
Coffman, Mary Magdalena 63
Coffman, Rebecca 59
Cole, Elizabeth App G
Cole, Mahalia Alice 44
Colgate, Charlotte 87
Colgate, Esther 62
Colgate, Robert 62, 87
Coleman, Elizabeth W 103
Coleman, Sarah App G
Coleman, Susannah 39
Coleman, William 25
Collins, James 25
Combs, Joseph H 100
Combs, Mary 100
Connell, Polly 12
Conrad, Elizabeth 37
Cooper, Frederick 26
Cooper, George 27
Cooper, John 26
Cooper, Michael 26
Copenhaver, Charlotte 51
Cordell (Kadel), Anna
Catherine 26
Cordell, Johann Adam 26
Corder, John 55
Corder, Mary 55
Cornett, David Canute 27
Cornick, Lemuel II 28
Cosby, Mary Ann 8
Cotton, Susannah 6
Cottrell, Charles Waddell 28
Coulling, James 28
Cowdery, Elizabeth 83
Cox, Armine Estes 14

Cox, David 28
Cox, James 28
Cox, Jane 53
Cox, Joshua McGowen 28 (2)
Cox, Ollie 77
Cox, Samuel 28
Craig, Margaret 9
Craig, Margaret 42
Craig, Patience 48
Craig, Robert 42
Cralle, Nancy 40
Crawford, Mary 24
Crawford, Sarah 9
Crawley, Lucy 19
Crawley, Robert 19
Crebs, Susan 27
Crider, John 29
Crittenden, Lemuel 24
Crittenden, Susan Lorinda 24
Crockett, Susanna 87
Crow, Hannah 21
Crowder, Mary Ann 29
Croxstall, Elizabeth 7
Crumley, Mary 33
Crutchfield, Malvina 74
Culton, Betsy 97
Culton, Robert 97
Cunningham, Betsy 76
Cunningham, James 76
Cunningham, Judith 22
Curlett, Frances 89
Currell, Mary 19
Current, Charity 88
Currie, Frances Hill 3
Curtis, George Sr 30
Curtis, Peggy 79
Curtis, Rachel 93
Custis, John Parke 30
Dabney, Elizabeth 64
Dabney, John 31
Dabney, Martha Burwell 23
Daffin, Betty M 31
Daffin, Vincent 31
Dalby, Hannah 85
Dalby, Joseph 69
Dalton, Rachel 45
Dalton, William 31
Dandridge, Dorothea
Spotswood 50
Daniel, Frances (Fanny) 75
Daniel, Leonard 31
Darby, Eliza Chruchill 39
Darne, Henry 31
Davenport, Elizabeth 99
Davidson, Agnes 87
Davidson, Benjamin 32
Davis, Betsy 55, 99

Davis, G. 31
Davis, John W 35
Davis, Martha Amelia 35
Davis, Mary 38
Davis, Mary Elizabeth 85
Davis, G 31
Davis, Nancy 31
Davis, Susannah 77
Davis, William 32
Day, Ann 77
Dean, Elizabeth 35
Dean, Mary 33
Dearing, John 33
Dearing, Nancy V 61
DeGougea, Charlotte Olympia
43
DeHaven, Peter 33
Delaney, Mary 74
Delong, Margaret 44 (2)
Dempsy, Milly 20
Derry, Elizabeth 97
Dew, Mary Ellen 47
Dew, Thomas Roderick 47
Dibrill, Anthony 33
Dick, Elizabeth 6
Dickerson, Jonathan 53
Dickerson, Nancy Sizer 53
Digges, Dudley Power 33
Digges, Martha 19
Diggfes, Sarah 45
Diggs, Charlotte 76
Dillinger, Rebecca 36
Dillon, Elizabeth 21
Dillon, Mary C 20
Dishman, Samuel 34
Dixon, John 34
Dodd, Sarah 69
Dodson, Elizabeth 10
Dold, Elvira A 17
Dold, William 17, 34
Donaldson, Mary 46
Donnan, David 63
Donnan, Jane 63
Doster, Rhoda 33
Doster, Thomas 33
Doughton, Irene Jane 28
Doughton, Joseph Bain 28
Douglas, Mollie 68
Dove, Catherine 35
Dove, George 35
Dove, Jacob 35
Dove, Levicy 35
Dove, William 35
Dovel, David 35
Dowell, Nancy M (or A) 46
Drake, Hannah 19
Drish, John 37

Nelson, Elizabeth 19
Nelson, Elizabeth 75
Nelson, Elizabeth 82
Nelson, Lucy 42
Nelson, Mary Ann 9
Nelson, Nathaniel 19
Neville, Rebecca 77
New, Anne 98
Newcombe, Nancy 31
Newcomer, Elizabeth 57
Newell, Margaret 20
Newton, Elizabeth 62
Newmann, Rebekah 32
Newton, Frances 80
Newton, Sarah 94
Newton, Thomas 94
Nicholas, Anna Maria 26
Nichols, William 79
Nickols, Edith 94
Nifong, Sarah 61
Norman, Edward 79 (3)
Norman, James S 79
Norman, Matthew 79
Northcraft, Verlinda 61
Nunnally, Mary 76
Nutty, Sallie 9
O'Bryhim, Ann 79
O'Bryhim, Clara 79
O'Bryhim, Joseph 79
Ogle, Elizabeth 56
Organ, Samuel 79
Orndorff, John 44
Orndorff, Rachel 44
Orr, Margaret 91
Osborn, Massey 23
Osborne, Ruth 28 (2)
Osburn, Virginia Jane 61
Otey, Frances 65
Otey, John Armistead 80 (2)
Overfield, Martin 19
Overfield, Nancy Ann 19
Overstreet, Barsheba 29
Owen, Mary 77
Owens, Margaret Susan 69
Pace, Amanda Polly 43
Page, Alice Grymes 33
Page, Eliza Nelson 82
Page, Elizabeth (Betsey) 42
Page, Lucy 19
Page, Mann 82
Palmer, Elizabeth 60
Palmer, Juliann 93
Palmer, Martha 66
Pamplin, Sarah 64
Pankey, John 80
Pankey, Nancy Branch 80
Pankey, Thomas 80

Parker, Elizabeth F 22
Parker, Polly 18
Parsons, Mary Ann 19
Parsons, Mary Polly 58
Pate, Elizabeth 41
Patterson, Elizabeth Tabb 103
Patterson, Louisa Gabriella 2
Patterson, Maria Booth 96
Patton, Lydia McIntire 52
Patton, Nancy McClung 61
Patton, Sallie 12
Paxton, Jane 59
Paxton, Mary Polly 54
Paxton, Samuel 34
Paxton, Sarah 34
Paxton, Thomas 54
Payne, Benjamin 80
Payne, Delilia 28
Payne, Elizabeth Barrett "Betsey" 84
Payne, Francis 30
Payne, Jemina 30
Payne, John 80
Payne, John 80
Peale, Amanda M Fitz Allen 59
Peebles, Agnes 50
Pendleton, Mary Maud 13
Pennington, Celica 27
Pennington, John Dees 61
Pennington, Lucy Ann 61
Pennington, Nancy Rebecca 32
Perreau, Mary 40
Perry, Agnes 21
Perry, Elizabeth 91
Perkins, William 83
Perkinson, Mary P 12
Peterson, Susan 1
Peyton, Catherine Storke 75
Peyton, Dorothy Smith 103
Peyton, Elizabeth 34
Peyton, Sarah 86 (2)
Phifer, Sarah 2
Phillips, Elizabeth 31
Phillips, Nancy 36
Phillips, Tobais 31, 36
Philpott, Elizabeth 21
Philpott, Julia B 21
Philpott, Sarah Ann 84
Plickenstalver, Catherine 65
Piggott, Sarah 17
Pinkard, John 82
Pitman, Lawrence 22, 82
Pitman, Nicholas 82
Pitman, Susan 22
Platt, Adline Sarah 27
Pleasants, Archibald 103

Pleasants, Elizabeth Rhodes 20
Pollard, Mary Anne Brook 57
Pope, Lucy Ann 31
Pope, William 31
Porch, Mary 93
Porter, Elizabeth 87
Porter, Nancy 77
Porter, Sarah 69
Porterfield, Eleanor 23
Powell, Elizabeth 36
Poythress, Agnes 70
Poythress, Susannah 12
Preston, William 82
Price, Catherine Fox 93
Price, Hannah Catherine 11
Price, Jane 71 (2)
Price, Thomas 83
Price, Margaret 27
Prill, Elizabeth A 26
Prosser, Elizabeth 64
Prosser, William 64
Purcell, Thomas Jr. 83
Purvis, Mary 64
Puryear, Mary 18
Quisenberry, Elijah 42
Quisenberry, Lucy Tate 42
Rader, Anna Marie 17
Rader, William 17
Ramey, Thomas 83
Randolph, Gabriella Harvie 16
Randolph, John 96
Randolph, Mary 7
Randolph, Mary Rebecca 40
Randolph, William 83
Rankin, Patsy 41
Rapley, Letitia Ann 8
Ratcliffe, Louisiana (Lucy) 47
Rahtuss, Salitha 41
Rayhill, Mathew 102
Rayhill, Sarah 102
Read, Hannah 22
Rector, Agnes/Agatha 44
Redman, Sarah 94
Reed, Elizabeth 90
Reed, Lucy 13
Reed, Peter 13
Reeves, Mary Ann "Polly" 28
Reid, Elizabeth 72
Reid, Elizabeth 72
Reid, Jeremiah 72
Reid, Sarah 1, 2
Renner, Elizabeth Rachel 44
Renner, Margaret 44
Reno, Lydia French 76
Repass, Elizabeth 78
Replogle, Margaret 22